WINNERS & LOSERS

KIERAN LEVIS

WINNERS & LOSERS

Creators and Casualties of the Age of the Internet

Atlantic Books
LONDON

First published in hardback and export trade paperback in Great Britain in 2009 by Atlantic Books, an imprint of Grove Atlantic Ltd.

The author and publisher would gratefully like to acknowledge the following for permission to quote from copyrighted material:

The Crack-Up by F. Scott Fitzgerald, published by New Directions Publishing Corp ©1945, reprinted by permission of New Directions Publishing Corp; The Future of Management by Gary Hamel, published by Harvard Business School Press © 2007, reproduced by permission of Harvard Business School Press; 'In Front of Your Nose' from Collected Essays, Journalism and Letters of George Orwell Vol 4 © George Orwell, reproduced by permission of Bill Hamilton as the Literary Executor of the Estate of the Late Sonia Brownell Orwell and Secker & Warburg Ltd; Capitalism, Socialism, Democracy by Joseph Schumpeter, published by Harper Brothers ©1942, reproduced by permission of Taylor and Francis.

Every effort has been made to trace or contact all copyright holders. The publishers will be pleased to make good any omissions or rectify any mistakes brought to their attention at the earliest opportunity.

1 2 3 4 5 6 7 8 9

A CIP catalogue record for this book is available from the British Library.

Hardback ISBN: 978 1 84354 964 2

Export and airside trade paperback: 978 1 84354 965 9

Printed in Great Britain by the MPG Books Group

Atlantic Books
An imprint of Grove Atlantic Ltd
Ormond House
26–27 Boswell Street
London
WC1N 3JZ

www.atlantic-books.co.uk

To Angela, my ever-fixéd mark

CONTENTS

1   Introduction

MARKET CREATION

17   1   Prodigious Partners
Apple, Sony

75   2   Capabilities and Vision

86   3   e-Merchants
Amazon, Webvan

113   4   Propositions and Discipline

125   5   Shooting Stars
Netscape, AOL

160   6   What It Takes

170   7   Network Models
eBay, Google

THE BIGGER PICTURE

215   8   New Markets and Networks

236   9   The Disruptive PC
IBM, Encyclopædia Britannica

263   10   Creative Destruction

291   11 Wireless Winners
           BSkyB, Nokia

329   12 Missing the Big Picture

355   13 The Right Stuff

386   Postscript

391   Sources and Bibliography

399   Acknowledgements

401   Index

INTRODUCTION

A Tale of Two Capitalists

Jeff Skoll was pitching a business plan to a partner at the Mayfield Fund, one of the top venture capital firms in Silicon Valley. Unlike most dot.coms in 1997, AuctionWeb was actually making money. In fact it had margins of 85 per cent and its sales were growing at more than 30 per cent every month. But the man behind the desk was baffled. 'Let me get this right, people are going to buy and sell antiques online? I gotta go.'

A few weeks later, Jeff's partner Pierre went to see two of the partners at Benchmark Capital, who were surprised he hadn't come with a slick presentation. He couldn't even give them a demonstration, because AuctionWeb's computer system was down, yet again, but they were intrigued and wanted to know more. They had no idea how big this business would become, but they started to get it. In June 1997 Benchmark placed what turned out to be the most profitable investment ever made by a venture capital firm. For $5 million it bought 21 per cent of the shares in a business that was now called eBay. Two years later its stake was worth more than $1 billion.

Nobody got eBay at first. It wasn't cool or sexy like Amazon or Yahoo. Its users were not the digerati but the kinds of people who went to yard sales. Pierre and Jeff were even having difficulty persuading programmers to come and work at such a strange business.

eBay had only become a business by accident two years earlier, when Pierre Omidyar had been toying with several website projects. He had always thought it would be cool if there were a level playing field, where people could trade with each other on equal terms. He started AuctionWeb so that computer buffs like him could buy and sell bits of old equipment to each other, with no idea of making money from it. Anyone in what he called the community could post items for sale, and anyone else could bid for them. The first item sold was a broken laser pointer, posted by Pierre himself, for which he was amazed to receive $14.

AuctionWeb was not Pierre's most important project and he paid little attention to it until his internet service provider told him that there was too much traffic on the site, and he would have to start paying the business rate of $250 a month. So in February 1996 Pierre asked the community if they'd mind contributing a percentage of the value of each sale they made. In March he was amazed to receive hundreds of cheques adding up to $1,000. By May they amounted to $5,000, and he had to hire someone to open the envelopes and take the cheques to the bank. In June, when $10,000 arrived, Pierre decided to give up his day job, and asked his friend Jeff to become his partner.

While Pierre concentrated on making the rather wobbly computer system more reliable, Jeff wrote the business plan, but he couldn't quite believe the numbers he was coming up with. Every month, without them doing a thing, revenues kept growing. It wasn't so much computer paraphernalia now – toys and dolls were selling like hot cakes. Then antiques, stamps and coins took over. There was apparently no end to the markets where eBay could make things work better.

Although nobody realized it, even in 1997, eBay was benefiting from a combination of what economists call network effects and engineers positive feedback loops – the more people who used it, the more attractive it became to others. It would continue to grow exponentially for years to come, like a giant snowball rolling down a mountain, gathering more and more buyers and sellers. By 1997 the total value of sales was on course for $100 million, and eBay's revenues were $4.3 million. Ten years later total sales reached $53 billion and eBay's share was $6 billion.

The picture fades

If eBay did not have the air of a winner in the mid-1990s, Kodak did not look like a loser. It was the biggest photographic company in the world, with revenues of $14 billion and a healthy balance sheet. It was one of only two companies from Forbes' list of the top hundred in 1917 that was not only still there but had out-performed the stock market index. It was a leader in research and development, with one of the best-known, most trusted brands in the world.

Its founder, George Eastman, had effectively created the industry a hundred years earlier, with his invention of roll film and the Brownie camera. He had also dreamed up a marketing message that had stood the test of time: 'You press the button, we do the rest'. One of his shrewdest insights was that there might be more money to be made from the production and processing of film than from selling cameras. Kodak became the only company that was successful at doing both, though film and printing made much more money. The high levels of research required in this industry created substantial barriers to entry, there were few global players, and fat margins financed investment in new areas.

Kodak had not rested on its laurels and had been a serious innovator, most notably in digital photography and imaging. In the late 1990s it was registering a thousand new patents a year in imaging technology and distribution. As it looked with bemused delight at the role photographic images were now playing on the World Wide Web, it concluded that a new, $200 billion 'info-imaging industry' was emerging and that Kodak was at the heart of it. In its annual report for 2000 Kodak's CEO, Daniel Carp, noted that 80 billion photos had been taken in the previous year, and 100 billion prints made. Growth was particularly buoyant outside the US and Europe. 'This is a very smart time to be in the picture business,' he declared.

It was certainly an interesting time. Kodak knew that digital photography was likely to replace film eventually and expected film sales to start to decline from around 2004. In 2000, only 8 million of the 400 million amateur photographers in the world had digital cameras, and most were not comparable with the best conventional models. A gradual transition was an entirely reasonable expectation, and Kodak was as prepared for it as anyone.

The pace and extent of change, however, took everybody by surprise. Consumers discovered that digital photography was a great deal easier, and a big part of the attraction was not having to mess about with film and prints. Despite an economic downturn, sales of digital cameras in the US rose sharply – 6.9 million in 2001 and 9.4 million in 2002. In 2003 digital sales overtook those of conventional cameras, reaching 12.4 million. The same crossover occurred world-wide in 2005, when over 42 million digital cameras were sold.

Kodak's film business's sales started to dip in 2001. At first its management attributed this to the gloom following the NASDAQ crash, and September 11. Even in 2003 Mr Carp insisted that 'our traditional film business is sound, as digital imaging continues to evolve'. But revenues from film plummeted, overall earnings fell by half, and in each of the next three years the company lost money.

Kodak could not be accused of hiding its head in the sand. It eagerly embraced a 'digital transformation' strategy, coming up with a raft of new products and services aimed at businesses and consumers. It particularly targeted the booming new market for digital cameras and in 2003 sold nearly 2 million of them, achieving the second-largest share of the American market. By 2005, it was the largest supplier in the world, earning revenues of $5 billion.

But cameras had always been a low-margin business for Kodak, and this market was to prove even tougher. New entrants flooded in – consumer electronics companies, personal computer suppliers and, astonishingly, even mobile phone makers. It seemed as though almost anybody could make these things – or easily outsource their manufacture in the Far East. The basic digital camera, a novelty a few years earlier, was becoming a commodity. The biggest supplier by volume was soon Nokia, churning out more than 12 million phones that could take pictures in 2005. These were not, of course, serious cameras, but they turned out to be good enough for many consumers. Apart from the top end of the market, digital cameras were going the way of much of the consumer electronics industry – rapid commoditization and wafer-thin margins. Although the Kodak brand provided some differentiation, the company made no money at all from its new consumer digital imaging business. In 2006 it closed it down and laid off 27,000 people, 42 per cent of its work-force.

Kodak had suffered a double disaster. The main new market into which it had plunged so boldly had turned out to be 'a crappy business', in the words of its new CEO, Antonio Perez. And the old market that had provided such reliable streams of income for so many years was disappearing faster than anyone had imagined. None of this was Kodak's fault. Nothing it could have done could have prevented the switch to digital – or turned digital cameras into a business as profitable as film. In a few brief years its world had been turned upside down.

Winners and losers

This book describes how a handful of businesses in the last thirty years were born and rose to enormous heights – and how others fell from them. None of the new businesses would have made sense either to customers or to investors much before 1980. It was not clear at first how many of them made money. Few of them actually sold products – and those who did sub-contracted their manufacture to others. Many of them provided information or services free, or at ridiculously low prices. Yet nearly all of them mushroomed from nothing to revenues of billions in just a few years.

They did this in a new economy – globally integrated, electronically networked and ferociously competitive. Thousands of new technologies emerged, millions of new businesses sprang up, but most failed to find a market or were quickly overtaken by others. Established businesses, accustomed to stability and continuity, found this brave new world distinctly uncomfortable, with unfamiliar competitors snapping at their heels and disruptive technologies and bizarre business models threatening to marginalize them. Illustrious names like IBM, AT&T, GEC and *Encyclopædia Britannica* came humiliatingly close to extinction. It seemed as though business life had become a brutal Darwinian struggle where no company had any security of tenure.

Yet a select few went from rags to undreamt-of riches in just a few years. In 1984 a married couple, moonlighting from their day jobs, started assembling networking equipment in their living room, and called their part-time business Cisco; for a few giddy weeks in 2000, Cisco was the most highly valued company in the world. In 1990 Nokia was an unwieldy 115-year-old Finnish conglomerate with 187 businesses, in deep financial trouble; ten years later it was the global leader of a new mobile telephone industry. Google only became a business in 1998 and earned virtually nothing for the next two years, but by 2007 it was grossing $16.6 billion.

How on earth did they do it? What was it about the winners described here that enabled them not just to define a new way of doing business but to dominate the markets they had created? What qualities did they have in common? And how did they manage to stay on top of the heap? What does it take to become the long-term leader of a new industry in today's economy? How were they able to fight off waves of challengers, when so many others were toppled?

The answers to these questions lie in the stories of some notable recent successes and failures. These show how and why a small number of very different organizations were so successful, and the unusual attributes they had in common, and why the absence of these qualities crippled those who failed. This does not pretend to be science, but these are not questions that rigorous, quantitative analysis can ever answer – it played little part in the birth of most of the businesses described here. Much of business, like much of life, is inherently unpredictable, untidy and uncertain. Managers can reduce the uncertainty but never eliminate it. In particular, as all these stories demonstrate, they can never know or control the future. All of the outstanding successes defied the conventional wisdom of experts – and enjoyed a considerable amount of luck.

Mortality, however, is unavoidable – not many businesses last more than a few years, and only a tiny number as long as a human life. In Joseph Schumpeter's sobering words, 'all successful businessmen are standing on ground that is crumbling beneath their feet.' The process of creative destruction he identified, whereby businesses are constantly challenged and eventually displaced by the innovations of others, has been going on for centuries, normally slowly and imperceptibly. In the age of the Internet the process has been speeded up and intensified. Globally integrated markets and the revolution in information and communications technology have led to vastly more innovation in new products, services and business models, but also to much more competition and disruption. Some recent financial innovations led to a crisis in 2008 that shook the global economy to its foundations. Creative destruction always produces more losers than winners, though in the long run most consumers are modest beneficiaries.

Winners and Losers aims to describe and explain, not to prescribe or predict. Most of the describing is done in the twelve detailed stories that make up the bulk of the book, most of the explaining in the chapters of ideas and arguments that follow each pair of stories.

In many languages, history and story share the same word. Human beings have been telling each other stories since prehistoric times, and sometimes learning lessons from them. We are still fascinated by Homer's heroes thousands of years after he brought them to life, because they are both extraordinary and recognizably human. Five hundred years ago Machiavelli showed how international politics really worked by describing how men like Cesare Borgia won power and exercised it. Peter Drucker and Alfred Chandler took a not dissimilar approach to describing the business corporations of the twentieth century. Their works are still widely read today because they explain so much.

The point of the stories here is to help us to understand how these markets work and, above all, why a few succeeded where so many failed. Ideas are particularly important in the new economy – knowledge and innovation are its lifeblood. All the new businesses described here started out with an idea that hardly anyone took seriously at the time, yet that was the seed from which not just a business but an entire market grew. Concepts like creative destruction, disruptive technologies and positive feedback are crucial to understanding the dynamics of these markets, but few executives are familiar with them, let alone general readers.

This book does not pretend to show business executives how to become winners themselves – there are more than enough of those already. It is intended for anyone who is interested in understanding how markets are created and disrupted, and why some winners seem to take all the prizes. Some ideas add less to our understanding than others. Catchy phrases like first mover advantage, winner takes all, content is king, get big fast and Web 2.0 contain a grain of truth, but have led some seriously astray. Change is confusing and it is tempting to clutch at soundbites for simple explanations. Unfortunately, they invariably oversimplify and encourage the belief that there is one big idea that can explain everything, that if we get one key thing right everything else will fall into place. These stories show that this is never the case – but that some patterns are almost universal.

Grand narratives

The fundamental problem all businesses face is how to adapt to apparently sudden changes in their environment. Making sense of them means taking a broader view than most businesses or business books do. Business is not a passion-free realm, separate from the rest of human life, and the humanities can shed as much light on it as economics, mathematics and evolution.

History offers not so much clear-cut lessons as perspective. Without some knowledge of the past, it is impossible to understand the present, let alone speculate sensibly about the future. History shows us that change is constant, though not always obvious, and rarely foreseeable. Though history never really repeats itself, as Mark Twain remarked, it does often rhyme. A globally integrated economy of sorts existed in the decades leading up to 1914, and steamships, railways and the telegraph in the nineteenth century shrank distances every bit as much as the telephone, the jet plane and the Internet in the twentieth. The fact that two world wars and the Great Depression shattered that era of peace and prosperity is a grim reminder that progress is far from inevitable.

Traditional economics, with its fixation on theoretical states of equilibrium, has its limitations in explaining how markets are created and disrupted, but it does provide 'a box of tools', a framework of near certainties: if costs and prices fall, sales will increase; successful business pioneers will attract competitors; competitive markets will eventually see prices fall close to marginal cost.

It is the economic consequences of technological innovation, particularly dramatic changes in costs, which transform markets, industries and societies. The biggest single factor in the growth of a global economy over the last two centuries has been the massive fall in the cost of transporting, first, people and goods and, latterly, information. Since the flourishing of the Internet, the marginal cost of transmitting information has fallen close to zero. The significance of Moore's Law (that the number of circuits that can be packed on to a chip doubles every eighteenth months) is not just that computers get ever more powerful, but that they become ever cheaper and more affordable by ever more people.

In the late 1990s some economists thought that digitization was creating a 'new economy' where old rules no longer applied. Capitalism is of course constantly mutating – that, as Schumpeter showed, is its nature. In the 1980s it went through two revolutions, one technological, one political-economic, leading to a radically different business landscape from the nationally protected, manufacturing-dominated economy of most of the twentieth century. The most striking differences are the abundance of new technologies and business models and the greater value attached to intangible assets like knowledge, brands and human capital. Less obvious is the ever larger role that networks of different kinds play in making this new business landscape more complex, more interconnected and more volatile, and how feedback loops speed up both growth and decline. eBay and Microsoft enjoy lucrative network monopolies, but most businesses find lasting competitive advantage more elusive than ever before.

To understand the transience of success, the tragic end that awaits so many businesses, and the limits to what we can really know, philosophy and literature may be our best guides. Isaiah Berlin has shown that in human affairs there are no universally applicable theories or formulae, no single right answer to complex questions. Shakespeare has more to tell us about the slings and arrows of outrageous fortune than any business book, and Homer more about hubris.

Some would say that trying to explain lasting business success is itself hubristic. Books like *In Search of Excellence*, *Built to Last* and *Good to Great* have been the targets of much academic scorn, partly because they relied on stories, but more tellingly because many of their winners later floundered. That, however, only goes to show that no-one is immune from the ravages of age and creative destruction, and does not make the quest futile or the conclusions entirely wrong. There is wisdom in all these books, notwithstanding Jim Collins's conviction that his research for *Good to Great* was scientifically rigorous and that he was discovering 'the enduring physics of great organizations'. Promises of 'timeless, universal answers that can be applied by any organization' are doomed to disappoint – if there were such answers, everyone could be a winner. Exceptional organizations have exceptional attributes and assets, the foundations of their competitive advantage. And even they cannot hold on to it for ever.

Winners and Losers aims simply to identify the qualities shared by all the market creators examined who went on to establish industry leadership that endured for more than a few years. (Most managed a decade or so.) These do not constitute a formula for lasting success, but the conditions necessary for it. External factors, notably customers and competitors, are crucial. And success, especially the lasting sort, is always relative.

For some people the problem is not so much business books as business itself, and the whole idea of winners and losers – one of the things they find so distasteful about capitalism. The term is used here mainly to describe businesses that had enormous wins or losses – in some cases both. Most of the losers described here either had reasonable hopes of hitting the jackpot or, like Kodak, had previously been long-term winners. Many of its workers were tragic losers from creative destruction – when an industry is wiped out, entire communities can find themselves on the scrapheap of history.

In a constantly changing competitive economy, nobody can be a winner for ever – incumbents are always challenged, and eventually displaced, by new forms of competition. None of the winners described here possessed all their attributes all of the time – many of them they learned, often from making mistakes, and some they forgot. Their stories show that contemporary capitalism is more diverse than its more sweeping critics might think, and that imagination, idealism and the search for excellence sometimes play a bigger role than avarice in the birth of businesses. In scarcely any of the new companies described here was financial gain the main goal of the founders – in some cases making shed-loads of money came as a delightful bonus.

There is no shortage of nasty businesses to confirm the suspicions of convinced anti-capitalists. Many companies as they get bigger put short-term financial performance above all other considerations, and rely more on locking customers in than on earning their lasting loyalty. Big companies may execute with chilling efficiency, but they frequently lack humanity and open-mindedness. Success and incumbency easily breed self-satisfaction, arrogance and hubris, but markets eventually bite back. In an economy where the skills and knowledge of employees and close relationships with customers and suppliers are critical to success, it does not pay to rest on one's laurels or to treat any stakeholder cavalierly.

The exceptional companies who establish lasting industry leadership are generally good at, among other things, nurturing human capital, cherishing customers and building mutually beneficial relationships with suppliers. The greatest single contribution to surviving the gales of creative destruction is the ability to continue to do better what the business already does well – enhancing organizational capabilities and learning new ones. This is what inspires the people who work in consistently innovative companies, and delights customers – and ultimately therefore shareholders.

Self-improvement and consideration for others do not of course guarantee success – unfortunately nothing does. But the long-term costs of complacency, of squeezing the last penny out of customers and suppliers, and the last ounce of effort out of employees, sometimes outweigh the short-term gains.

There is no overriding key to success in new markets, apart perhaps from luck. The combination of attributes required is highly unusual, which is why very few become big winners. Lasting competitive advantage and industry leadership call for an even more demanding set of qualities. A few organizations manage to master all of them, for a time at least. This book is mainly about them.

Heroes and villains

The book is organized around the stories. Each odd-numbered chapter, from 1 to 11, contains a pair of detailed stories and is followed by a generally shorter, even-numbered chapter that explains the ideas the stories highlight, and why qualities like distinctive capabilities and disciplined entrepreneurialism are critical to the success of all market creators.

The first half of the book concentrates on how new markets are born and on the success factors for those who created them. In between the stories of Apple and Sony, Amazon and Webvan, AOL and Netscape, eBay and Google, we examine, in chapters 2, 4 and 6, the eight attributes shared by all market creators.

The second half looks at the bigger picture. Chapter 8, 'New Markets and Networks', considers what makes new markets different from mature ones and how networks, both physical and virtual, make recent ones even more different. Chapter 9, 'The Disruptive PC', describes how IBM and Encyclopædia Britannica, Inc. were almost destroyed by the PC revolution and the failure of their leaders to understand the challenges they faced. Chapter 10, 'Creative Destruction', discusses the constant evolution of the business landscape and the ways in which markets are disrupted by new technologies, new forms of competition and other discontinuities. Chapter 11, 'Wireless Winners', tells the stories of two masters of creative destruction, BSkyB and Nokia. Chapter 12, 'Missing the Big Picture', considers why it is so difficult to recognize, understand and come to terms with radical change.

Finally Chapter 13, 'The Right Stuff', considers the really big question – what does it take to achieve lasting competitive advantage in a new industry. It examines the eight qualities shared by those organizations who held on to long-term industry leadership – significantly different ones from those for market creators.

Our protagonists are organizations, but they are made up of people and share several qualities with Homer's heroes – audacity, ingenuity, determination – and often hubris, wishful thinking and refusal to face uncomfortable truths. They are also remarkably diverse. The values and culture of an aggressive, hard-selling company like Dell or Rupert Murdoch's buccaneering BSkyB could scarcely be more different from those of the idealistic founders of eBay or the high-minded mathematicians at Google, intent on doing no evil and organizing the world's knowledge. Yet these four shared some very rare qualities, as did all the outright winners. These attributes, and other patterns found in many of these cases, cannot be shoehorned into anything resembling a scientific theory. There is never a single explanation for success or failure.

The criteria for selecting subjects were that they could genuinely be said to have created a new market (or tried to) or that they lost leadership to a new form of competition, and that there were several independent, reliable sources of information about them – companies' own accounts are invariably selective and bland. Successes do not lack chroniclers, but failures tend to be glossed over. Fortunately the early years of Netscape, Webvan and AOL have been almost as well documented as those of Amazon, Google and Nokia. The failures of heroes are particularly instructive, and I have rather unkindly concentrated on the unhappier episodes in the mostly illustrious histories of Apple, IBM and Sony.

Of the twelve organizations described in depth, five – Amazon, BSkyB, eBay, Google and Nokia – succeeded both in creating new markets and in establishing long-term leadership of an industry or sector. Three – Apple, IBM and Sony – were both winners and losers: each created several new markets and made themselves lasting leaders of most of them, but they also knew bitter defeats. Four – AOL, Encyclopædia Britannica, Netscape and Webvan – were ultimately losers. The stories of many other organizations are told more briefly.

Microsoft, one of the outstanding winners of the era, often through invading markets others had created, is not treated comprehensively, but plays an important supporting role in several of the company stories, and not always as the villain. There are few out-and-out villains here – none of these ventures was entirely without merit. And none of the heroes is flawless or immortal.

Most of the markets discussed took off during the 1990s, and with the benefit of hindsight we can make better judgements than were possible at the time. In a few, very recent, cases judgement needs to be particularly provisional. The stories of Google and the iPod thread in Apple's go up to about 2007, because it was in this decade that these markets took shape. The others concentrate mainly on earlier periods, where more perspective is possible.

The stories are not assessments of the current competitive positions of the companies profiled. Indeed, many of the winners are now mature, and inevitably have lost some of the qualities that made them great market creators. None of them is invulnerable to new kinds of competition or to the perils of rigid thinking, and we can expect several of them to lose leadership over the next few years. Dell appeared to have done so at the time of writing, and others have had setbacks. They will not be the only ones – that is the nature of creative destruction.

MARKET CREATION

1

PRODIGIOUS PARTNERS

Apple and Sony are the most talented market creators of modern times and their stories make an intriguing contrast. Like Google, which resembles them in other respects, each was founded by two talented partners, passionate about technology. Apple was inspired by Sony in its early days, but many years later became a formidable competitor to it. These stories concentrate on these companies' failures, but their achievements were truly heroic.

Apple

Woz and Jobs

Apple started with a friendship that became a business. The business took off like a rocket, but the friendship sadly died. The two young men who founded Apple Computer first met in 1972, when Stephen Wozniak, invariably known as Woz, was twenty, and Steven Jobs was sixteen. They had grown up in the area south of San Francisco that was to become known as Silicon Valley. Both were outsiders, with few friends or shared interests with their schoolmates and limited social skills. Woz had already dropped out of college and Jobs was to do so later.

They were fascinated by electronic circuitry, in Woz's case to the point of obsession. From an early age he had shown an extraordinary talent for designing gadgets, and a taste for juvenile practical jokes. One of their wheezes was a device for making free (i.e. illegal) phone calls, which Woz produced and Jobs sold.

Early in 1976, mainly to impress other members of the Homebrew club of amateur electronics enthusiasts, Woz designed and built the circuitry for what was to become the first Apple computer. It was a characteristically maverick solo achievement, based on a micro-processor whose chief merit was that Woz could obtain it for $20. Like all his designs it was stunningly simple and elegant, so it could be manufactured easily and cheaply, but economics were not upper-most in Woz's mind. He proudly gave away the schematics to other members of the club.

Steve Jobs turned out to be equally brilliant, but in very different ways. His genius was for superb design and for developing and sell-ing ideas, for getting others to share his extraordinary, egotistical vision, which some later called his 'reality distortion field'. Self-taught, he would become the most brilliant, intuitive marketer of his generation. His boundless self-belief and energy bulldozed the unworldly, reclusive Woz into turning his invention into a business.

Without each of them it would never have happened. Wozniak single-handedly designed the first two Apple computers, but it was Jobs's drive, vision and hustling that pushed the business forward. They both liked to attempt the seemingly impossible. Woz delighted in leaving everything to the last minute and then hurling himself into frenzied bursts of creative activity, going without sleep for days on end. Yet he produced astonishingly stylish designs with fewer parts than anyone else had previously imagined. Jobs revelled in chivvying others into pursuing his dreams of perfection, in never taking no for an answer, not least from his partner. It took months of pressure to convince Woz to commit himself to the business and even longer before he would give up his modest day job at Hewlett Packard. His family also had reservations about his going into partnership with this bumptious, manipulative and none too scrupulous youngster.

Jobs justified himself by winning their first big order. Their original idea was to make a hundred boards at $25 each and sell them to their fellow hobbyists for $50. Much to their amazement the Byte Shop offered them $25,000 for fifty machines. This was the spur that made it all suddenly much more than just a sideline. Jobs's problem then was to persuade sceptical suppliers to give them thirty days' credit for the necessary parts, and to organize 'manufacturing'. When they were thrown out of Woz's family house, they decamped to Jobs's parents' home. Running out of time and money, he got his pregnant sister and friends to help with the assembly and, working day and night, they managed to fulfil the order on time. When Jobs proudly took the first ones to the Byte Shop, he discovered that the proprietor had expected fully assembled computers, with keyboards and software, not just boards. Somehow Jobs managed to bamboozle him into making payment in full.

This early success enabled Jobs to hire more people, but he soon had to borrow more money, and the business existed largely hand to mouth. In their first year, 1977, they only sold 150 machines. While Woz's enhancements, like a colour display that he integrated into the microprocessor, were major innovations that paved the way for a whole PC on a single board, they did not do much for immediate sales.

Jobs realized that if Apple was to survive as a business, let alone advance, it needed a new model – and to be a serious, properly financed business. If their new machine, the Apple II, was to get beyond the tiny hobbyist market, it would have to be a professionally produced and marketed product. The machine itself would need to be completely self-contained, with its own operating system, power supply and keyboard, and housed in an attractive case. As he put it subsequently, 'The real jump of the Apple II was that it was a finished product. It was the first computer that you could buy that wasn't a kit… You didn't have to be a hardware hobbyist with the Apple II. That's what the Apple was all about.'

Meeting the grown-ups

At this stage, Jobs knew little about marketing, but he had a nose for who did. He wanted Regis McKenna's agency, which was handling a stylish advertising campaign for Intel, the creator of the microprocessor. McKenna was the top marketing strategist in the Valley and would normally not consider a tiny, start-up outfit as a client.

Undeterred, Jobs made dozens of calls until he finally got them a meeting with the great man, but when McKenna suggested that Woz avoid making a magazine article too technical, he told him he wasn't having any PR guy mess with his copy. Jobs managed to smooth things over, but it took a lot more phone calls to persuade McKenna to take the account.

What swung it was McKenna's conviction that the future of electronics lay in applications aimed at non-technical customers, not raw technology. He had campaigned hard within Intel for the potential of the microprocessor and a shift towards selling devices as products.

McKenna was to play a big part in Apple's early development, as was another experienced businessman he introduced to Woz and Jobs. Don Valentine was a tough venture capitalist, who had been an executive at leading semiconductor companies, and who would later fund Cisco, the maker of Internet infrastructure. Initially Valentine, like McKenna, was none too impressed by these two naive, arrogant, scruffy kids, now working out of a garage. They did not look remotely like the kind of management team VCs liked. As he put it to McKenna, 'Why did you send me these renegades from the human race?' They did not even know what a business plan was and had no strategy at all. But he felt they had something interesting and told them that he would consider investing if they brought in someone with marketing experience. After a week of Jobs making three or four calls a day, Valentine introduced him to a former colleague, Mike Markkula.

Markkula had been a marketing manager at Intel who had picked up several million dollars in founders' stock. He had retired, aged thirty-three, and was enjoying a relaxed, civilized lifestyle which he did not intend to change too much. He had, however, long been convinced that there was soon going to be a mass-market break-through with microprocessors. The Apple II, he decided, could be that breakthrough, and a fascinating sideline: not only would he mentor the two Steves, but he would guarantee Apple a loan, reducing the immediate need for seed capital. In return he would take a third of the equity. However, Markkula was not interested in running the company. He would be the chairman and advise on marketing. He persuaded another old friend from the semiconductor industry, Mike Scott, to become president and effective CEO.

Scott was a tough, hands-on manager of manufacturing operations, someone who made things happen. Jobs was always ambivalent towards him but knew that he was not yet ready for the top job. As Scott owned much less stock than the other three principals, the dynamics were always rather strange. Jobs, as heir apparent, felt free to get involved in any aspect of the business and nibbled constantly at Scott's authority. His main role was evangelizing and achieving distribution, which he did brilliantly, ensuring that the Apple II had what was then the best distribution system in the young PC industry. Tensions, however, were frequent.

Scott respected Jobs's vision and intellect: 'The great thing about Jobs was that you always understood where you stood with him. He never said what you wanted to hear. His positions were well thought out, and he always told you where he was coming from. It could be stressful, but the trick was not to take it personally.' However, it was also clear that 'he cannot run anything. He doesn't know how to manage people. After you get something started he causes lots of waves. He likes to fly around like a hummingbird at ninety miles an hour. He needs to be sat on.'

This was the view of most people who worked with Jobs – apart from those who simply adored or loathed him. Although Jobs's vision and charisma were inspirational, he was cordially disliked by many engineers for his arrogance, insensitivity and passing off of other people's ideas as his own. On the other hand, his intuition and business judgement were often excellent. Scott thought that the ninety-day warranty standard in the electronics industry would be good enough for personal computers. Jobs argued passionately that a year would be needed to build trust with customers. After bursting into tears, he finally won the argument.

Initially, though, Scott's biggest problem was Woz. The Apple II could sell by the thousand, but they were totally dependent on Woz to complete its designs. 'Woz was very, very creative, but it came in spurts. And he would cover himself during the in-between times. He'd never say that he wasn't making progress on a project. Instead, he'd say everything was coming along just fine, and meanwhile he would be waiting for the little light in his brain to switch on and save him.'

Woz was the idol of the technological buffs, particularly the many talented engineers who joined Apple, and created, virtually single-handed, the first two Apple computers. He was not just a brilliant designer of elegant circuitry – he also wrote a new programming language for the Apple II. Knowing nothing about disk drives, he designed a completely new approach to the interface: ignoring the work of dozens of IBM engineers on synchronicity, Woz dispensed with the problem by simply holding data in a cache. His overall design for the Apple II, with an unheard-of 62 chips on the board, was applauded by the entire industry as an engineering masterpiece. *California* magazine proclaimed him 'King of the Nerds'.

The techie view of Jobs, shared to some extent by Woz and by many commentators repelled by his often appalling behaviour, was that he was a parasite on the real inventors, his main contribution to the Apple II being its elegant design and casing. With the benefit of hindsight, it is clear that his role was infinitely more important. Jobs imprinted his personality on Apple and its products like a Hollywood producer, with monumental ego and screaming tantrums to match. He gave the company its stylish, hip, rebellious image. His passion for great design and his perfectionism can be seen in virtually every Apple product in which he had a hand. He was obsessed with how customers would see the product and feel about it. The engineers who did more of the supposedly real work resented his appropriation of their creations, but there seems little doubt that, in his irritating gadfly way, he inspired Apple to produce 'insanely great' products.

McKenna carefully burnished the image of Jobs as the personification of Apple and created a climate of approval in the press and among opinion-formers. Apple's audience became young professionals more interested in how easy the computers were to use than their technical wizardry. They loved Apple's coolness and identified with Jobs.

However, the person who actually built the organization, created the administrative, manufacturing and financial infrastructure, and ensured that products got out of the door was Mike Scott, who imposed a modicum of order on Apple's often chaotic culture. Valentine, never lavish in his praise, called it one of the best pieces of execution he'd ever witnessed.

The perils of stardom

Launched in 1977, the Apple II was the company's greatest commercial success until the arrival of the iPod more than twenty years later. By 1978 it had become the leading personal computer in a market until then made up mostly of hobbyists. In its first year Apple sold 8,800, in the next, 35,000, and by 1980, 70,000, taking Apple's revenues to $118 million. Eventually it sold more than 2 million and was Apple's cash cow for most of the 1980s.

Its take-off was considerably helped by the appearance in 1979 of VisiCalc, the first spreadsheet program, produced by a small software developer. This gave the first, tentative business customers a reason for buying one of these strange new toys. Spreadsheets were not just aids to preparing budgets and business plans, but dynamic models that could almost instantly show the effect of changes which previously would have taken hours of laborious calculations. VisiCalc was also the first serious personal computer software designed for non-engineers. There were soon hundreds of independent software developers producing programs to complement the Apple II.

Another organization was to have an even more decisive influence on Apple's future – and on how we all use computers. In 1979 Jobs visited Xerox's legendary research centre, PARC, and what he saw there 'blew his mind'. PARC had developed some concepts for personal computing – a 'graphical user interface', icons and the mouse – that we now all take for granted. These were far too strange for Xerox's management to pursue, but Jobs saw at once that they could transform how people used computers. Apple signed a licensing agreement with PARC and recruited several of its people.

The concepts were first incorporated into Apple's most ambitious new project, the Lisa, which Jobs championed personally. No fewer than three rival projects began in 1979, when Apple spent $7 million on research and development, bumping this up to $21 million in 1980, and $38 million in 1981. Jobs, however, had eyes only for Lisa, and loudly disparaged other teams, notably that of the II.

The Apple III was rushed out hurriedly in 1980. Intended as the successor to the Apple II, it was severely limited by the need to use the same Motorola processor and to be able to run all the software used on Apple II. There were big conflicts between marketing and engineering and major compromises on quality and reliability. It was Apple's first failure.

Fortunately, it attracted comparatively little public attention. The triumphant success of the Apple II and the enormous media interest in the young company led directly to an early stock market flotation. Valentine had previously organized some private placements of capital and sold his own stake for a large profit in August 1980. The Initial Private Offering (IPO) in December 1980 was the most oversubscribed in twenty years. In August the stock had been valued at $5.44, but on 12 December it opened at $22 and closed the day at $29. The company, which had only been incorporated three years earlier, was valued at more than a billion dollars. Over a hundred of the now 1,000 employees became millionaires overnight.

The excitement leading up to the IPO, and the euphoria after it, went to many people's heads and contributed directly to some of Apple's subsequent misfortunes. Its timing had dictated the rushed launch of the Apple III earlier in 1980. It was quickly followed by some disastrous decisions, and the loss of Woz and Scott.

Most of the new millionaires satisfied themselves with buying a

Mercedes or a Porsche, but Woz was one of those who decided on a plane. In February 1981, before he had obtained his full pilot's licence, he crashed it and was in a coma for a week. He did not come back to work for nearly three years and never produced another computer for Apple.

It was also clear by now that the Apple III had been a disaster and Scott decided that heads must roll. He was probably right to conclude that Apple had acquired too many people in its headlong expansion and had got into sloppy ways, but his handling of the purge grated badly. He first demanded eighty dismissals, including the head of engineering. His colleagues baulked at this but finally agreed on forty. Black Wednesday caused almost universal outrage – Apple was not supposed to be that kind of company and Scott became intensely unpopular. A month later, while he was away on a long weekend break, Markkula, Jobs and other senior managers took what was probably a worse decision – to fire Scott himself. Markkula took over as nominal CEO, but much of the power was now with the 25-year-old chairman. The children had revolted against the grown-ups and turned their backs on disciplined management.

This kind of coup was to become a pattern at Apple for the next sixteen years. Every subsequent leader, from Jobs himself in 1985 to Amelio in 1997, was ousted in similar fashion following plots by their closest colleagues. Most of the firings were justified, but they never produced better leaders. On each occasion Apple failed to find a truly professional CEO with a coherent strategy for the company and the ability to get it to pull together. Scott may not have been the man to lead Apple to the next stage but neither were any of his successors until the return of the older, wiser Jobs in 1997.

Many years later, McKenna acknowledged that the defenestration of Scott had been a terrible mistake:

> Mike Scott was a tough and demanding boss. But he was also intent on putting a systematic decision-making process into place at Apple. And Apple got rid of him. Looking back, he was Apple's last chance to institute some kind of order. After that, the culture

became so overwhelming that even the toughest manager would come in to shake things up – and instead find himself two months later lounging on a beanbag... The mistake everyone makes is assuming that Apple is a real company. But it is not. It never has been.

Markkula and Jobs, however, were so pumped up by the IPO and all the adulation the company was receiving that they took another terrible decision. They decided that all these third-party developers were parasites making money out of Apple's brilliant innovations. Instead of recognizing that the relationship was genuinely symbiotic, they decided to discourage the parasites. Apple would from now on try to produce everything it could itself, starting with its own spreadsheet program, and moving on to disk drives and keyboards.

Fortunately Apple was not consistent about this, and later actively recruited developers for the Mac, but it gradually lost much of the army of independent developers that had played such a big part in making the Apple II so attractive to customers and was one of its most valuable assets. (A much larger army would soon do the same for the PC.) Trying to do everything in-house added yet more to Apple's rising costs, since it could never be the most efficient producer of every component. Most dangerously of all, it reinforced the belief that nobody outside the company could do anything better than Apple.

Big Brother

For its first four years, Apple had faced little serious competition. All the other personal computer makers made machines that for non-technical professional users were not real alternatives to the Apple II. This brief and entirely atypical period fostered the dangerous illusion of invincibility.

The arrival of IBM in the PC market in 1981, with its enormous financial and marketing muscle, changed everything. Apple cheekily ran a full-page ad in the *Wall Street Journal* saying, 'Welcome IBM. Seriously.' Privately it scoffed at the ordinariness of the new PC. Bill Gates was visiting Apple when the PC was announced and remarked that 'they didn't seem to care. It took them a year to realize what had happened.'

It was IBM who opened up the corporate market. It sold 240,000 PCs in 1982, and in 1983, with the more powerful XT in its portfolio, 800,000. This meant it had displaced Apple as the leader, though in reality the two companies were addressing different markets. IBM had the best sales force, the greatest reputation, the closest relationships with its customers of any company in any industry, not just computing. It was at the peak of its powers – nobody then foresaw its abrupt decline a few years later. All that Apple had were some nice products and early adopter customers. Its zaniness and anti-corporate attitudes, while good for its image among the young at heart, positively repelled organizations dubious about the very idea of purchasing PCs at all. IBM represented security and reliability.

For Apple, IBM represented the bad old corporate world that it was going to topple. As its own sales topped a billion dollars in 1983 and profits reached $79 million, it was still riding on the crest of a wave. It had not the slightest doubt that its brilliantly designed products would prevail over dull mediocrity and conformism and that its main weapon against Big Brother would be the flagship product it had been working on since 1979, the Lisa.

The Lisa was undoubtedly an immensely innovative machine, the precursor to the powerful workstations that Silicon Graphics and Apollo were soon to produce – but for engineers, not businessmen. The problem with the Lisa was that it was like a Ferrari in a market of Model-T Fords. Computer buffs and aesthetes loved it, but for corporations it was too unusual and, at a minimum price of $10,000, far too expensive. It flopped badly and Apple's profits in the quarter to September 1983 slumped from $25 million to $5 million.

Fortunately it seemed that a new champion was at hand, and one that was now Jobs's own baby.

A dent in the universe

The Mac had initially been developed on the fringes of Apple. Its original creator was Jeff Raskin, another technical genius, whom even Woz acknowledged as a peer. Raskin had long dreamed of a computer as easy to use and as inexpensive as a Swiss Army penknife, a People's Computer. Sadly for Raskin, Jobs, who had previously disdained the Mac, became a convert after he had been pushed off the Lisa team. The Mac, he realized, could be the Apple II of the 1980s, a 'friendly' computer. He worked his way into the development team and soon worked Raskin out.

His vision for the Mac was more ambitious and owed not a little to his earlier dreams for the Lisa. The Mac 'would make a dent in the universe' and be acclaimed as a superb piece of design. However egotistically Jobs may have behaved, there is no doubt that it was his leadership that made the Mac the product it became, more powerful than Raskin had intended (though not enough) and incorporating the first mouse. His silly slogan, 'it's better to be a pirate than join the navy', struck a chord with most of the team and working on the Mac was the most intense, traumatic and exhilarating experience most of them ever had. They thought of themselves, with some justification, as artists. John Sculley remarked later, 'It was almost as if there were magnetic fields, some spiritual force, mesmerizing people. Their eyes were just dazed. Excitement showed on everyone's face. It was nearly a cult experience.'

The Macintosh was a truly outstanding creation, worthy of its appearance in the Museum of Modern Art. It was announced with one of the most arresting television commercials ever seen, made by Ridley Scott, the director of *Blade Runner*. Shown only once, in January 1984 in the middle of the Superbowl, it depicted a thinly disguised IBM as the soulless, totalitarian Big Brother of Orwell's novel, proclaiming 'the unification of thought' to the massed ranks of grey, sleepwalking prisoners. Suddenly a dashing young woman runs through their ranks, throws a hammer at the screen and smashes it. 'On January 24, Apple Computer will introduce Macintosh. 1984 won't be like *1984*.' Quite what this meant was not clear to everyone who saw the ad, but it certainly caught their attention, was replayed on other programmes, and seen by 43 million people.

Two days later at the official launch, Jobs added his own dramatic flourishes:

It is now 1984. It appears that IBM wants it all. Dealers, initially welcoming IBM with open arms, now fear an IBM-dominated and controlled future. They are increasingly turning back to Apple as the only force that can ensure their future freedom.

Will Big Blue dominate the entire computer industry, the entire information age? Was George Orwell right?

To massed shouts of 'No', he unveiled the Mac.

The Macintosh was completely different from any PC that most people had ever seen. It embodied ideas that visionaries like Douglas Engelbart had expressed in the 1960s for a machine 'for the augmentation of man's intellect', complete with the now familiar mouse and windows. The Mac was the first real product inspired by the prototypes that the engineers at Xerox's PARC centre had been developing for years. Its graphical operating system was breathtakingly intuitive and user-friendly. It was to be another eleven years before ordinary PCs had a desktop interface that was comparable, when Windows 95 finally arrived, incorporating many features from the Mac operating system. Everyone who played with the Mac fell in love with it – including Bill Gates.

Unfortunately, not nearly enough people bought it. Sales for the year were 250,000, only half of the target Jobs had boldly set, and a sixth of the number of PCs IBM sold in 1984. For the corporate market, the Mac was too expensive, initially not powerful enough, with woefully inadequate memory, no hard disk and a feeble software library. Most of these flaws had been pointed out before, but Jobs was adamant that it was ready. As his lieutenant, Bill Atkinson, acknowledged later, 'In our efforts to change the world we were a little arrogant and unwilling to listen.'

Over the next two years, the problems were gradually fixed and subsequent versions of the Mac attracted passionate devotion. However, the cult that developed around it pointed to a strategic weakness for a product aimed at knowledge workers: it was simply too different, too quirky to appeal to a mass market quickly. It was a great product for early adopter consumers who could afford it and for niches like desktop publishing and education, but not remotely appropriate for taking on IBM and winning corporate customers. It seemed to IT managers more like a toy than a real man's machine.

Apple took a long time to recognize both this and the fact that selling to businesses was about a lot more than having great products. By 1984 there were more than a hundred manufacturers of clone PCs, most of them slashing their prices every few months, as Moore's Law kicked in. The vast majority never made a profit and quickly disappeared. Soon, not only was IBM itself to be eclipsed by Compaq, Hitachi and the rest of the pack, but Apple was up against a vast new, low-cost industry. Its chief adversary, however, became the owner of the operating system on all these new machines, Microsoft. The ever-growing number of people and organizations using first DOS, then its successor, Windows, and the number of developers producing software for the new standard, represented a tide of network effects that became irresistible.

The fact that the Mac's operating system was vastly more user-friendly than DOS and Windows counted for little. It was becoming the Betamax of the computing industry, stuck with the discriminating early adopters but missing out not just on the corporate market but also on the emerging mass consumer market. DOS and then Windows were the dominant standard, just as VHS had become in videocassettes.

For Apple to penetrate corporate markets, it would have needed either to build new capabilities or to form a partnership with someone who had them. The alternative, and more realistic, strategy was to concentrate on markets where it had real competitive advantage. Instead it spent years fighting an unwinnable war, first against the most powerful corporation in the world, then against the most formidable strategist of the information age, Bill Gates, and most hopelessly of all against the overwhelming momentum that developed behind ever-cheaper PCs.

Nobody anticipated at the time how much the PC market and industry structure would evolve, but McKenna gave Apple some sensible advice:

> You need to make a decision. Whether you want to be a Sony or an IBM. You can't be both... This path goes towards systems, and systems are different things than stand-alone computers. They need heavy support, infrastructure. You can't just sell *things*. If you want to sell things, then go the other way. Be like Sony... But you've got to decide.

It would take Apple another sixteen years finally to decide what kind of company it would be, and indeed could be, but in the mid-1980s it wanted to have it all, to dominate both corporate and consumer markets.

The Dauphin dethroned

It was during this critical time that there was a protracted, traumatic regime change. Markkula and Jobs had known since they fired Scott that they needed a real CEO. Their first choice was Don Estridge, who had brilliantly championed the PC at IBM, but he turned them down. He had no desire to leave the company he regarded as the best in the world, least of all to work with Jobs. They then went to Heidrick & Struggles, the leading head-hunters, with a curious set of requirements: they wanted an experienced CEO with consumer marketing experience, who was 'interested in' high tech, would serve as a mentor to Jobs, adjust quickly to the Apple culture and act as a visionary in the industry. What self-respecting CEO would want to be a mentor to the heir apparent, particularly one as capricious as Jobs? How could someone without anything more than an *interest* in computers be a visionary? Conspicuous by their absence in the specification were the strategic ability to chart a course in constantly changing terrain, a thorough understanding of technology, or strong leadership qualities.

Apple unfortunately was to get exactly what it asked for, a marketing tactician with little knowledge or understanding of the new industry, but one rather taken with the idea of being a visionary. After several top executives had predictably declined Heidrick & Struggles's approaches, the one big name that remained in consideration was John Sculley, President of Pepsi. Sculley's experience was entirely in consumer marketing in a mature industry, where he had been very successful at climbing the corporate ladder and in winning market share from Coca-Cola. However, he had no experience of devising and launching genuinely new products or of selling to corporations, and knew next to nothing about computers. He was also distinctly uncharismatic.

Jobs brushed aside these objections:

> What we're doing has never been done before. We're trying to build a totally different kind of company, and we need really great people... My dream is that every person in the world will have their own Apple computer. To do that, we've got to be a great marketing company.

This was classic Jobs hot air, which he almost certainly believed at the time. Not only was the dream unattainable, but if Apple were going to attack the corporate market the marketing skills needed would be business-to-business. Sculley, however, as his autobiography was to reveal, believed himself to be a marketing genius. Jobs courted him fervently, closing the deal with the irresistible line, 'Do you want to spend the rest of your life selling sugared water, or do you want a chance to change the world.'

Sculley, having secured a financial package that made him the best-paid executive in Silicon Valley, accepted. At his first public appearance in 1983 he unctuously declared, 'If you can pick one reason why I came to Apple, it was to have the chance to work with Steve. I look on him as being one of the really important figures in our country in this century. And I have the chance to help him grow.'

Deliberately or not, Jobs had got himself a mentor he could dominate. For much of the next year he largely got his own way, and the two indulged in frequent displays of mutual admiration. Sculley himself was happy to employ the prevailing Apple rhetoric – 'leveraging Apple's critical mass' and 'selling a total systems environment' – without questioning too closely what any of it really meant.

Privately, however, he was dismayed by the 'many competitive fiefdoms' he discovered. When *Business Week* ran a story saying that IBM had won the PC race, he was struck by the fact that nobody in Apple took much notice. 'Apple people didn't really read business magazines. So absorbed in what they were doing at the company, many had no touch [*sic*] with the outside world.'

At an early stage he raised the legitimate question of whether Apple shouldn't consider bringing out a product that was compatible with the new IBM standard. This heresy was quickly overwhelmed by a torrent of counter-arguments and he did not have the technical knowledge to sustain the debate. Jobs ended it by declaring categorically that Apple was all about being different. 'Compatibility is the noose around creativity.' Sculley went back to being a mentor, though he seemed to be doing more of the learning than Jobs.

One useful thing that the CEO did manage to do in these tentative early months was to restore some of the battered morale of the Apple II team. Starved of resources and regularly berated by Jobs as bozos, they were bitterly angry. Sculley made himself the head of the division, increased the advertising budget, cut prices and offered dealers new incentives. In December 1983 the II had its best-ever month and saved the company from the disaster of the Lisa: 110,000 were sold, helping overall revenues reach $1.5 billion. The II was to continue to outsell the Mac for some years to come.

Sculley's contribution to the launch of the Mac was less fortunate. Raskin had originally had in mind a price of $995 for his People's Computer. Jobs and his team had decided their Mac should sell at $1,995. Sculley believed that new products should be priced high to skim the market and recover development costs. He insisted at the last minute that it should be $2,495. This made the Mac a premium product – one that everyone liked to play with, but few could justify buying.

As the scale of the sales challenge became clear, Jobs's behaviour became ever more erratic. He launched himself into a frenzied promotional campaign, flying around the country to demonstrate Macs to celebrities like Mick Jagger and Andy Warhol. In a further flight from reality, he devoted an enormous amount of time and more than $5 million to the production of what he declared would be the greatest annual report of all time, with endorsements from Ted Turner, Kurt Vonnegut and Stephen Sondheim. Apple's advertising budget, which had been $4 million in 1983, reached $100 million. Much of it was brilliant and stylish, but a disastrous TV ad showed businessmen as lemmings walking blindfold over the edge of a cliff. This insult to IT managers, the customers Apple needed to win over, was resented for a long time.

For some months, nobody dared to tell Jobs the real sales numbers, continuing to use estimates made in January. Senior executives who questioned the figures were removed. Yet by October, when it was clear that sales were massively short of target, Sculley meekly accepted Jobs's business plan that predicted a million Mac sales in 1985 – four times higher than those for 1984 – and a doubling of company revenues to $3 billion. As the real sales figures emerged, depression spread throughout the Mac team, and many of its key members left the company. Woz himself, furious at the constant disparagement of the Apple II team, resigned angrily in February 1985, declaring that 'Apple's direction has been horrendously wrong for years'. Sculley finally plucked up the courage to confront Jobs. In April, with financial disaster looming, the board gave Sculley the authority to act like a real CEO and decided that Jobs should be removed from the Mac division.

Jobs's reaction was to plan a coup against Sculley. This was not an entirely ludicrous idea. The CEO was clearly no more up to the task of leading Apple than Jobs, but he was a more astute politician. Presented with evidence of the plot, he immediately challenged Jobs. At a meeting of top executives, he forced them to choose between him and Jobs. Jobs pointed out what many suspected but dared not acknowledge, that Sculley was completely out of his depth.

Nevertheless, faced with the alternative of the devil they knew only too well, everyone, including Regis McKenna, voted for Sculley.

Jobs hung around for a few more months in brooding semi-exile. He then announced that he was leaving the company to start NeXT, taking with him several Apple people, and selling all but one of his shares in Apple. As NeXT aimed to produce what was effectively a top-of-the-range workstation for the education market, it would compete directly with Apple. This looked to all concerned like a final parting.

The Trojan niche

Sculley's immediate task was to deal with the financial crisis. This was due not just to the Mac's failure to penetrate business markets but to a general slump in demand for personal computers. In June 1985 he laid off 1,200 people, a fifth of Apple's workforce. Apple's direct sales force disappeared almost entirely. At the same time, Apple announced its first quarterly loss, of $17 million; but by the end of the year, with Apple's highest-ever quarterly earnings, the crisis appeared to be over,

Sculley's priorities, now that he was free at last from Apple's sacred monster, were to make the company an orderly, marketing-led operation and to maximize profitability. In this he had considerable success, ending the civil war between the Apple II and Mac divisions, supervising the roll-out of urgently needed extensions to the Mac range, and seeking every opportunity to improve margins:

> The heroic style – the lone cowboy on horseback – is not the
> figure we worship any more at Apple... Apple is a more
> disciplined, grown-up company today. People have grown
> to recognize that discipline is not a threat to innovation...
> I know what it takes to be a success.

Although not acknowledged explicitly, Apple was gradually but significantly modifying its strategy. In 1984 Jobs and Sculley had articulated a new mission statement: 'Produce high-quality, low-cost, easy-to-use products that incorporate high technology for the individual. We are proving that high technology does not have to be intimidating for non-computer experts.'

Implicit in this and in the winding down of the sales force was concentrating on consumer and professional markets, where it was making some headway. The Apple II had always done well among schools, but the Mac discovered a vital new market thanks to three innovative products, only one of which originated within Apple itself. The critical one was Pagemaker , the creation of Paul Brainerd. He had realized that PCs would one day be capable of doing the design and layout for newspapers that then required immensely expensive machines. When he discovered the Mac, he could see at once that it was ideally suited – it was infinitely easier to use than PCs, already had graphical capabilities and it was the only computer where the screen display was the same as what appeared on the printed page – What You See Is What You Get.

Pagemaker was complemented by LaserWriter, a printer that actually incorporated a computer, but one with more memory than the Mac and which ran much faster. Most people in Apple had thought that customers would baulk at paying $7,000 for a peripheral device, but Jobs had seen its potential. His really smart move though was to buy a stake in a company called Adobe, which had produced another stunning innovation that made use of LaserWriter. PostScript was a page description language for printers, developed by John Warner, an alumnus of Xerox PARC and the founder of Adobe. PostScript, among other things, made it easy to use many different typefaces and fonts. It transformed the way documents were produced and was the forerunner of the now ubiquitous Adobe Acrobat.

The combination of these three with the Mac produced the revolution known as desktop publishing, where Macintosh remains the standard platform today. Like the spreadsheet, Pagemaker's real value lay in its ability to model – suddenly it became incredibly easy to try out different layouts. What had once taken hours of work against tight deadlines could now be done in a flash. This would transform newspaper publishing and enabled virtually anybody to become a publisher on a modest scale.

Sculley could see that Pagemaker would be a brilliant way of differentiating the Mac, and Apple supported its launch. As publicity departments started to buy Macs to produce elegant documents in-house, other parts of the organization started to place orders – it became known as the Trojan niche. These sales were not enough to halt the ever-rising market share of DOS-based machines, but they did help to save the Macintosh, and hence Apple, as sales of the Apple II finally started to fall off. Customers who bought Macs to use Pagemaker were happy to pay a premium, but most others were not.

There was a steady roll-out of versions that rectified the deficiencies of the 1984 model. The more powerful Mac Plus came out in 1986 and the Mac II in 1987 was an even greater advance, described by its 28-year-old developer, Mike Dhuey, as 'a market-driven Mac, rather than what we wanted for ourselves'. By 1990 the Mac family included a low-cost model, the Classic, that realized Jeff Raskin's dream of a Swiss Army penknife computer, selling at $1,000, and in 1991 there was finally a laptop version, generally acknowledged to be the best on the market.

For three years sales grew at an annual rate of over 40 per cent from $1.9 billion in 1986 to $5.3 billion in 1989. It had taken five years to sell the first million Macs, but in the next five years, ten million would be sold. The product had clearly established itself, but in 1990 sales stopped growing and it became clear that the company was in big trouble.

Decline and drift

The immediate cause of the problems was pricing. For most customers the difference in cost between ever more affordable IBM clones and what had once been called the People's Computer, but was now downright expensive, was simply too great to justify. Pricing, however, was part of a wider strategic muddle. Sculley simply did not understand the dynamics of the industry and had no strategy for dealing with a constantly changing marketplace. If the personal computer market had been a stable, predictable one like soft drinks, his segmented approach aimed at maximizing margins would have made sense. But in a market that was in constant flux it became an artificial exercise that was nearly always out of date.

In the short term, facing no direct competition, Apple was able to raise prices and improve profitability without introducing new products. This made Sculley popular with the stock market and boosted the share price. Strategically, however, it accentuated Apple's marginalization from the main PC marketplace, as its overall share got smaller and smaller. In the eyes of Apple devotees he did something far worse – he squeezed the soul out of the company and made it dull.

Apple's top management team contained few serious, practical technologists: Sculley preferred to surround himself with marketing and finance professionals. However, he shuffled them around constantly, and anyone who began to look like an heir apparent quickly disappeared. Many of the most talented people, like Bill Campbell, who had made a real success of Claris, Apple's semi-autonomous software developer, departed in disgust. There was a major exodus of engineers to the emerging workstation industry, the successor to the Lisa. Apple itself had nothing that could compete with the products of Silicon Graphics and Sun, so that new market was closed to it.

Sculley had been caught by Regis McKenna's curse of the bean-bag. He was captivated by the idea of revolutionary new products and of himself as a technological visionary. His big idea, to which he devoted considerable energy, was a hazy concept called the Dynabook or Knowledge Navigator. This would be 'a discoverer of worlds, a tool as galvanizing as the printing press. Individuals could use it to drive through libraries, museums, databases or institutional archives... converting vast quantities of information into personalized and understandable knowledge.'

Such a device, of course, still does not exist, though the Internet now provides many of the elements. There was nothing wrong with the ideas, as daydreams, but they had little to contribute to the strategic challenges facing Apple – ferocious competition from IBM PC clones and steady marginalization. By 1990 Windows had incorporated many of the features that had made the Mac OS so distinctive. The Mac's share of the overall market was now less than 10 per cent, and was to fall much lower. By 1993 Microsoft had sold 10 million copies of Windows.

Sculley compounded his failings by talking publicly about his 'visions' as if they were something that might result in real Apple products. He even got George Lucas to help him make some documentaries about them. This generated good media coverage initially but little respect in Silicon Valley. When Sculley announced in 1990 that he was now Apple's Chief Technology Officer, he made himself a laughing stock. In 1993 he launched the Newton with an enormous build-up of publicity. This 'Personal Digital Assistant' sounded to many like a Dynabook, but its capabilities were so feeble that not only did it flop in the marketplace, it almost ruined Apple's reputation as an exciting innovator.

The Newton was the final nail in Sculley's coffin. After a slump in the share price, the board fired him.

Renaissance

It would be tedious to relate Apple's history over the next few dismal years under Sculley's successors, Mike Spindler and Gil Amelio, when sales were static at best, profits non-existent and morale rock-bottom. Although Amelio did a lot to clean up Apple's finances and initiate new product ideas, its market share was down to 3 per cent and in 1997 it lost $1.6 billion. To most Apple watchers and insiders it looked as if the end was nigh. As an act of apparent desperation in 1997 the board turned to the former *enfant terrible*, and invited the infuriating, but extraordinarily gifted, Mr Jobs to come back as an adviser.

In the twelve years he had been absent, Jobs had matured considerably and become even more creative. He had made an enormous success of Pixar, his computer animation company. With hits like *Toy Story*, it had made him a movie mogul and was later to be sold to Disney for $7.4 billion. His computer company, NeXT, had been a critical success but a business flop: Tim Berners-Lee had used a NeXT to develop the World Wide Web. However, before he signed on with Apple, Jobs showed that he had not changed all his spots. Apple desperately needed a new operating system, and NeXT had a good one. He persuaded Amelio to purchase not just the operating system but the whole company for what many observers thought a very generous $400 million.

Almost his first piece of advice to the Apple board was to fire Amelio and for the adviser to become the interim CEO. He then replaced the entire board of directors, but this time brought in some people whose advice he was prepared to listen to. He cut a pragmatic deal with Microsoft to support its browser in return for a commitment to developing new versions of Office for Mac and a $150 million investment. He continued the cost-cutting that Amelio had started and ended the licensing of the operating system to third parties. Being the only company with 'the whole widget', combining hardware and software, was what made Apple different in his eyes.

Jobs had two priorities: the first, ironically, was to reimpose some of the discipline that had been lost since Sculley's departure. There was a myriad of confusing, mainly mediocre, products. Jobs's mantra now was simplicity and focus and he ruthlessly discarded the 'crap products' and 'the bozos' who'd designed them. The second was 'giving Apple back its soul'. He quickly identified the truly talented among Apple's ranks of engineers and designers and urged them to press ahead with all speed on a new computer, aimed strictly at young-at-heart consumers – the iMac. It was hard to believe that this tear-shaped, transparent, brightly coloured marvel was comparable to a dull Dell box: its main function was fun. This time Jobs's many suggestions were mostly welcome and he did not seek to take most of the credit himself. A new advertising campaign reinforced Apple's radical image, with the theme 'Think Different' emblazoned on portraits of iconaclastic, Jobs-ean heroes from John Lennon to Picasso: 'The people crazy enough to think they can change the world are those who do.'

Jobs's stated intention was to put Apple back on its feet as quickly as possible and then get back to Pixar full-time, but the more he got into rescuing his first love, the more clearly he heard destiny calling. Destiny was the word he used in May 1998 when he unveiled both the startling new iMac and his strategy to the world. It was actually not that different from the 1984 mission statement, except that it was aimed primarily at discriminating consumers. What differentiated Apple would be not just making 'the whole widget' but being by far the most innovative company in the computer industry. Innovation and great design would be the key to its renaissance.

The iMac became a smash hit – 2 million were sold in the first year. It was followed in 1999 by a family of laptops that were generally acknowledged to be the best available. Powerful new desktop machines for designers also had a warm reception. In 2000 sales revenues rose a third to $8 billion and profits hit a record $786 million. To massed applause Jobs announced that he was dropping the 'i' from his 'iCEO' title. In 2001, the long-awaited OS X emerged, designed with new users in mind, and set a new benchmark for user-friendly operating systems. The first of a number of bold new Apple stores opened. These were designed to offer 'the best possible buying experience', partly by selling and encouraging visitors to play with all the goodies. Within three years they were generating revenues of more than $3 billion. The annual gatherings at MacWorld Expo trade shows became evangelical events once again, the faithful whipped into a frenzy by the prophet of cool.

Although Jobs had achieved an astonishing revival in Apple's image, self-confidence and share price, none of this was going to be enough to make Apple a reliably profitable company in the long term. Indeed, in the next financial year, when trading conditions were tough, sales slumped back to $5.4 billion, profits disappeared again, and there was to be precious little growth in sales over the next two years. The fundamental problem was that however stunning Apple's products were, they still had to compete with utilitarian PCs whose performance kept rising and prices dropping in remorseless synchronicity with Moore's Law. Apple could charge a premium, but it too had to keep cutting prices every few months or it would get out of line with the main market. The glory from being so innovative was gratifying, but Microsoft could get away with shamelessly and clumsily copying innovations like OS X, in Vista, its replacement for Windows, and make far more money from it than Apple ever could.

Apple needed to find more radical ways of being different.

A musical revolution

Jobs's main strategy for tempting customers away from Windows conformity was the Digital Hub: the best possible portfolio of consumer software for people who mainly wanted to have fun with their computers. The first item, and the one Jobs expected to be the biggest hit, was iMovie, a simple editing tool for users of camcorders. iPhoto was to prove another hit with the rapidly growing number of digital photographers. In 2000, work started on an easy way to organize digitally stored music, and in January 2001 iTunes was born. iTunes was a delightful program, but it lacked a satisfactory physical platform. Most if not all MP3 players were, in Jobs's words, crap. Their capacity was limited, they were difficult to load and use, distinctly un-cool, cheap plastic gadgets for geeks. Jobs decided that Apple could do very much better and the iPod was conceived.

The Apple team was fortunate in finding some excellent off-the-shelf technology, in particular a tiny Toshiba hard disk with enough capacity for a thousand songs, but what they produced was as different from any previous MP3 player as a swan from an ugly duckling. The man who made the iPod a masterpiece of minimalist design was Jony Ive, who had been responsible for the startling appearance of the iMac. He was egged on by a perfectionist CEO with his customary maniacal attention to detail, endlessly pressing for more simplicity, speed and elegance. It was at Jobs's insistence that the iPod had no specific on-off switch, but the really radical decision was to dispense with plus and minus keys for making choices, and use a wheel to whiz through selections. Jobs had long believed that the really important issue in design was deciding hat to leave out. According to Ive the task is 'to solve incredibly complex problems

and make their resolution appear inevitable and incredibly simple.'
The most complex problem on the iPod was navigating all that
content, which the wheel solved brilliantly. The atmosphere on the
project was reminiscent of that on the original Mac eighteen years
earlier, except that it was not so much revolutionary fervour that
drove them on as a strong desire to have one of these things themselves and a strong hunch that other people would too. Financial
analysts who got wind of the iPod thought it was a typically zany Jobs
distraction from the serious business of competing with Microsoft –
how could there be any money in this?

What emerged in October 2001, just in time for Christmas, was
the iconic Apple product for the twenty-first century – brilliant technology, incredibly easy to use and simply beautiful. As with the Mac,
people fell instantly in love with the iPod, but in this case they bought
it. It was mostly about being able to carry all your music with you in
one tiny device, but according to one survey 13 per cent of purchasers
did so primarily on aesthetic and sensual grounds. They just could-
n't stop touching that wheel and revelling in the sheer coolness of
the thing. Over the next twelve months 300,000 were sold, and in
2002–3 nearly a million, generating revenues of $345 million. This
was becoming much more than just a satellite to the Digital Hub –
it was a business in its own right.

One thing the iPod lacked was a way of loading music other than
just copying CDs into iTunes. It needed something as easy to use as
Amazon, but would people pay for music they could get free? Up to
now, online music had been a disaster for the record industry – the
young saw no stigma in sharing their songs, and were doing so on a
frightening scale on online networks like Napster. First there was the
even knottier problem of persuading the major labels to allow Apple
to create a legitimate online music store. The music industry's attitude to technological change to date had been a blanket refusal to
consider any accommodation with it, and to sue anyone or anything
that might pose a threat to its copyright. In 1998 it had tried, and
failed, to prevent the first manufacturer of MP3 players even from
selling such dangerous devices.

As a bona fide media player in his own right, and someone who understood both technology and the creative process, Jobs was well placed to convince them that allowing him to sell their songs online was more of an opportunity than a threat. To sweeten the pill, he allowed the companies to keep 66 of the 99 cents charged for each song – a much better margin than they got from conventional retailers. He just wanted the store to break even – its function was to feed demand for iPods. He reluctantly accepted restrictions on how often songs could be copied, using the 'protection' of Digital Rights Management, and only asked for a year's commitment initially. Jobs emphasized that the experiment would be confined to Mac users, the only ones currently using iTunes and iPods, and told the majors, 'If a virus gets out, it's only going to pollute five per cent of the garden.' (Actually an overstatement of Apple's market share.) They decided that Apple couldn't ruin the record business in a year and bit the bullet.

The iTunes Music Store opened in April 2003 with 200,000 songs initially. In its first week it made a million sales, more than had ever been legally downloaded before. In its first year it sold a hundred million and within three years a billion. The site worked particularly well because it was integrated with the software all Mac users had. Apple also had many years' experience of downloads through its popular QuickTime movie trailer service. But the key factor was that buying music in this way, instantly and by the song, rather than a whole album, was so convenient.

To make this, and iPod, a real mass market, iTunes needed to be opened up to the Windows barbarians, but it took some while for the iPod team to convince Jobs, as the original point of the Digital Hub had been to differentiate the Mac. Eventually he accepted that the scale of the opportunity justified some dilution. Persuading the music industry was easier – iTunes was the only new source of revenue they had at a time when record sales were collapsing. Releasing what Jobs mischievously described as 'the best Windows application ever written' opened the floodgates. In the last quarter of 2003, 730,000 iPods were sold. Then something extraordinary happened – instead of falling off in the quarter after Christmas, sales actually rose. In 2004 as a whole, 8 million iPods were sold, and in 2005, 38 million. By 2008 a total of 160 million had been sold.

This of course transformed Apple's finances. Sales revenues reached $8.3 billion in the year to September 2004, but net income of $276 million was still only 3 per cent of this. In fiscal 2005 sales hit $13.9 billion and profits $1.3 billion; by 2007 they were $24 billion and $3.5 billion respectively. The halo of the iPod and the immensely popular Apple stores were helping sales of computers, but it was one particular satellite of the Digital Hub that was doing all the pulling. By 2007, Apple's share of the PC market had doubled to 63 per cent.

What was most extraordinary about this was that Apple faced no serious competition. Other makers of MP3 players churned out machines that matched some iPod features but none came close in customer appeal. Even with a substantial price premium and a not exactly stellar reputation for durability, the iPod still had more than 60 per cent of the market in 2007. While the real competition at the low end of the market was coming from mobile phones with enough memory for music, Apple trumped this in 2007 with the iPhone, built on an iPod platform, and a much easier, distinctly cooler, way to use the mobile Web than anything Nokia had devised.

One explanation for Apple's dominance may be that most consumer electronics manufacturers just didn't get it – they did not understand what was so special about these little things, and, if they did, had no idea how to make something comparable. In 2005 Michael Dell, who seven years earlier had publicly advised Jobs to close Apple down and give the money back to shareholders, announced what it called the Dell DJ, smugly declaring that 'Style is nice but function and value are what matter to customers.' Apparently not – a year later the DJ was withdrawn from the market.

Apple had effectively created an entirely new market in which it alone had the capabilities to compete effectively. The only serious challenges were coming from new models it brought out itself. It stunned the market in 2005 with the tiny iPod Nano – only 0.27 inches thick, yet still with a capacity for 1,000 songs.

Nobody could accuse Jobs of resting on his laurels:

> Our revenues have doubled in the last two years, our stock price is high and our shareholders are happy. And a lot of people think it's really great, we've got a lot to lose, let's play it safe. That's the most dangerous thing we can do. We have got to get bolder, because we have world-class competitors now and we can't just stand still.

Insanely great

This turnaround would amount to a fabulous success story by any standards. For a company that had been on the ropes for years it seemed miraculous. Jobs could not take all the credit, and was happy to share it with Ive and other colleagues. His distinctive style of leadership and creativity, however, have clearly been crucial.

It is perhaps too soon to make an objective assessment of this renaissance and its durability. The digital music market is still very young, and it remains to be seen whether Apple can retain leadership – it still only makes money from the devices rather than the music. What is clear is that it is a different company from the one that was on an endless rollercoaster ride for its first twenty years, one rather more like the Sony that Regis McKenna had suggested should be its model. It is not so different, however. What has remained constant, even if somewhat stifled during the dark days, is a matchless ability to be creative and a compulsion to be different. Those qualities have sometimes represented terrible weaknesses. They have also been its greatest strengths.

Even without the brilliant successes of the iPod and iTunes, there are grounds for qualifying the conventional wisdom on Apple's 'failure' to hold on to leadership of the PC industry it played such a big part in creating. That industry grew massively and changed beyond recognition, and for most of the 1980s Apple grew with it. It lost market share and made dreadful mistakes, but no company has been able to dominate the PC industry in the way that IBM did the mainframe world, and no one company ever could have. The leadership of Microsoft, Intel and Dell is of a very different nature. Apple never had the capabilities to challenge them in the mass markets that emerged, but it did carve out its own premium segment.

Many have argued that if Apple had licensed its operating system software to other manufacturers, it could have mounted a more serious challenge to Microsoft. That may be true, but it is far from certain that the benefits to Apple would have outweighed the costs and the risks. Microsoft is a formidable adversary, especially when threatened, and could well have crushed Apple, which was in no position by the end of the 1980s to mount a serious challenge. The network effects and feedback loops that developed behind Windows were irresistible, just as they had been for VHS in the videocassette war of the 1970s.

But the Mac did not go the way of Betamax. In fact, of all the pioneers of the PC industry, Apple is the only one whose staying power compares with Microsoft's. Even when the company itself was badly led, it held on to that segment of the market it articulated in 1984: individuals who value 'high-quality, easy-to-use products'. Unlike nearly all of its rivals, most of whom faded away, it was never simply selling technology, but great products that incorporated brilliantly innovative hardware and software. It redefined on several occasions the concept of what a personal computer could be. Though it never really penetrated corporate markets, and probably never could have, it has been consistently good at attracting the most talented designers, visionaries and marketers, and has produced time after time products that amaze and delight consumers. Its 'niche' has never become commoditized, as most of the PC market has. It has the highest brand loyalty and rates of repurchase of any computer manufacturer. And in 2007 its supply-chain management was ranked second only to Nokia's, ahead of Cisco and Wal-Mart.

The crucial quality it lacked in its early years was discipline, but it was so gifted in other respects that it got away, just about, with breaking rules that others disregard at their peril. Scarcely any companies have the talent to model themselves on Apple, and in later chapters there are frequent references to its weaknesses, but there is little doubt that the world is the richer for Apple's existence. It has come as close as any business has to making a dent in the universe.

Sony

The founders

Japan in the years after the Second World War was even more devastated by crushing defeat than Germany. Nearly half of the population of Tokyo were homeless and 20 per cent suffered from tuberculosis. Many died in the streets from under-nourishment and hypothermia. It was against this dismal background that the 39 year-old Masara Ibuka founded Tokyo Tsushin Kogyo (Tokyo Telecommunications Research Institute), known as Totsuko. He had been an inspirational leader to other engineers since he had won the Gold Prize for invention at the Paris World Fair in 1933 and had gone on to lead the wireless communications arm of the Japan Measuring Instruments Company, which was diverted into military work during the war. As soon as the Emperor announced the stunning news of Japan's defeat, Ibuka decided to go to Tokyo to start anew. Seven of his colleagues followed him.

Totsuko had little money, not much of a business model and no product to speak of until 1950. It survived mainly by selling voltmeters and repairing radios that the military police had disabled during the war. Ibuka's sights, however, were set on achieving the highest levels of innovation, never imitating and creating products 'from the heart'. In January 1946 he wrote a ten-page 'Founding Prospectus', which is still proudly shown to visitors today. His goal was to create 'an ideal workplace, free, dynamic and joyous, where dedicated engineers will be able to realize their craft and skills at the highest level'. Profit was 'a secondary motive. Our service commitment should be pure and total.'

The young man who became his partner modified this vision only slightly when they turned Totsuko into a business. Akio Morita was the 26-year-old eldest son of a line of sake merchants that went back fifteen generations, and was expected to join the family business. However, he had been fascinated with gadgets, especially phonographs, since he was a boy. In the closing years of the war, as a naval lieutenant he had briefly worked with Ibuka, who was developing an amplifier that could detect submarines. As soon as he heard that Ibuka was forming a company, he knew that he must persuade his father to allow him to join. This was an exceptionally bold step for a Japanese son and heir to take, the first of many by Morita. But he won his father's consent and even persuaded him to invest in the fledgling business.

Morita became the business and marketing brains of the company, and the partnership proved an exceptionally harmonious one: nobody ever saw the two men quarrel. As important to them as innovation was the idea of the company as a family, where disagreements were healthy but loyalty paramount.

Several things marked out Totsuko and its founders for future greatness. They were exceptionally fast and thorough learners, quickly mastering new capabilities in a variety of technologies, in manufacturing processes and in marketing and brand building. They were repeatedly willing to stake everything on one big idea, even when they had virtually no previous experience of it. Many of these ideas came from Ibuka initially, but the vision was a shared one, and was always focused on eventual products rather than on technology as an end in itself. They were obsessed with making superb products and building a great company, pursuing their own path even in the face of considerable opposition, and attracting the most talented and idealistic engineers of their generation.

Overcoming seemingly impossible odds and defying conventional wisdom was to become their hallmark.

Building new capabilities

The first big new idea to capture Ibuka's imagination was the tape recorder. He only saw one for the first time in 1949 and decided immediately that Totsuko must build one. They had no model to work from and worked out virtually everything from first principles. He and Morita also decided that they should manufacture their own tapes, to generate a continuing revenue stream. Unfortunately the right kind of plastic was not available in Japan at the time, so they persuaded a paper manufacturer to produce especially smooth paper, strengthened with hemp, which they then coated by hand with magnetic paste. Considering obstacles such as these, it is remarkable that they produced a working product at all. The first model, however, weighed 75 pounds and was too expensive to attract bemused customers. Morita realized that they had to develop design and marketing skills to match their engineering expertise and make much smaller, lighter and cheaper models. Subsequent ones became good enough to compete internationally with the best German and American brands.

In 1952 Ibuka discovered transistors and lost most of his interest in tape recorders. This was an entirely different technology, recently invented at Bell Labs, the research arm of AT&T. When he requested funding for a licence, MITI, Japan's immensely powerful Ministry of Trade and Industry, laughed at him. How could a tiny company with no previous experience of electronics hope to master something this complex? Nonetheless, Morita persuaded Western Electric to grant Totsuko a licence, and for the next two years Ibuka's engineers pored over the two volumes of technical guidance issued by Bell Labs – their only source of information. Bell had advised Morita that the transistor would only be good enough for hearing aids and never produce the frequencies necessary for good sound quality. Ibuka, however, was determined. Not only did his team succeed, they took transistor radios to a level of quality and miniaturization unequalled by any competitor.

They launched their first radio in 1955 and, in 1957, the TR-63, a 'pocketable' model. Dealers had told them that there was no market, especially in the US, for such small products, but the transistor radio hit a wave of social change. A distinct new category of consumer had arrived, teenagers, who no longer sat next to a set in the living room, listening to the same music as their parents. The radio was gradually ceasing to be a stationary object, like the television set, and music started to become portable and personalized. The TR-63 was an instant hit in the Christmas season in the US, sold 1.5 million worldwide, and took company revenues to $2.5 million that year.

Morita set his sights seriously on international markets on an extended business trip in 1953. The scale of the US and the power of its economy daunted him somewhat. In Düsseldorf he was deeply depressed when a waiter pointed out that the tiny paper parasol on his ice-cream was from his country: were trinkets like this all that Japan stood for in the West? He took comfort when he moved on to Eindhoven, and to Philips. Here was an engineering company, with values similar to his and Ibuka's, that had established a well-earned international reputation. 'If Philips can do it, perhaps we too can succeed,' he wrote to Ibuka. The two companies were to go on to have a long collaborative relationship, with Ibuka and Morita generally providing the creative spark.

An essential step towards building a global company was to find a name that would work globally. The name 'Sony' was first applied to the radios and in 1959 to the company as a whole. Morita, who was passionate about music and sound quality, deliberately chose a word with a connection with sound ('sonus') that also had friendly, youthful connotations in the English language. It also had the advantage of not seeming obviously Japanese. However, an important part of what he and Ibuka were trying to do was to change the image of Japanese products.

The most important market was the US, but it needed to be penetrated in the right way. In 1955 Morita persuaded an Amercian company, Bulova, to take an order for 100,000 radios, but it insisted it would only market them under its own name. He told them, 'Fifty years from now, I promise you that our name will be just as famous as yours.' Sony's board thought this was too big an order to turn down and instructed Morita to accept Bulova's terms, but he defied them. For him the overriding goal was to establish the company name and reputation for premium products. No deal, however lucrative, should compromise this.

Inititally Sony worked through an American distributor but by 1960 was ready to defy conventional wisdom again and establish a direct subsidiary, Sony Corporation of America. The overriding goals were to build the brand and develop market understanding. Most Japanese businessmen in the US clung together, which Morita thought was like the blind leading the blind. He engaged with Americans and appointed savvy, experienced executives like Harvey Schein to build a strong marketing capability, though he had difficulty persuading them that short-term profits should not be the main criterion of success. He opened a showroom on Fifth Avenue and moved his family into a fancy apartment on the Upper East Side, where they lived for several years, and he networked energetically.

Trinitron

Sony started applying miniaturization techniques to television sets as early as the 1950s but was slow to develop a colour model. By 1961 this was a major weakness in the product range and dealers pressed for the company to obtain a product from another supplier. This was anathema to Ibuka – nor did he favour licensing RCA's technology, then the world standard. (RCA was at that time the leading American maker of televisions and audio equipment.) A Sony product, he insisted, must be innovative and distinctive. His first approach was to license a colour tube called Chromatron that had been developed for military purposes, and developing a Sony product based on this became his top priority. However, it proved impossible to mass-produce these reliably and the project turned into a disaster. Only 13,000 sets were ever sold, at a fraction of their cost, and by 1966 the company was in deep financial crisis.

Even Morita now favoured cutting their losses, but Ibuka was resolute. Despite the dire financial situation, he insisted on looking for an alternative. The company had to borrow ¥650 million ($2 million) to finance the project. One of their young engineers, Senri Miyoake, came up with the idea of using a single gun to produce three electronic beams, rather than three colour guns. This produced an exceptionally bright though blurred picture, but Ibuka was convinced that this approach would produce a reliably sharp picture and devoted every available resource and virtually all his own time to developing a single-gun tube. By 1968 he had a product and announced that 10,000 units would be ready by October. This was an almost impossible deadline – Miyoake requested an extension and was promptly taken off the team, but it was achieved, even if some sets had to be recalled subsequently.

The Trinitron 12-inch television set instantly established itself as by far the best product on the market, vastly superior to anything else available. To meet the demand, Sony established the first Japanese manufacturing facility in the US in 1972 when it built a TV assembly plant in San Diego. The Trinitron was Ibuka's proudest achievement and added to the company legend of never giving up. What it needed to complement it, Ibuka believed, was a video recorder.

Bitter defeat

Ibuka had been passionate about video recorders since the mid-1950s. Sony produced one of the first commercial products in 1963, when they were still enormous reel-to-reel machines and only afford-able by TV professionals. In 1971, Sony transformed that market with U-matic, a more compact approach using videocassettes, which quickly became the near-ubiquitous standard. It then modified U-matic technology to develop Betamax, a simpler videocassette recorder (VCR) for the mass consumer market. When it launched the first Betamax product in 1975 it looked like a sure winner. Its biggest problem looked like being the long legal battle started by Universal City Studios and Disney, who alleged that the VCR was 'a tool of piracy'. Sony was to win that battle in the Supreme Court in 1984, but by then it had lost the more important war. Its rival, Matsushita, and Matsushita's subsidiary, JVC, had comprehensively outmanoeuvred Sony and made VHS the dominant standard.

Yet Betamax had started with enormous advantages – it had Sony's now stellar reputation, its picture quality was superior, and it had a product on the market a year before its rival. VHS had rather more features than Betamax and initially had longer tapes that could record an entire movie. What made the big difference though was Matsushita's aggressive licensing strategy – it encouraged dozens of other manufacturers to produce VHS machines, all competing with its main subsidiary, JVC, and some of them very cheaply. It did its best to flood the market. Sony, as a premium brand, disdained this approach, though eventually felt obliged to follow it. There were soon many more VHS than Betamax recorders in homes. This in turn had a big impact on a crucial ancillary market, that for pre-recorded cassettes.

Contrary to what everybody had anticipated, consumers proved much less interested in recording television programmes to watch later than in using their machines as players for pre-recorded material. When Hollywood got over its intense suspicion of the new medium and started releasing movies on cassettes at affordable prices, it uncovered a major new source of revenue and stimulated the rapid growth of the VCR market. Within a few years sales of VCRs were exceeding those of television sets. And an ever-growing majority of them were on the VHS standard, which was enjoying economies of scale and scope, and charging consistently lower prices than Sony. In the new video rental stores that mushroomed in the late 1970s and early 1980s, there were always more titles available on VHS. The slight superiority of Betamax counted for much less in playing pre-recorded tapes and by 1983 VHS was outselling it by four to one. The following year Sony dropped out of the American market and in 1988 started to manufacture VHS machines itself. VHS remained the universal standard until DVD players started to replace VCRs in the 1990s.

This defeat created a financial crisis for Sony in the early 1980s. However, Alfred Chandler has argued that because Matsushita's victory had been due to its functional (marketing and distribution) capabilities, it did not give it major advantages in other markets and it did not dislodge Sony from leadership in commercializing technical advances in consumer electronics. That only made failure all the more galling.

Morita drew two lessons from the debacle. One was that Sony

should not try to set standards on its own again but should cooperate with its competitors. The other was that content had clinched it. If Sony had owned a library of films and TV programmes, instead of fighting with the Hollywood studios, the VCR battle would have turned out very differently. It was content that really drove demand, not just for VCRs but for most of Sony's products. If it was to avoid further defeats like this, it had to have a stake in the media industry. It had to be a player in content.

Ohga to the rescue

In 1968 Sony had taken what at the time seemed a radical step in this direction, when it formed a joint venture with CBS, the owner of the Columbia record label. Within two years, CBS/Sony Records was earning annual revenues of $100 million and quickly became the leading record company in Japan. It benefited from a rapidly growing Japanese music market that made it second in size only to the US. CBS/Sony quickly became the most profitable business of either of its parent companies, generating returns each year that greatly exceeded their original investment. Although the company sold large numbers of Columbia titles in Japan, it also developed Japanese talent and popular music and was run entirely by Japanese staff. The start-up business was headed by Norio Ohga, who was to become almost as heroic a figure in the Sony saga as the two founders.

Morita had gone to great lengths to persuade the young Ohga to join Sony and had long had him in mind as his likely successor. Ohga was a polymath – a shrewd businessman with a musician's sensibility and deep technical understanding. In the 1950s he had pursued a career as a classical singer, while advising Sony in his spare time on ways to improve the quality of its recording and audio products. When he eventually joined the company in 1959 at the age of thirty he rose rapidly. Within two years he was responsible both for the tape recorder division and for company-wide product planning, design and advertising.

He was also largely responsible for refining the look and design of Sony products. 'I'd been telling Mr Morita for years that what we needed to do was create products that looked smart, stylish, international, and start advertising stylishly... Mr Ibuka and Mr Morita were great visionaries and entrepreneurs, but they didn't necessarily have much sense of style.' Design, of course, became one of the keys to Sony's differentiation. Ohga probably did more than any other single person to make the Sony brand one of the most valuable in the world.

Ohga's success with the record business and his passion for music played a big part in Sony's next great technological and product breakthrough, the compact disc. He became managing director of Sony in 1972, while remaining chairman of CBS/Sony Records, and made himself the champion of digital recording within the company. In his view digital was 'like removing a heavy winter coat from the sound'. This was a revolutionary position to take in Sony, as most audio engineers, led by Ibuka, were utterly convinced of the superiority of analogue sound. Some of the development work had to be kept secret from Ibuka. The other-worldliness of his engineers sometimes exasperated Ohga as much as the prejudices of the analogue ideologues. In 1976 they presented him with a laser disc prototype that carried a 'platter' of thirteen hours of music, which he laughed at as technology for its own sake. 'I love technology and I love technical details, but inside Sony I was always first and foremost a businessman.'

In the late 1970s, fighting a bitter war in the VCR market, Sony desperately needed a great new product. The challenge of making compact discs a success was that they had to overcome the massive investments that both record companies and consumers had made in LPs, a standard that seemed to suit everybody. Consumers had spent a lot building their library of records, so they would need compelling reasons to switch. And the music industry resisted bitterly. Dislodging the LP would be much more difficult than opening up the market for VCRs.

In 1979 Sony started what proved to be an immensely successful programme of collaborative research with Philips on an optical CD. Ohga had himself laid the groundwork for this in 1966 with a cross-licensing deal on laser technology, and both companies had done a great deal of work on their own since then. This project, however, was an intensive joint product development which resulted in a specification for CDs known as the 'Red Book'. Ohga's personal contribution was resisting Philips's proposal that the disc should have a capacity of sixty minutes. In his eyes, this was 'unmusical': it would not be long enough for works such as Beethoven's Ninth Symphony or an entire act of an opera. They eventually agreed to make the capacity of a CD 700 MB, equivalent to a duration of seventy-two minutes.

The 'Red Book' was published in March 1980 and the Sony–Philips design was so clearly the best one proposed that it was adopted by most other manufacturers and quickly became the industry standard. Just about everybody in the music industry, however, including CBS, bitterly resisted this disruptive technology, just as the movie industry had execrated the videocassette. They were adamant that the LP platform was the mainstay of their business, in which they had made enormous investments, and that this threat must be resisted resolutely. Ohga tried to persuade a gathering of industry executives that the future was almost certainly going to be digital but was shouted down by massed chants of 'The truth is in the groove. The truth is in the groove.' Reasoned arguments were clearly hopeless. Ohga decided that if no-one else would produce music for the new platform, Sony would go it alone.

Sony itself was in tight straits in the early 1980s following the Betamax disaster, but the joint venture with CBS had created substantial reserves, and now was the time to reinvest them. Ohga had frustrated CBS in the past by refusing to distribute all the profits as dividends. He now convinced his partner that CBS/Sony should invest $100 million in factories in Shizuoka, Salzburg and Indiana to manufacture CDs, telling CBS that he was presenting it with a brand new business for nothing. CBS subsequently asked Sony to buy out its interest in these plants and also tried to persuade Ohga to postpone the launch to avoid upsetting the rest of the industry. Ohga accepted the first suggestion and ignored the second. In 1982 he proudly announced not just the first CD player but the first fifty CD titles, all produced by CBS/Sony Records.

Ohga felt that a significant proportion of the first customers would be his kind of music lovers, so the fifty CDs included a high proportion of classical pieces. This proved to be a shrewd move, not least because only serious audiophiles could afford the $700 price of the first player. These early adopters, however, were enthusiastic and generated considerable word-of-mouth recommendations. It gradually dawned on the music industry that if this new format succeeded not only could it charge more for a CD than an LP, but that many consumers would want to replace their entire libraries. By 1984, when Sony launched a simpler, cheaper player, there were thousands of music titles available, and sales took off. Since scarcely anyone else had invested in CD mastering plants, there was a serious shortage of capacity, and Sony's three factories were working flat out. The one in Indiana accounted for Sony Corporation of America's entire profit that year – CBS of course missed out on this revenue stream. By 1986 total CD sales in Japan reached 45 million and overtook those of LPs, and the same crossover took place in the US two years later.

The inspired hunch

Sony's other saviour in this period, and a totally unexpected one, was the result not of innovative engineering or of far-sighted product development, but of the indulgence of one of the company's founders and the sudden hunch of the other.

In February 1979 the 72-year-old Ibuka was about to make some long flights and asked if the company had something that would work as a portable music player. The tape recorder division was happy to oblige: in four days, they modified a product devised for journalists, the Pressman, taking out the recording mechanism and speaker and replacing them with a stereo amplifier and circuitry – plus of course headphones. Everyone was surprised at how good it sounded.

Ibuka was delighted and on his return showed it to Morita. After playing with it incessantly over the weekend, he decided that this should be a Sony product. He had seen the lengths young people went to in order to have music with them at all times, lugging radios around with them on hikes and at the beach. Sony should launch a real product – and do so by June, when students started their summer vacation.

Virtually everyone, including Ibuka and Ohga, was distinctly dubious, but Morita was convinced it would be a winner. He took personal responsibility, specifying just two modifications to the original version – twin outputs to allow couples to listen together and, at his wife's request, a fader button so that conversation would be possible.

There was no time for market testing and little for planning. Morita fixed the price at ¥33,000 ($125), to make it accessible to young consumers. Given that the Pressman sold at $400, the accountants calculated that the Walkman could only be profitable at production runs of 30,000 a month – twice the level of Sony's most popular tape recorder. Retailers were universally negative – why would anyone buy a tape recorder that did not record? Morita's response was that Sony was going to create a new 'headphone culture'. If they did not sell 30,000 a month, he would resign as chairman.

The Walkman was promoted as a fashion item and came out in July 1979 with ads showing young people cycling, skateboarding and exercising while listening on their headphones. For a month nothing much happened, then sales exploded. The first 30,000 went by mid-September, the next month they doubled, then tripled. Plans to launch in Europe and the US had to be postponed, as factories could not keep up with domestic demand. The Walkman became Sony's fastest-ever hit and its best-known product, with eighty different models. By 1998, 250 million had been sold.

It was in some respects not a typical Sony product, cobbled together quickly from bits and pieces, and innovative primarily in marketing terms. What was characteristically Sony about it was the boldness and panache with which Morita followed his hunch. The next time Sony based a strategy on the chairman's gut feeling the results were less happy.

Seriously seeking synergies

Sony, not surprisingly, does not like talking about its acquisition of Columbia Pictures in 1989, and the saga has a strong air of unreality even now. Fortunately we have two detailed and largely consistent accounts: John Nathan's 'personal history' of the company, based on lengthy interviews with all the key players apart from Morita, and *Hit and Run*, a racy description of Hollywood shenanigans by Nancy Griffin and Kim Masters. This vividly describes the farce the acquisition became, even by Tinseltown standards, and how Sony allowed itself to be duped by greedy opportunists.

The idea of seriously diversifying into music and movies emerged gradually from Morita's conviction that if only Sony had controlled content it would not have lost the Betamax–VHS war. This was reinforced by a big, but fuzzy, idea that took hold of much of the media, IT and communications industries at this time – 'convergence'. Clearly the fates of these industries were increasingly intertwined, but quite how they and their products would converge or, according to some scenarios, become indistinguishable, was never spelled out. On one thing, however, most pundits were agreed: in this multimedia world, 'content' was the El Dorado. The problem with this line of reasoning was that nobody was entirely clear what all the abstractions meant. 'Content', like 'convergence', could mean many different things. People in technology-driven industries tended to see it as an elixir that would magically enhance the appeal of their platforms and pipes. But there is an enormous difference between a library of old films and a studio producing new ones, and managing them requires completely different capabilities.

Sony's vision for combining software and hardware, as it liked to put it, was not exactly a clearly articulated strategy. There is little to suggest that the pros and cons of moving into show business were thoroughly evaluated before Sony committed itself. What fuelled the move more than any rational business considerations was Morita's dream of owning a movie studio, combined with Sony's conviction that it could overcome any obstacle, as it had done so many times in the past, and master anything it put its mind to. Vanity and personal ambitions played a big part in how the acquisition proceeded.

The principal actors in the drama on Sony's side were Ohga and Mickey Schulhof. Morita was involved from time to time but it was these two who cut the deals. Mickey Schulhof was now Sony's principal American lieutenant, and regarded by Ohga almost as an adopted son. The smart, streetwise wheeler-dealers Morita had recruited to build Sony Corporation of America had been a little too tough and blunt for Japanese sensibilities, but Schulhof was different. Cultivated, sophisticated and with a wide range of interests, he made himself part of the Sony family and won Ohga's complete trust. A physicist by training, he seemed able to grasp the essentials of technical and business issues as quickly as Morita and Ohga themselves and had established himself as the most versatile trouble-shooter in the company. However, he had never actually run a business – he was a strategist rather than a manager, and he had no experience at all of the media industry. Nobody, least of all Schulhof, felt this disqualified him from creating a media empire on Sony's behalf.

An important supporting role was played by Walter Yetnikoff, the colourful CEO of Columbia Records. 'The Führer of records', as he liked to call himself, led a rock-and-roll lifestyle, but had formed an unlikely friendship with the straight-laced Ohga in the early years of the joint venture. He certainly knew the music business well, and had close relationships with artists like Michael Jackson, Bruce Springsteen and Barbra Streisand. Yetnikoff was personally keen to break free of CBS and, like most of those who got involved, saw Sony as a once-in-a-lifetime opportunity to make his fortune and to take his career in new directions. Ohga's dream was to be a major player in the classical music world.

The first act opened in 1986 when Laurence Tisch, CBS's latest chairman, toyed with the possibility of selling off Columbia Records. Yetnikoff tipped off Schulhof, and Sony quickly decided to bid. There were compelling business arguments for this move – CBS/Sony Records had proved to have real strategic value for the company, as well as being a money-spinner; Sony had shown that it could run a record business; owning Columbia's libraries could help the forthcoming launches of DAT and minidiscs; no longer having CBS as a partner would be a big relief; and Ohga certainly did not want Columbia to fall into other hands. Schulhof, however, signalled such eagerness that Yetnikoff realized he could have extracted even more generous personal terms than the $50 million he had asked for himself and his close colleagues for delivering the management team and artists. Tisch drew similar conclusions. To everyone's intense disappointment, though, the CBS board overruled Tisch and decided against selling.

The following year Tisch contacted Schulhof again. The company might be for sale, but the price had gone up from $1.25 billion to $2 billion. Almost immediately, and against Yetnikoff's advice, Morita and Ohga authorized Schulhof to agree, but CBS deferred its decision. Then, on 7 October 1987, the stock market crashed, and CBS suddenly needed to sell. The asking price, however, remained $2 billion. Again Sony acquiesced without demur, and in February 1988 the deal was finalized. In fact the price, at five times earnings, turned out to be not unreasonable. Yetnikoff delivered on his promise to keep the management team intact and the major artists loyal, and the music business delivered good profits for the next few years.

However, Ohga was determined to make Sony 'the most important classical label by the end of the century' and to spend whatever it took to do so. He poached Gunther Breest from Deutsche Grammophon – an unusual step in the cosy classical world – and encouraged him to outbid all rivals to sign conductors like Claudio Abbado and Riccardo Muti. This spurred the previously frugal classical sector to increase A&R budgets recklessly. Peter Alward of EMI said later that Sony 'almost bankrupted the industry'.

This, however, was merely a rehearsal for a much bigger spending splurge.

Hollywood or bust

Almost immediately after the acquisition, Morita, Ohga and Schulhof decided that the next step in the synergy strategy should be to look for a movie studio. What kind of research they conducted or how they evaluated their options is not recorded. Certainly only the most cursory consideration was given to how Sony might manage such an acquisition and what it would take to do so well. Almost all the advice the company appears to have taken was from people who had a vested interest in Sony proceeding down this perilous path. Yetnikoff, the closest they had to a media figure, had ambitions, of which he made no secret, to emulate his friend Steve Ross of Time Warner and become the head of a giant entertainment empire. He now made the first of his fateful introductions.

Michael Ovitz was the head of the Creative Artists Agency, and one of the most powerful figures in Hollywood. Like Yetnikoff, Ovitz saw Sony's venture as a major opportunity for self-aggrandisement and happily set up a series of meetings with movers and shakers in the film industry. It soon became clear that only MGM and Columbia might be for sale. Morita was a big fan of MGM, but it had recently sold most of its library, the single most important reason for making an acquisition. So that just left Columbia, and in November 1988 Sony opened negotiations to purchase it.

Columbia Pictures, which had only an historical connection with the record label, had been going through a troubled period, with frequent changes of management and few recent hits. Most of its money now came from television productions at its Tri-Star studio, which also made movies. In the past it had collaborated frequently with independent producers and directors like David Lean, Otto Preminger and Sidney Lumet. This had enabled it to build up a large film library, with prestigious titles like *Lawrence of Arabia*, *On the Waterfront* and *Close Encounters of the Third Kind*, and an even bigger stock of TV programmes. Coca-Cola had acquired Columbia in 1982 but, like almost every other normal business that had invested in Hollywood, had not found the experience a happy one. It had sold off some of its shareholding and was interested in disposing of the rest. It is indicative of Sony's state of mind that the fact that Coca-Cola wanted to get out was seen not as a warning but as an opportunity.

The man who handled the negotiations on behalf of Columbia

shareholders was Howard Allen, whom Ohga subsequently described as the shrewdest businessman in America. Allen quickly
concluded that Sony was so hungry for the deal that he could extract
a very good price. He started by asking $35 for shares that had been
trading at between $7 and $17, and quadrupled his own holdings in
Columbia in anticipation of a large profit.

The Blackstone Group, who had handled the financial side of the
Columbia Records purchase, had little idea of how to value a movie
studio, so Sony hired Ovitz to advise Blackstone. For the month's
work that this took him, they agreed to pay a fee of $11 million.
Between them, Blackstone and Ovitz produced professional-looking
but meaningless financial projections which could be used to justify
an offer of around $20 a share. Neither adviser was exactly disinter-
ested – Ovitz had his eye on running the new studio and Blackstone
stood to earn big fees if the deal went through.

Negotiations dragged on for most of 1989, with Sony not apparently willing to go higher than $17 a share, and had to be suspended
for a month after Ohga had a heart attack. Ironically this precipitated
a hardening of Sony's resolve to go ahead at almost any price. In
August, at a meeting of the executive committee, Morita expressed
concern, in the light of Ohga's fragile health, about Sony entering a
business it did not know well. Ohga agreed with Chairman Akio and
a decision to abandon the acquisition was minuted. That evening at
dinner, however, Morita remarked on how sad he was to see his
dream disappear. Nobody commented, but the Emperor had spoken.
When the meeting reconvened the next day, Ohga announced that he
had reconsidered and proposed that they should go ahead. Morita
graciously concurred. They also concluded that in view of what they
saw as the uniqueness of the opportunity they should not haggle too
much over the price. The die was cast.

Everything now hinged on how they would manage this business
which they knew practically nothing about, and who would actually
do the managing. The first to put himself forward was Ovitz. His
demands, however, were so preposterous – as well as enormous
financial rewards he expected to have almost unfettered control –

that Sony quickly rejected them. They had little luck in identifying anyone more suitable who was actually available until Yetnikoff suggested his friends, the independent producers, Peter Guber and Jon Peters.

At first glance, Guber looked promising. He and his partner had a string of recent hits to their names, from *Batman* and *The Color Purple* to *Rain Man*, which had just won four Oscars. And Guber was charming and charismatic – he even expressed modest doubts as to whether he was quite up to the job. Schulhof quickly became convinced that he was and sent him off to meet Morita and Ohga. If they liked him too, Schulhof recommended that Sony make its big offer to Allen within the next two days. Morita asked him no questions about the business, and Ohga too was bedazzled. He told Nathan that 'his eyes sparkled and he really forgot time... I could feel that this man really loved film. What I bought was his passion... I felt that this really was Mr Movie Man.'

And so they took an enormous business decision on the basis largely of their personal impressions of a plausible stranger. They wanted to believe that Guber was the answer to their problem, and convinced themselves that he was. Schulhof took out only the most perfunctory of references. It was as if all normal ways of doing business no longer applied. Nothing, including the most elementary forms of due diligence, could be allowed to prevent the realization of the 'vision'.

Having secured, so he told himself, an executive team, Ohga authorized Schulhof to make a formal bid to Allen for Columbia at $27 a share – $10 higher than their previous tentative offer. Allen did not hesitate to accept. These were better terms than anyone had dreamed of a year earlier. Coca-Cola's exit from the movie business made its once problematical investment a highly profitable one.

For Sony the whole venture now depended on Guber and on his ability to make a success of managing Columbia. But there was a great deal more, and less, to Peter Guber than first appeared.

Preposterous partners

Only to star-struck outsiders did Peter Guber look like a heroic Hollywood chief executive. He was not even an executive but a smart Hollywood deal-maker, good at spotting and acquiring promising properties for movies and at marketing them once they had been made. His experience of production was limited and he had an active aversion to hands-on management. He and Peters were well known in Hollywood for seeking to get their names associated with as many movies as possible and their desire to make 'more money than anyone in the history of motion pictures'. In most cases they played little part in making the films in their portfolio, other than as packagers. They delegated script development and production to others, and mainly got to work on promoting films when they were completed. Steven Spielberg insisted on a clause in his contract that neither of them should ever visit the set while he was shooting *The Color Purple*. The idea of Guber and Peters as heads of a major studio was greeted in Hollywood with stupefaction.

They also had a history of ingratiating themselves with naive foreign technology businesses seeking to diversify into software. Polygram, a record company jointly owned by Philips and Siemens, and much admired by Sony, had branched out into other media ventures in the 1970s, mostly disastrously. In 1977 Guber persuaded Polygram to invest in his company and proceeded to spend its money liberally. They had hits like *Midnight Express* but ran up large losses, which did nothing to curb Guber's enthusiasm for expenditure on chartering yachts at the Cannes Film Festival and other essential items.

In 1980 Guber had formed a partnership with the even more extravagant Peters, best known as the former hairdresser and boyfriend of Barbra Streisand. Peters was a flamboyant wheeler-dealer who occasionally came up with a brilliant idea but was also prone to wild outbursts. One of the mildest of his excesses was charging his large grocery bills to Polygram as expenses, as he 'spent most of his time at home thinking about movies'. Polygram also paid them a producer's fee of $750,000 for every film made, so they had a strong incentive to make as many as possible, regardless of commercial or artistic merit. As Peters explained, 'It doesn't matter if the movies make money. We make a fortune.'

After running up losses of $220 million on its various media ventures in the US, Polygram decided to call a halt in 1982. It took another three years to return to profitability, which might have been expected to give pause for thought to other companies seeking synergy with software. Polygram executives talked freely to the authors of *Hit and Run* about their resentment of Guber and Peters and would almost certainly have done so with Sony, if they had been asked. Schulhof, however, was not looking for reasons why proceeding might not be advisable.

What enthralled Ohga and Schulhof was the run of success the two partners had enjoyed with Warner Bros., after Polygram's withdrawal. They had had an enormous hit with *Batman*, and several other films did well. Their other business dealings had been less illustrious and their own company, Guber–Peters Entertainment, was a troubled one. Guber hated the responsibilities of being chairman and longed to be free of his troublesome fellow shareholders. They had targeted Sony as a possible purchaser before Yetnikoff approached them – everyone in Hollywood had identified Sony as an easy mark. Making that sale was Guber's main goal when the negotiations with Sony started.

Given his distaste for management and for the nitty-gritty of film-making, Guber must have known that he was not cut out for the job of managing Columbia Pictures, and that Peters would be a major liability. Indeed Peters was kept firmly in the background throughout the discussions for fear of alarming Sony. But the enthusiastic attention he got, first from Yetnikoff, then Schulberg and Morita, and finally from Ohga when he went to Japan, was immensely flattering. More importantly, this represented the best opportunity he had ever had to make the enormous financial killing he had long dreamed of. Guber's seduction of Sony's management was his finest hour as a salesman. He struck exactly the right note with each of them – visionary, charismatic, but slightly hesitant. He cultivated a friendship with Schulhof that lasted until the business unravelled.

Once Ohga and Schulhof had convinced themselves that Guber was their man, every obstacle only reinforced their conviction that he was now indispensable to the whole deal. They agreed to buy Guber–Peters Entertainment in parallel with acquiring Columbia. The company had just lost $19 million on revenues of $23.7 million and its main asset was an exclusive contract with Warner Bros., which the acquisition would soon make worthless. Yet Sony agreed to pay a total of $193 million for it, and accepted $30 million of debt. It took on trust Guber's assurances that Warners would release them from their obligation. Each partner would receive a salary of $2.75 million, plus large profit shares, and would share a virtually guaranteed bonus pool of $50 million. This might conceivably have been justified if Sony had been buying a proven management team of outstanding ability, but nothing could have been further from the case.

When the overall deal was publicly announced, it became apparent that Warner Bros., despite Guber's previous assurances, would not release the partners from their exclusive contract. Time Warner, too, saw an opportunity to relieve Sony of some of its cash and sued it for a billion dollars. Against the advice of its lawyers, Sony agreed to indemnify Guber and Peters, and in a final bizarre twist paid $400 million to agree terms with Warners.

And so Sony became committed to paying close to $6 billion to realize Chairman Akio's dream: $3.2 billion for the shares in Columbia plus $1.6 billion of debt; $223 million for Guber–Peters Entertainment; $400 million to settle with Warner Bros, plus assorted fees and expenses. Even with the yen high against the dollar this was a fabulous amount. Apart from the library, which was not worth a fraction of this, it was not clear what exactly Sony had got for its $6 billion, other than a licence to spend a lot more money. Columbia at the time was practically moribund, with scarcely any projects in the pipeline. And Guber and Peters, it soon emerged, were totally unqualified to run a major studio, let alone rescue one.

Debacle

The debacle that ensued was all too predictable. Sony's management involvement was minimal other than writing cheques. Ohga wanted a hands-off approach, partly in reaction to the media furore that greeted the news that another great American institution was falling victim to the rapacious Japanese. Only in Hollywood was it clear that if anyone was being raped it was Sony. Ohga blithely told the *New York Times*, 'If we act like the American occupying army that controlled post-war Japan, we will be bashed, but if we keep it totally as an American company, everything will work out fine.'

Guber and Peters's brief, when they started work in November 1989, was to 'jump-start' the studio to build market share. No expense was spared. One of their top priorities was to order a grandiose rebuilding of the Culver City studio Sony had acquired as part of the deal with Warners, at a cost of $110 million and much of the new chairmen's attention. While waiting for it to be completed they moved into a temporary penthouse, which they first fitted out with a Japanese motif – pools of water, bamboo blinds and mud walls – for a mere $250,000.

Since word was out that money was no object, Peters and Guber had no difficulty attracting movie projects. They bid recklessly high to buy unproven properties, most of which never turned into films. Several independent producers, like Danny De Vito, were enticed to come and work on the new lot, but many of them complained of a lack of follow-up. Guber and Peters were more interested in chasing the next deal or futuristic project to devote much time to nurturing talent and developing a steady flow of new material. Some modestly successful films were made, like *Prince of Tides* and *Hook*, but there were many more flops. The criteria applied to green-lighting projects were confused, and Guber frequently washed his hands of decisions, showing more interest in Sony theme parks and acquiring football teams.

So many senior executives were hired that Columbia became known as 'the elephants' graveyard'. Wives and girlfriends were not neglected – there was an expensively fitted office for each of them and funding for their personal projects. Guber's chief delight, however, was the fleet of corporate jets, which he used intensively to travel to meet Schulhof in New York and between his various homes and yachts. Peters also made sure they were kept busy, on one occasion sending a plane filled with flowers to a Swedish model on the East Coast.

Guber's top priority was managing the relationship with Schulhof, and keeping him happy with frequent star-studded parties and joint family vacations. He changed the company name to Sony Pictures Entertainment in 1991, in order to consolidate the parent's commitment. He insisted that nobody else should communicate with Schulhof and completely marginalized Yetnikoff. As Schulhof achieved something similar in his relationship with Ohga, the chain of command was a curious one. All three of them were hands-off, each of them thousands of miles away from each other, and all hopelessly out of their depth.

Peters lasted less than eighteen months. His behaviour became too outrageous even for his partner, but characteristically Guber asked Schulhof to do the actual firing. Peters could console himself with a pay-off of $20 million and a contract as an independent producer.

Sony Pictures did achieve a market share of 20 per cent by 1992, but at enormous cost. Its overhead of $300 million a year was significantly higher than that of all the other studios and another $500 million was needed just to service debt. The average cost of its movies, $30 million, was much higher too, but their average grosses were the lowest – $24.8 million for Columbia and $22 million for Tri-Star. The losses were publicly masked by profits from the music division.

This mess was soon compounded by even more serious problems. Sony's sales in consumer electronics hit a plateau in the early 1990s, just when it had to write off its large investment in high definition television. In 1993 the company made its first-ever loss. To cap it all, in November 1993 Morita suffered a stroke, from which he never fully recovered, and found himself briefly in the same hospital as

Ibuka, who had had a heart attack. The two old partners sat hand in hand for hours on end, while the company they had created came close to falling apart.

To stem the losses, Schulhof was instructed to look around for outside investors in the movie business. Despite the perceived value of a studio at a time when convergence hype was at its peak, nobody was prepared to put up any money. Part of the problem was that Sony did not own a television network. A bigger one was that nobody who understood anything about the business would consider investing in it with Guber at the helm.

For Sony Pictures Entertainment, 1993 turned out to be the worst year so far. After the massive flop of *Hudson Hawk* came that of *Last Action Hero*, one of the most expensive movies ever made. In 1994 Schulhof finally fired Guber, to the latter's evident relief, while announcing that he looked forward to investing in Guber's future ventures. Either Schulhof had learned nothing from his experience or he had acquired Hollywood depths of insincerity. Market share was now back to 9.4 per cent, the level it had been at when Sony had taken over the studio.

In November, the company announced one of the biggest write-offs in corporate history, $3.2 billion, equal to the amount it had paid for the equity in Columbia. This was as close as Sony came to acknowledging that the whole venture had been a disaster. In 1996, Ohga's replacement as chief executive, Idei, fired Schulhof. All the main characters in the drama had seen their careers end ignominiously.

Where did the synergies go?

We will not take the Sony story any further. It did pull itself back from the abyss, and found new and competent management to run the movie business, but it lacked Morita's brilliant intuition about new products and created few new markets. Sony's minidisc in the nineties was not a success, and the company did not stamp its mark on the new DVD standard. It did, however, triumph in the market for games consoles Nintendo had created, launching PlayStation in

1994 and going on to make itself the market leader, selling 100 million units over the next ten years. It had a modest success with its Vaio range of portable computers, particularly in Japan, though this was innovative only in terms of design. Overall sales revenues, which had languished at or below ¥4 billion from 1990 to 1994, bounced up to 6 billion by 1997, but the consumer electronics side of the business was soon in trouble again, and the brand has never quite shone with the radiance it once had.

How does the synergy strategy look today with the benefit of hindsight? Has it, or could it yet, yield long-term benefits? Might it have made sense if it had been better executed? The short answer to the last question is that the reason it was executed so badly is that Sony did not know what it was doing. And that is why it was such a misconceived strategy. It did not have the capabilities even to be competent at managing a movie studio, let alone derive synergy from doing so.

Sony certainly had good reasons for seeking a different strategy in the 1980s. Its core consumer electronics business was becoming very much more competitive, in some respects commoditized, and this trend has intensified since. The days are long gone when it could expect to reap decades of competitive advantage and high margins from its innovative products. An ever-growing host of rivals were baying at its heels, offering 'good enough' substitutes at lower prices. While content undoubtedly drove demand for Sony's platforms, it did so indirectly. The blind hostility of the media industry to technological change did pose a serious problem, and deploying its own content had helped Sony to win some wider battles, notably in the launch of the CD.

However, this strategy, which Sony has stubbornly pursued ever since, has yielded precious little synergy so far and created a fresh set of problems, which may make resolving those of its core business even more difficult. Having a foot in so many camps creates conflicts of interest. When Sony joined the rest of the music industry in viscerally opposing Internet file sharing and downloading, it lost its leadership in the market for personalized, portable music players. Refusing for many years to develop an MP3 player itself, it has seen Apple usurp the Walkman's throne with the iPod, a disruptive product worthy of Sony at its best.

Instead of being a potential partner to the entertainment industry, Sony has become a rival to other media companies. In some respects it has less influence over them than if it were neutral, especially now that they are more realistic about adapting to, rather than bitterly resisting, technological change. Its presence as a media player did not help it to establish the minidisc or its DVD videodisc in the 1990s and may have been a hindrance.

It is virtually impossible for Sony's top management to have the same understanding of the media business as of consumer electronics. The capabilities and markets are completely different. Two of its biggest problems now are overstretch and coherence. Few other organizations in recent times have attempted to span so many different activities. Sony's most recent response to the challenges of running such an extended enterprise was to make a distinguished media executive, Howard Springer, its overall boss in 2005. It remains to be seen how well he can cope with the challenges facing the core consumer electronics business and with making such an unwieldy group cohesive.

From the perspective of new and changing markets, one of the most important lessons from Sony's story is the danger of taking abstractions like convergence too literally. Up to a point this is a useful metaphor for the ways in which telecommunications, television, entertainment and technology markets interact and overlap, but only up to a point. Likewise, using words like 'software' to talk about different kinds of 'content' invites even more conflation of different ideas, and distracts attention from the new products that real customers might want and the capabilities needed to devise and produce them. It is ironic that Sony started to try to unite software and hardware in the 1980s, just when most of the computer industry had divided sharply on these lines.

Consumers may use an ever-wider range of platforms to receive content and communicate, but they have no desire to watch or listen to Sony-produced entertainment just because they are using Sony devices. They want the widest possible choice of artists, movies and music. And the 'converging' industries themselves remain fairly distinct. The creativity behind a Walkman or mobile phone that is also a music player is completely different to that which conceives and produces entertainment. Likewise the cable TV industry has something in common with telephony, but almost nothing with television production.

Convergence does present opportunities for versatile, entrepreneurial companies to carve out new areas of competitive advantage, as we shall see in the BSkyB story. Sony's capabilities complement those of Ericsson, which has made their partnership in mobile phones a success. Its strategy for PlayStation 3, a hub for all kinds of home entertainment, could be the route to making its high definition approach to DVDs the standard.

All these ventures are built on what really counts in these markets – capabilities. No company, however gifted, can be good at everything. Sony was as talented a company as there has ever been, but repeated success against the odds led it to attempt the impossible. However, one should never underestimate its ability, like that of Apple, to bounce back and to dazzle us with fresh innovations.

2

CAPABILITIES AND VISION

Always be the best and maintain your superiority over others.
Homer, *The Iliad*

It is the customary fate of new truths to begin as heresies.
Thomas Huxley, *The Coming of Age of the Origin of Species*

Two of the most striking things about Apple and Sony are how radically different their products were and how consistently innovative they have been. Innovation sometimes appears to be the result of a sudden flash of inspiration, like Morita's idea of making a product of the Walkman, but it invariably builds on knowledge and abilities developed over a long period.

Few companies are able to build a lasting success on the basis of a single innovation – typically these are imitated or improved upon by others. Most technology pioneers never reap the rewards of their inventions – they are quickly bypassed by others with more capabilities and assets. In industries where product innovation is critical, the winners are those who can produce a steady flow of them. This requires rather more than a few brilliant flashes.

Apple and Sony at their best were organizations optimized for innovation. They did not just depend on the genius of a Wozniak or an Ibuka, vital though they were. Innovativeness and engineering excellence became organizational capabilities. They were engrained in the DNA of these companies.

Organizational capabilities

Capabilities, in particular what business gurus call distinctive capabilities (or core competencies – there are lots of terms for what is essentially the same idea), are the most important and lasting source of competitive advantage for most firms. They are what the organization does better than its competitors, what differentiates it from them. They are part of the architecture of the firm, part of its culture. They are more than just the talents of individuals, they are what the firm does so well that it is almost second nature. In cases like Sony and Apple they define the company.

The Apple II redefined the market for personal computers. Nobody else at the time could produce anything remotely as good for those early adopter customers. Apple had assembled such an array of talent, so dedicated to developing insanely great products, that for a while no competitor could touch it. One of its distinctive capabilities, which never entirely deserted it, was the ability to innovate, to produce yet more beautifully designed, easy-to-use products. Even during the darkest days, it had enormous competitive advantage in markets like design, publishing and education, and among its dedicated following of idiosyncratic individuals, prepared to pay a premium for great products. Since its renaissance it has been universally acclaimed as the most innovative company in the world.

Innovation, engineering excellence and originality were key principles in Ibuka's founding charter for the organization that became Sony. He had been pursuing these goals most of his life and the founders never deviated from them. Although Morita became, like Jobs, a brilliant self-taught marketer, great engineering was in his bloodstream almost as much as his partner's. Like Apple, for which it was long a model, Sony achieved consistent excellence in innovative engineering, in design and in marketing. It had an unequalled capacity to acquire new capabilities and to master one technology after another.

An organization that understood the importance of capabilities for competitive advantage long before the invention of the modern corporation was the General Staff of the German army. For most of the nineteenth century and up until 1945, the Prussian, then German, army was the most consistently effective military force in the world. In the Franco-Prussian War of 1870 it swept aside the armies of Napoleon III. In almost every month of the First World War, German soldiers, man for man, killed or captured more French, British and Americans than their opponents. Even in 1944, when Germany was clearly losing the Second World War, facing overwhelmingly greater Allied forces, they consistently out-planned, out-manoeuvred and out-fought the Allies. Trevor Dupuy has calculated that 100 German troops were the equivalent of 130 British or Americans.

This superiority has been popularly attributed to a militaristic mentality or to Nazi fanaticism, but even if there were some truth in this, it could not account for such a persistent disparity. Much more plausible are the analyses of those historians who have shown that the German army was simply better organized, better trained and better led. Its officers were competitively selected, and if they did not perform were quickly removed. Contrary to popular legend and unlike the rank-conscious British army of 1914, junior German officers were trained to use their initiative in the chaos of battle, not to await orders and obey them blindly. According to Dupuy, the army was successful for so long because it had 'institutionalized military excellence'.

That is very close to the achievement of outstanding companies like GE, Shell, HP and IBM, who have performed consistently better than their rivals over decades. The excellence they have institutionalized is in management, technology, logistics and the areas where they have strong specialist capabilities. They have also, of course, developed other sources of competitive advantage, notably strategic assets like strong brands and binding relationships with their customers and suppliers. Ultimately, however, it is their organizational capabilities that have been their greatest strength.

Capabilities in new markets

In new markets, capabilities are the single most important factor in success or failure. It is capabilities more than anything else that enable a company to make a new market its own.

Every firm that succeeded in creating a new market started with at least one unusual capability that gave it a unique edge and generally developed more. Between 1999 and 2001 Google redefined Internet search, because its radically different approach was simply so much better than anybody else's. Southwest worked out a way, which eluded conventional airlines, to keep its aircraft in the air for twelve hours a day and full of paying customers. It was this more than anything else that enabled it, and its imitators like Ryanair, to create an enormous new market for cheap air travel. Amazon's rapid early growth was primarily due to the careful design and meticulous professionalism of its overall operation. It developed highly distinctive capabilities – in user-friendly software, in direct marketing, in customer service, in the logistics of storage and delivery, in the presentation of information about books and other products and in the processing of payments. Everything from the ordering process to the delivery was honed endlessly to maximize customer satisfaction. It was years before competitors like Barnes & Noble could catch up.

Companies who rely for competitive advantage on a single innovation are vulnerable to attack from others who may have other capabilities and assets. This is why it is so important for capabilities to be truly distinctive. Those that are easy to acquire attract many suppliers and the markets quickly become crowded – it is not difficult to learn how to become a fast-food outlet, a bicycle courier or a minicab driver. For most manufacturers of consumer electronics it was easy enough to produce an MP3 player, and there were soon dozens on the market (though only one iPod).

If, however, the capabilities are complex, entering a new market is much more of a challenge. It is even more so if the would-be supplier does not understand what the challenges are.

Go-getting universities

In the late 1990s many American universities and business schools formed partnerships with entrepreneurs who persuaded them that the Internet would change education fundamentally and represented an enormous opportunity to market their courses more widely.

The management guru Peter Drucker believed that 'higher education is in deep crisis... Already we are beginning to deliver more lectures and classes off campus via satellite or two-way video at a fraction of the cost. The college won't survive as a residential institution.' Merrill Lynch, then an investment bank, declared in 2000 that 'the digitization of education has made the university ripe for the kind of rationalization that took place in the health industry in the 1990s.'

New York University started NYU Online as a wholly owned profit-making corporation offering corporate education and training. Columbia University invested $25 million in 2000 in a wholly owned company called Fathom to offer a variety of humanities and other courses online in collaboration with other institutions. Western Governors University was set up in 1996 with a big fanfare as a broker of courses developed and delivered by a mixture of universities and private enterprises. It was going to provide a new kind of 'competency-based' education. Pensare was a venture capital-funded 'aggregator' of business school content into online courses aimed at large companies.

The most ambitious, best funded and best connected of the Internet start-ups was UNext Inc, which founded Cardean University. It raised over $100 million in funding, much of which it quickly spent on developing cutting-edge technology and courseware, and guaranteeing substantial revenues to its academic partners, the universities of Chicago, Columbia, Stanford and Carnegie–Mellon and the London School of Economics.

Although online learning grew significantly during this period and several American universities developed outstanding courses, none of these ventures found many paying customers. NYU Online at its peak never managed to enroll more than 500 students. After spending $20 million on the venture, NYU closed it in December 2001. Columbia and Cornell did likewise. Four years after its foundation, Western Governors had only enrolled 200 students on degree courses. Pensare went broke in 2001. Cardean only survived thanks to the support of Thomson Learning.

There were many reasons for these failures – greed, naivety, ignorance of customer needs, weak value propositions, muddled thinking about the value of their brands. In most cases all that was on offer was a heavily diluted version of a conventional course. The critical stumbling block though was capabilities. Most of these would-be entrepreneurs had little idea of what it would take to be successful in these very immature markets. What is more remarkable is the narrow view they took of what constitutes a university education and how people learn.

The great myth of the e-learning boom, in both the corporate and the academic worlds, was that the key ingredients in the magic formula were content and technology: essentially all that was needed for a learning revolution was for technology to deliver more content to more learners at lower cost. Some even interpreted this to mean simply sticking lecture notes on a website, with little consideration of what good teaching and effective learning entail.

Distance learning, whether online or not, has been shown to work well mainly with highly motivated, self-disciplined adults, but even they need a degree of guidance, encouragement and social support. The outstanding successes like the Open University have developed a rich set of capabilities to meet the needs of their students: an intensely loyal, committed community of teachers, part-time tutors, and present and former students; a multidisciplinary team approach to developing courses and producing course materials; and expertise in supporting distance learning on a large scale, through a dedicated organization.

Capabilities like these take time to develop and the immense dedication of large numbers of people.

Knowledge vs know-how

Most of the e-commerce companies that sprang up in the late 1990s – and disappeared when the stock market collapsed in 2000 – lacked even the basic capabilities in processing orders and prompt delivery at which Amazon excelled. Indeed, many of them could not even operate websites capable of handling large numbers of orders. They relied mainly on the buzz surrounding the Internet and the willingness of investors to fund their half-baked ideas. Well-established mail-order companies like Lands' End, however, found it comparatively easy to move into web marketing, because they already had the most important capabilities and assets: customer knowledge, reputation, direct-marketing skills, proven product range, experience in prompt fulfilment of orders. It was comparably easy to acquire web management skills. Indeed, website operation, which was regarded in the Gold Rush days as a mystery only a technical elite could handle, is now a routine function easily outsourced to specialist providers. In itself it is therefore unlikely ever to be a distinctive capability for any company, other than in exceptional cases.

When traditional telephone companies moved into cellular telephony, they assumed they would have a head start over new outfits. But in almost all countries it was specialized new businesses like Vodaphone, Orange and McCaw that performed better. Instead of treating mobile networks as a sideline to serious telecommunications, these companies recognized that this was an entirely new industry, requiring new capabilities and approaches. Quite apart from the different technologies and infrastructure employed, mobile telephony was from the outset in most countries a competitive industry. Typically, there were two or more roughly equally matched licensed operators at the outset, one of whom was an offshoot of the local telephone monopoly. Marketing and responding quickly to competitive initiatives were therefore critical but unfamiliar capabilities.

It is not just monopolies that are uncomfortable in competitive markets. The most intriguing might-have-been story of the personal computer revolution is that of Xerox PARC, the legendary research and development arm of the photocopying giant. As we saw in the Apple story, it developed many of the seminal and now ubiquitous concepts for personal computing, like windows, the mouse and local area networks, producing brilliant prototypes of all of these which the company never took forward commercially.

PARC's scientists and technologists and Xerox's executives lived in different worlds and conducted a dialogue of the deaf. The scientists seemed arrogant and unintelligible to the managers, who were dismissed as stupid and blinkered by PARC. The organization as a whole did not have the capabilities to turn brilliant ideas into real products. It was Apple in the 1980s who incorporated many of PARC's ideas into the Macintosh, and Microsoft who later copied them.

Shifting sands

When a market evolves rapidly, so do the basic capabilities. This happened most dramatically in the case of the PC industry in the 1980s, described in detail in Chapter 9. IBM, which had played a big part in creating the new market, rapidly lost not just its position as leader but eventually the ability to compete effectively in PCs. The capabilities that had given it enormous competitive advantage in the world of mainframe computing were not relevant in the mass market that emerged for PCs. Once the hardware had been standardized, the PC became a commodity and competition focused on price. The winning combination of capabilities in the hardware side of the industry turned out to be Dell's: mass customization, direct selling, virtual integration and very low costs. However nimble and far-seeing IBM might have been in the early 1980s, there was no way it could have succeeded in this marketplace.

Until about 2005, Dell was the only mass market PC supplier with substantial competitive advantage. It had started learning very early on how to deal with customers directly and to meet their needs effectively, making itself a master of direct marketing, rapid fulfilment and minimal inventories. Not only could other computer makers not catch up on these capabilities, but they were inhibited by their anxiety not to 'cannibalize' their existing distribution channels.

The recent decline in Dell's competitive position is a reminder that leaders can never rest on their laurels for long. Capabilities that were once highly distinctive can gradually become boringly normal, as competitors learn how to imitate and sometimes surpass the pioneers. That is why, in the long run, really successful businesses do what Sony and Apple have been so good at – they extend their capabilities and learn new ones.

Radical visions

Few organizations can rival Sony and Apple for the boldness of their ambitions, but every one of our market creators started out with a radical vision that most sensible people at the time either ignored or dismissed. Their businesses were all based on an original idea that, unlike most corporate mission statements, could be expressed clearly and simply, that defined what was new and different about the business and the direction in which it would move. It was rarely a comprehensive long-term strategy at the outset, and certainly not a detailed blueprint, but it did suggest a long-term goal.

Jeff Bezos at Amazon saw in 1994 that the Internet could become a powerful platform vehicle for e-commerce and that selling books online would work particularly well. Sergey Brin and Larry Page believed that Internet search methods based on the structure of the Web would enable Google 'to organize the world's information and make it universally accessible and useful'. Pierre Omidyar at eBay dreamed of a perfect marketplace, where all buyers and sellers could participate on an equal basis. Klaus Heymann saw that CDs represented an ideal platform for high-quality, low-cost recordings of classical music, and that there could be a substantial global market for the business he named Naxos.

The idea is a combination of creativity and business judgement: creativity to envisage a significantly different way of doing things; judgement to assess whether the idea offers real benefits to customers and can be put into practice.

The vision is generally inspired by the insight that there is something missing from, or wrong with, existing markets. In some cases, there are people whose needs are not being well met: coffee drinkers who could not find a decent cup of espresso; anyone wanting to obtain information quickly; music lovers who would like to listen to their choice of songs while on the move. In others, the present methods are clearly flawed: occasional buyers and sellers of low-value items cannot easily hook up with each other or get a fair price; telephone conversations confined to fixed lines were inconvenient for people on the move and for the young; search engines produced misleading results.

In many of these cases, customers were paying more than they needed, either because of cartels, as with CDs and flights, or because customary sales and marketing methods were unnecessarily cumbersome and labour-intensive. Simpler methods, cutting out salespeople, improved choice and offered lower prices.

Almost invariably, the idea is counter-intuitive, in some cases heretical. The experience of eBay's founders with prospective investors, described in the introduction, is entirely typical. Many new technologies, of which the PC and the Internet were the most notorious, appear at first to be unimpressive, irrelevant to established companies and their customers, attractive only to insignificant markets. Sony's retail distributors and many Sony executives told Akio Morita that there was no market for a portable tape recorder that could not record. He was convinced that the Walkman would create a new headphone culture and insisted on launching it. Apple was written off by most of the business world for most of its history until 2001, since when it has been almost universally acclaimed.

The visionaries are those who are brave enough and sometimes crazy enough to defy conventional wisdom. With the benefit of hindsight, the merits of these ideas are very clear. They were not at the time. The maverick nature of new business ideas is one of the main reasons why established companies, especially large ones, rarely develop them successfully. They tend to favour initiatives in areas they understand and where the evidence is clear-cut – and where many others are already looking.

The vision is not just a clever idea – it is a source of inspiration to those working in the new organization. Whether it is Starbucks or Sony, Apple or Amazon, the people working there are fired up like evangelical missionaries. The conviction that they are doing something vitally important and new gives them enormous energy. It can sometimes blur their judgement: Apple's notion of being 'insanely great' came painfully close to the literal truth in the mid-1980s.

Over time the vision evolves, and the strategy is elaborated considerably, but most of our market creators have remained fairly faithful to their original vision. The virtue lies not in the faithfulness but in the sharpness of the vision.

3

E-MERCHANTS

The vision of Amazon and Webvan was to use the Web to revolutionize retailing. Amazon succeeded spectacularly; Webvan lost a billion dollars.

Amazon.com

Jeff Bezos became the most celebrated of all Internet entrepreneurs, partly because his business was one that most outsiders thought they could understand. In the late 1990s his face was on the cover of countless magazines and he had a ready supply of memorable quotes for the media, suddenly obsessed with the Net and the instant accumulation of wealth. He described himself as a nerd, but he had grasped the commercial potential of the Internet long before most people, and acted quickly and systematically to seize the opportunity it presented.

He was an outstanding student with a flair for quantitative analysis, but at Princeton University he realized that, although he was among the top twenty-five students in physics, 'there were three people in the class who were much, much better at it than me'. He switched to engineering and computer science and graduated, summa cum laude, in 1986. Bezos was offered jobs by blue chip companies like Intel and Andersen Consulting, but decided to join a small Wall Street start-up, Fitel, where he developed an information service for cross-border trading by brokers, investors and banks. He then moved to Bankers Trust, where he became its youngest-ever vice-president at the age of twenty-six. He led a programming team that developed a communications network which enabled clients to monitor the performance of their investments. Traditionalists at Bankers Trust disliked PCs, and in 1990 Bezos decided he would move to a company where he could conduct 'second-phase automation – which would allow him to fundamentally change the underlying business process and do things in a completely new way'.

D. E. Shaw & Co was a quantitative hedge fund, founded only two years earlier and described by *Fortune* as 'the most intriguing and mysterious force on Wall Street today... a place where the avant-garde meets arbitrage, and intellectualism and profit-seeking mix harmoniously'. It was an ideal environment for Bezos and he was put in charge of exploring new markets.

Opportunity

Early in 1994 he conducted a study of ways to make money from the Internet and was staggered to discover that Web usage was growing at 2,300 per cent per annum. Clearly this was going to create enormous business opportunities, though their nature and scale were difficult to grasp. 'Human beings aren't good at understanding exponential growth... it's invisible today and ubiquitous tomorrow.' He drew up a list of twenty types of product that he thought could be sold efficiently online and decided that books and music were the most attractive. They were standardized products that customers understood, so they would know exactly what they were buying. Of the two, he favoured books, as there were more titles and 'no eight-hundred-pound gorillas' in the fragmented publishing and distribution industry.

It was clear to Bezos that an online business could offer customers a wider selection than the biggest stores and a more convenient way of buying: the very largest stores stocked 175,000 titles, yet there were 1.5 million books in print in the US. An online store could offer very much greater choice, and without the real estate and staffing costs of physical stores it should be possible to achieve both lower prices and higher margins. 'So that became the idea... easily find and buy a million different books.' The inefficiency of the industry was exemplified by the fact that 35 per cent of the 460 million books shipped in 1994 were returned to publishers. Bezos came to realize that this was 'not a rational business. The publisher takes all the return risk and the retailer makes the demand predictions.'

D. E. Shaw decided that it did not want to get into this business, so Bezos decided to do it himself. Characteristically, he constructed what he called a 'regret minimization framework' to decide whether to jump. 'I knew that when I was 80 there was no chance that I would regret having walked away from my 1984 Wall Street bonus in the middle of the year... But I did think there was a chance that I might regret significantly not participating in this thing called the Internet that I believed passionately in. I also knew that if I tried and failed, I wouldn't regret that. So, once I thought about it in that way, it became incredibly easy to make that decision.'

Start-up

Bezos decided to locate the business somewhere where there was a large pool of technical talent and near a large book wholesaler. According to Amazon legend, he wrote the business plan on his laptop, while his wife, Mackenzie, drove them West. In fact, they flew from New York to Texas, and drove from there, while the business plan was not written for another year. He did, however, choose a house with a large garage, in the hallowed tradition of start-ups.

He chose the name 'Amazon', so that it would come up early on alphabetical lists and the brand could stand for anything. From the outset he had the idea of moving beyond books at some stage. He insisted that the company should always be called 'Amazon.com', which later helped to differentiate it.

His first recruits were computer specialists. Sheldon Kaphan, who was to become Amazon's CTO, was an experienced software engineer. He was initially reluctant to commit himself to another start-up, but Bezos eventually convinced him that Amazon had a chance of being a big winner. He was later to say that finding Kaphan was one of his luckiest breaks. Paul Burton-Davies was younger, had experience of the Web, but like Kaphan was chosen primarily for his intellectual ability. Neither of them had experience of writing business software for users. This became a pattern – Bezos insisted on getting the brightest possible people, regardless of their specific experience. Scarcely any early recruits came from the book business. Together Kaphan and Burton-Davies constructed the first website over the next year, working on two Sun workstations in the converted garage. The fourth member of the team was Mackenzie Bezos, who handled the accounts and administration until 1996.

While they beavered away, Bezos took himself to a four-day course on book selling and was inspired by the emphasis on customer service, above and beyond the call of duty. He decided that he must make customer service 'the cornerstone of Amazon.com', the best possible shopping experience that could be provided online. This idea subsequently crystallized into the goal of making Amazon 'the most customer-centric company on the planet'.

Amazon was not the first online bookstore. Computer Literacy Bookshops had been selling technical books by email since 1991, when only the technically adept were using the Net, and overt commerce was widely abhorred. In 1992 Books.com started up, using bulletin boards for customer communications, and in 1994 launched a rather crude website with 400,000 titles at discounted prices.

The Amazonians felt that they could do better, but were happy to adopt many of Books.com's features such as reviews, excerpts and attempts to cultivate community. They also hedged their bets, building an email-based system for ordering books as well as one for the Web. In 1994–5 however, nobody knew whether many consumers could be persuaded to make purchases online. Amazon anticipated that many would prefer to give credit card details over the telephone.

Bezos played an active part in defining the look of the website, the operating system interface and the database. His own software experience made him a demanding but realistic client. Burton-Davies said that one of the reasons for Amazon's success was 'Jeff's insistence on everything being done right'. The standard software packages used by mail order companies would not be good enough for them.

The database was constructed mainly from Bowker's *Books in Print*, a catalogue of 1.5 million titles. Converting it from CD-ROM was a lengthy, labour-intensive process. Amazon's operating system made it possible to store details on the thousand most popular books in computer memory, which made most responses very fast. When the site was launched, the database contained more than a million titles. Each customer entering the site was assigned an ID so that browsing and buying habits could be analysed.

Amazon knew that it had to be seen to have a secure system for dealing with credit cards, but had no direct experience or knowledge of how card transactions were processed. Security was something they had to work out from first principles, a long learning process. A crucial, much imitated innovation was the idea of the 'shopping basket'. In 1995 the very simple, five-step ordering process was a big breakthrough. Scarcely any other early sites were as well planned and easy to use.

Bezos insisted on keeping transactions as simple as possible, both to make it easy for customers and because scarcely any of them had high-speed connections. Fancy graphics for users with slow dial-up access made many early websites painfully unusable. Amazon pages, consisting mainly of text, were designed to be downloaded in seconds. Amazon's site stood out by its stripped-down approach, part of its rigorous focus on the customer experience. For many customers its main advantage was speed and convenience. Bezos believed that 'in the late twentieth century, the scarcest resource is time. If you can save people money and time, they'll like that.'

Launch

After a lengthy and surprise-free beta test, the site was made public in July 1995. The logo was a large 'A' with a river running through it and the tagline 'Earth's Biggest Bookstore'. The page mainly consisted of text and navigational bars. 'Spotlight' was a daily feature on a book with a 40 per cent discount. The top twenty best-sellers carried a 30 per cent discount, and about 300,000 books a discount of 10 per cent.

They were expecting a trickle of orders in the first few weeks. However, three days after launching they received an email from Yahoo, asking if they could include Amazon on their 'What's Cool' list, then the busiest page on the Web. Kaphan wondered if accepting might be like taking a sip of water through a fire-hose – in its first week, Amazon took hundreds of orders, but was only able to send out a handful of books. During the first month, they shipped books to every state in the Union and to forty-five countries.

These were very much bigger numbers than Bezos had planned for. 'I think one thing we missed was that the Internet was exclusively made up of early adopters at that time. So all the people online, even though it was a relatively small number compared to today, were those who liked to try new things.' Bezos had believed that it might take years for large numbers of people to feel comfortable about buying books online and that sales would build slowly. In fact Amazon's challenge became that of managing very much greater demand than anyone had dreamed of. It was one that it handled brilliantly well, by a mixture of inspired improvisation, systematic analysis and competitive drive.

In the launch period, scarcely any provision had been made for packing, so Bezos and everybody else had to work until midnight every night getting packages ready. They were mostly working on their hands and knees until they had the bright idea of getting tables. Returns were another thing they had not really planned for. Bezos wanted to be generous in order to develop customer loyalty and gave customers thirty days to change their minds, and the first returns created accounting chaos – yet another set of operational skills and processes that had to be learned from scratch.

Partly for reasons of frugality, and even more to symbolize it, Bezos made himself a desk from a door. He wanted to demonstrate that the company was only spending money on things that mattered to customers. He had to be talked out of putting stickers on all the furniture to show how little money had been spent on it. His desk, however, proved a sound investment. It featured in nearly every media interview and in a *Vanity Fair* photo shoot. Soon everyone at Amazon had door desks.

Funding

Initially Bezos paid his colleagues out of his own pocket, putting $60,000 dollars into the company in 1994. His parents invested another $250,000. Money was very tight – the loss for the year would be $303,000, and Bezos anticipated several more like that. He desperately needed more funding and decided to raise it from 'angel' investors in Seattle, rather than from venture capital.

Bezos subsequently said that raising this first round of just short of a million dollars, mainly in lots of $30,000, was the most difficult task he ever had. He had to pitch to sixty different people to find twenty investors. He told them, 'I know nothing about the book industry, nothing… But I know that I can get the books here, and I can get them to the customer and forget about bricks and mortar.'

One investment adviser did extensive research on the industry and decided that Barnes & Noble would crush Amazon when they moved online. Most investors raised the objection that people liked going to bookshops. Bezos's initial response was that books were one of the few categories where computers were already an aid to selling. But his main argument was that an online bookstore offered customers choice and convenience that conventional stores could not. A survey by the American Booksellers Association showed that book buyers were almost twice as likely as average shoppers to use the Net and that half of book buyers under the age of fifty were disposed to buy books online.

Amazon's business model was also very much more efficient. Conventional American bookstores turned over their inventory less than four times in a year – often much less. Amazon would achieve 150 turns on much of its stock. Furthermore, it would receive payment from credit card companies fifty-three days before it had to pay its suppliers. Conventional stores had to pay for their books on average seventy-nine days before they got their money, mainly because each book sat on their shelves for 161 days before it was sold.

This was a major advantage to Amazon in terms of working capital. The extent of Amazon's enormous economies of scale is more debatable: distribution and customer service were hardly fixed costs. And as Amazon grew, it ceased to be a virtual organization, needing to build large warehouses of its own.

Bezos's concern was that customers would require a long period to be 'educated'. His business plan in 1995 showed two scenarios. The moderate one projected sales of $11.5 million by 1997, with a profit of $50,000. His fast-growth scenario was for sales of $17.7 million and a profit of $143,000. (Sales would in fact almost reach the larger of these two estimates in the first quarter alone of 1997.)

By November 1995, Bezos managed to convince twenty-one individual investors to stump up a total of $981,000. A few months later he was approached by General Atlantic Partners, a large venture capital company on the East Coast. With sales revenues now heading for $5 million, GAP valued the company at around $10 million.

Bezos and his colleagues, however, were realizing that Amazon had the chance to dominate not just a large online book sales market but many others. It could quickly become a multibillion-dollar business, and it should not be too difficult to raise the finance to achieve this. Eric Dillon, one of the early investors and now a financial adviser, saw this as 'a crossover point... We changed our whole focus to one that was driven by momentum. We had the momentum and we had to keep the momentum. We needed to bring in money. We needed to be the first ones to use national advertising.'

In May 1996 Amazon was effectively granted this wish, free. The *Wall Street Journal* ran a front-page story under the headline 'How Wall Street Whiz Found a Niche Selling Books on the Internet'. The article could have been written by Amazon's PR department. 'Its site on the World Wide Web has become an underground sensation for thousands of book lovers around the world, who spend hours perusing its vast electronic library, reading other customers' amusing online reviews and ordering piles of books.' Orders doubled overnight and continued to accelerate.

The article also triggered a swarm of calls from the venture capital industry. John Doerr of Kleiner Perkins, who had previously been ignoring Eric Dillon's calls, decided to come to Seattle. His was the firm that Bezos most wanted, for the prestige it would bring, and its contribution to momentum: 'Kleiner and John are the gravitational center of a huge piece of the Internet world. Being with them is like being on prime real estate.' Doerr not only invested $8 million for 13 per cent of the company, valuing it at $60 million, he agreed to join the board.

Get big fast

The goal now was to grab market share as quickly as possible, regardless of short-term profitability. The strategy was expressed in three words emblazoned on company T-shirts: 'Get Big Fast'. Bezos wanted to make Amazon's brand as big and as valuable as Disney's, and make it difficult for later entrants to online retailing to dislodge it. When Barnes & Noble announced a deal with AOL, George Colony of Forrester Research pronounced Amazon 'dot.toast' – as accurate a judgement as Forrester's many inflated forecasts.

Amazon used Kleiner Perkins's money to launch an advertising campaign in the *New York Times*, *Wall Street Journal* and *USA Today*, and did deals with the *New Yorker*, *Atlantic Monthly* and *Wired*. It also continued to get lots of free, almost entirely uncritical publicity. Everyone wanted to write about the Web, and Bezos made Amazon into a story everybody could understand and admire. In December 1996 *Fortune* declared, 'Amazon is truly virtual. Though it has become a multimillion-dollar business that employs 110, there's still no storefront and little inventory.' In fact, the month before, Amazon had quietly opened its own 93,000-square-feet warehouse in Seattle to store copies of popular titles. Having got pretty big very fast, it no longer made any sense to wait to order every book in response to customer orders.

Amazon appointed Deutsche Morgan Grenfell as its financial advisers and in March 1997 filed a prospectus. This positioned Amazon as a glamorous technology company, rather than a low-margin retailer. It talked of 'virtually unlimited online shelf space' and 'lack of investment in expensive retail real estate and reduced personnel requirements'. It also disclosed losses to date of $6 million and said that it expected losses to continue for the foreseeable future and at significantly higher levels. It was paying on average $16 to buy each book it sold for $20, and spending $8 on advertising. It acknowledged that barriers to entry were low and that it expected competition to intensify. It could not say when the company would become profitable.

In normal times this would have put off most investors, but these were not normal times. There was no shortage of experts like Mary Meeker of Morgan Stanley to convince investors that market share was much more important than profits, that companies like Amazon would enjoy big 'first mover advantages' and that they would quickly come to dominate their markets. At that point they would be able to extract good profits. Many people were convinced by the 'winner takes all' and 'land rush' theories and that was what was driving stock market sentiment.

In the first quarter of 1997 Amazon's sales reached $16 million, more than for the whole of 1996. Losses were almost $3 million, but enthusiasm for the IPO enormous. It raised $54 million, valuing the company at $548 million in May. By September, the stock had more than doubled, and Amazon secured credit of $75 million for further expansion. The publicity from the IPO rebounded directly on sales – revenues in 1997 hit $148 million, almost ten times their level the previous year.

Bezos was not just thinking big about book sales. 'Our strategy is to become an electronic commerce destination. When someone thinks about buying something online, even if it is something we do not carry, we want them to come to us. We would like to make it easier for people online to find and discover the things they might want to buy online, even if we are not the ones selling them.'

In July, Amazon made its long-planned move into music, with 130,000 CD titles available, divided into fourteen genres and 280 sub-genres. As usual, great care had gone into the software and the support systems. Customers could sample any of 225,000 songs online. The *New York Times* rated the search system for songs the best on the Web. Within three months, Amazon had overtaken CD-NOW, the established leader in online music sales, incidentally disproving the universality of the first mover advantage and land rush theories. Almost certainly, this success owed much to Amazon's customer base and high public profile, but its ability to deliver on the promotional promise was critical.

In 1997 Amazon also started a programme of acquisitions that was later to reach a frenzy. It started with IMDB, an online database of information about movies, in preparation for a move into videos. PlanetAll was a 'contact management service' that collected personal information from consumers. Junglee Corp was an early search engine that enabled consumers to make price comparisons. All of these were complementary to Amazon's main business. It also formed alliances with most of the main players on the Internet at the time – Yahoo, Excite, AOL, Netscape and @Home.

Inside Amazon

The new funding enabled Amazon to take on a lot more people. However, Bezos remained extremely picky – he only wanted the very best, people who had 'been successful in everything they had done', who would commit themselves to the intense corporate culture he was trying to create. He liked to tell potential recruits, 'You can work long, hard, or smart, but at Amazon.com you can't choose two out of three.'

John Doerr encouraged him in this, though it frustrated Kaphan and Burton-Davis, who were often desperate to get people on board. Dillon recommended four friends, whom Bezos rejected as not quite good enough. Everyone was interviewed several times, with Bezos always having the last word in the early days. 'Jeff would find things wrong or right with people that no-one else spotted', according to

Glenn Fleishman, the catalogue manager. When he joined in 1996, he felt 'it was like going to the world's best college. I was surrounded by smart people.'

Bezos thought a lot about culture. He told recruits he wanted Amazon's to be 'intense and friendly. In fact, if you ever had to give up "friendly" in order to have "intense", we would do that. So, if we needed to be "intense" and "combative", we would do that before we'd be "not intense".' He was modelling himself on his neighbour, Bill Gates, but hoped to avoid the ferocious internal competition that was rife at Microsoft. Two key executives came from the Beast of Redmond: David Risher became Vice-President of product development and Bezos's right-hand man; Joel Spiegel became VP of engineering.

We have a fascinating account of what it was like to work at Amazon from James Marcus, who joined in 1996 and spent five years there as an editor, mainly writing book reviews. His rather precious perspective is very different from that of the business journalists, breathless with admiration for Jeff Bezos. Despite his distaste for the 'Culture of Metrics' and his difficulties maintaining the 'ever more precarious balance between art and commerce', he paints a largely sympathetic portrait of Bezos, certainly in the early days:

> I had already succumbed to his brand of anti-charismatic
> charisma, which would have mortified a Great Man of a century
> ago, but which seemed just right for our nerd-driven meritocracy.
> Jeff it should be stressed was a likeable and normal person... the
> habit of humorous self-effacement kept his Napoleonic side
> under wraps.

Bezos did not just want a metric for customer enjoyment, he felt that they should eventually be measuring customer ecstasy. He believed that 'content' would be an essential tool for Amazon and Marcus was one of dozens of professional writers he hired to create it. Marcus liked the 'delightful uncertainty about what the company was, when it could be taken for a bookstore or bulletin board or electronic agora, a revolutionary enterprise or (as Jeff continued to say) a destination'. He graphically describes the feverish pace, the constant improvisation, the endless interviews of prospective staff, the boundless ambition. He reports with gloom the arrival of droves of MBAs at the end of 1997, bringing with them a plague of jargon: 'Pulling on revenue levers meant making more money. If we leveraged our verbiage correctly, the division would soon reach an inflection point (we would make more money). The main thing in any case was to monetize those eyeballs.'

By 1999 Marcus felt that 'monetizing those eyeballs had become a lynchpin of Amazon's corporate philosophy', and as time went on he was observing the company from the outside, where he clearly intended to be as soon as he could exercise his stock options.

Customer-centric

Bezos frequently declared his ambition to make Amazon 'the most customer-centric company in history'. Certainly he made it more responsive to customers than any other e-commerce business, none of whom ever actually met any of their customers. He believed that the Web required a completely different approach. In the physical world a merchant might spend 30 per cent of his time creating a good customer experience and 70 per cent shouting about it. On the Web, the proportions had to be reversed.

He set out to engineer the customer experience, systematically planning every step, every element, from the time to download Web pages and the taking of the order through to delivery. He wanted the brand above all else to stand for a great customer experience. Precisely because the early Internet was primitive and slow, the value proposition needed to be 'overwhelming', to persuade potential customers to change their habits. Hence the metric for customer enjoyment and the musings about ecstasy.

He acknowledged that:

> We will never make Amazon.com fun and engaging in the same
> way as the great physical bookstores are. You'll never be able to

hear the bindings creak and smell the books and have tasty lattes and soft sofas at Amazon.com. But we can do completely different things that will blow people away and make the experience an engaging and fun one.

In the early years, editors like Marcus tried giving customers the kind of advice they might find in an independent bookstore. Marcus's boss, Rick Ayre, said he wanted users to 'get a sense of the quirky, independent, literate voice, and that behind it all you're interacting with people, and that it's people who care about these things, not people who are trying to sell you these things'. The emphasis had changed by 1999, but there is little evidence that many customers were put off. Given that the vast majority were using Amazon for the choice and convenience, the sophisticated editorial offerings were a nice bonus for some, rather than an essential feature. What was important was having enough information about the book, and Amazon was happy to get this from any reputable source, notably the publisher, and reviews written by customers. While forthright critical reviews came to be discouraged, as they generated hundreds of complaining emails, Bezos himself stirred up a controversy in 1999 when he refreshingly summed up a book on Internet business planning as 'Stupid book... don't waste your time.'

Increasingly, Amazon started to make suggestions to individual customers, based on previous purchases, part of its shift towards personalizing their experience. These new approaches made use of relational software and collaborative filtering, which told customers of books chosen by other customers who had bought the same title. In 1998 Bezos declared that most e-commerce was just about 'findit, buy-it, ship-it'. What Amazon was doing was 'e-merchandising'. 'We can use advanced technology to not only understand our products... but to understand our customers.'

Amazon invested heavily in market research, particularly focus groups, to get a detailed picture of customer likes and dislikes, and before introducing new features like '1-click shopping'. This was another significant step towards making the process of placing an order as simple and easy as possible, indeed too simple for some, who did not believe that they had actually completed it until Amazon made the confirmation more explicit.

The term 'e-merchandising' never quite caught on, but the idea of systematically organizing the approach to the customer experience did clearly distinguish Amazon from most of its competitors, direct and indirect, in the early years of e-commerce. Like Dell and Starbucks, it made its marketing operations almost scientific in their precision. In the early days, it took four days to ship books that distributors had in stock, but they aimed to get that down to twenty-four hours. To ensure that every book arrived in perfect condition they 'over-packed'.

The brightest people in Amazon worked in customer service. They were trained to know every step in the company's operation, and to rectify the impersonal aspect of buying books online. If there was a problem, Bezos was determined that Amazon would not only be responsive but demonstrate an exceptional commitment to helping customers. Patricia Seybold recounted receiving a package with a handwritten note: '"We know you ordered the softcover version of this book, but it's out of stock so we sent you the hardcover version for the same price." That little handwritten note cemented my love affair with Amazon.'

Delusions of grandeur

The remarkable growth of the Internet in the late nineties and the even more extraordinary rise in share prices went to the heads of most entrepreneurs, and Bezos was no exception. Given that he was not the most modest of men to begin with and that he was being hailed as a business genius by almost the entire media industry, this is hardly surprising. In November 1999 *Business Week* wrote breathlessly:

> When we try to comprehend something as vast, amorphous and
> downright scary as the Internet, it's no wonder we grope for
> familiar historical precedents – the railroads, the interstate

highway system, the telephone network. But none of these really captures the Internet's earthshaking impact on the business world. For that, we must take the advice of Internet commerce pioneer, Jeffrey P. Bezos.

It was in this year that he had started to implement his dream of making Amazon 'the Wal-Mart' of the Internet, and dominating not just sales of books and music but just about anything. He believed that eventually 15 per cent of all retail sales would move online, and he intended Amazon to have a large chunk of that $500 billion.

In March 1999 Amazon attacked the rising star eBay, with its own auction site and much better customer service. The previous year, eBay executives had gone to Seattle to propose collaboration on cross-promotions. To their astonishment, Bezos informed them that, with 8 million customers, Amazon would soon crush their little operation. Amazon's auction site was positioned as a service that helped consumers to find products and was aimed rather more at buyers than sellers.

We want to build a place where people can come to find anything they might want to buy online. You realize very quickly that you can't sell everything people might want directly. So instead you need to do that in partnership with thousands and indeed millions of third-party sellers in different ways. To try to do that alone, in strictly a traditional retailing model, isn't practical.

In June, following eBay's acquisition of Butterfield's, the US auction house, Amazon bought a stake in Sotheby's. They set up a joint auction site for upmarket items, which was launched in November, but it was not a success. Traditional Sotheby's customers had less need of an online service than the smaller-scale dealers and buyers who had been flocking to eBay. Their collectors valued the atmosphere of live auctions in swish locations.

With its stock riding high, Amazon embarked on a shopping spree of acquisitions and investments. In February 1999 it bought 46 per cent of Drugstore.com and in March 50 per cent of Pets.com. Both had business models clearly inspired by Amazon's, but with far less competitive advantage. In April, Amazon acquired outright e-Niche, which ran two marketplace sites, and Accept.com, which developed software that enabled person-to-person Web transactions. In May it acquired 35 per cent of Home Grocer.com for $42.5 million in cash; in July, 49 per cent of Gear.com; and in September, 20 per cent of a wedding gift registry.

Its biggest move towards becoming the 'electronic commerce destination' was the launch of zShops in November 1999. It had been working all year on building a network of thousands of affiliates: zShops was to be an online shopping mall, where customers could find everything bar firearms, live animals, pornography and tobacco. 'Sixteen months ago, we were a place where people came to find books,' said Bezos. 'Tomorrow, we will be a place to find anything, with a capital "A".'

This particular venture, unlike most of the others and the core business of books and music, really did enjoy minimal marginal costs, since Amazon itself, like eBay, did not have to handle or deliver anything, but simply collect commissions on every transaction. The underlying logic of the new strategy of getting many times bigger than anyone had previously dreamed of was increasing returns to scale. Yet, in 1999, Amazon opened five new warehouses in addition to the two it had in Seattle and Delaware. These employed thousands of people, not all of whom were counted as employees. It was a logical step in providing better service to customers, but it showed that Amazon was not as virtual as it had first appeared and nor was its business model anything like as radical as eBay's.

This paradox was highlighted by the financial results. In the third quarter of 1999, sales rose 130 per cent from the same quarter of 1998, to $356 million. Losses, however, rose more than fourfold to $197 million – customer acquisition costs had risen to $35. For the first time, several prominent analysts started to question Amazonian economics. Henry Blodget, whose bullish pronouncements normally made Mary Meeker sound conservative, downgraded Amazon's rating to 'accumulate', and several others marked it down. The chinks in the armour were clear to many now.

Nonetheless, the year ended gloriously, with Bezos's face on the cover of *Time* magazine, as their 'Man of the Year', and the share price at a new peak of $113.

Crisis

The crisis that hit Amazon in 2000 was primarily a general collapse of confidence in Internet business models and valuations. Ominously, in February, Pets.com's IPO flopped and Amazon announced losses for the previous quarter of $323 million, seven times their level a year previously. Bezos described this as a 'high water mark'. From now on he would be leading a 'drive towards profitability' and specifically to reduce the percentage loss on sales to single figures by the end of the year. For a few weeks the market seemed reassured by this, but these were the last days of the bubble.

In March, NASDAQ, the stock exchange where most Internet stocks were traded, reached its peak and then started its rapid fall. Amazon's share price plummeted to $52 in April, but there was much worse to come. In June, Ravi Suria of Lehman Brothers wrote that Amazon 'shows the financial characteristics that have driven innumerable retailers to disaster throughout history'. He predicted that it could run out of working capital by the end of the year. Amazon called the report 'hogwash', but clearly Suria was not entirely wide of the mark. However, Bezos had been serious about improving profitability. In the third quarter, losses were reduced from 22 per cent of sales to 11 per cent. The books, music and video division earned a profit of $25 million on $400 million sales. In the new mood of deep investor pessimism, this brought only temporary relief. The share price ended the year at $15.

The rate of attrition was frightening. Pets.com, which was totally dependent on external funding to cover its huge losses, was the first publicly traded Internet company to close down, taking with it the $60 million Amazon had invested. HomeGrocer.com was offloaded at a loss.

However, Amazon was slowly turning the corner. Although losses for the year as a whole were $1.4 billion, they were reduced in each quarter and there was still $1.1 billion in the bank. In January 2001, the company closed down two of its distribution centres and laid off 1,300 people. The era of land grabs and get big fast was over, but the core business was now quite healthy.

The company achieved the highest-ever score for a service business in the University of Michigan's Customer Satisfaction Index in 2000 and maintained the same score in 2001. Sales growth stagnated for a while in the US but grew by 72 per cent internationally. Losses were reduced to $567 million, but the share price continued to drop, to $6 by the end of the year. That was as much as anything else a reflection of general business confidence in that traumatic year. In fact, online sales did much better than conventional retailing in the period after September 11.

Postscript

Amazon's former pre-eminence as *the* e-commerce business has waned slightly with the many other retailers who now have an online presence, but it remains dominant in its core media business of books, music and DVDs, where sales reached $9.2 billion in 2007. However, intense competition in this sector will make it difficult ever to achieve high levels of profitability.

Since 2001 sales have grown steadily, particularly internationally, in non-media products and through third parties, reaching $14.8 billion in 2007. The company achieved a tiny operating profit in 2002, and in 2003 its first-ever net profit, of $35 million, less than 1 per cent of sales. By 2007 profits had climbed to $588 million, 4 per cent of sales. Margins would have remained considerably lower had it not been for sales through third parties, where Amazon's variable costs are negligible. By 2004, there were more than 100,000 third-party sellers – Amazon Marketplace, for part-timers, and Amazon Merchants for other retailers – and these accounted for nearly a third of total revenues.

For a while it appeared to be eclipsed by the growth of eBay, which became the largest and most profitable e-commerce business. However, Amazon's networked model, with its emphasis on buyers rather than sellers, began to look more durable than eBay's rather grasping one, which was alienating many of its traders. Certainly Amazon's brand shines more brightly, commands strong customer loyalty and wins high levels of repeat business.

In 2004, Amazon started to reinvent itself in a way that surprised even Bezos. It realized that its enormous computing infrastructure, built to cope with Christmas peaks, could be used by other firms. Starting with its retail partners, by 2008 it had signed up 370,000 customers, from tiny websites to the *New York Times*, who rent computing capacity instead of building their own data centres.

Judged by most standards, Amazon's achievements are astounding. It is only in relation to the over-ambitious goals it set itself in the late nineties, and the ridiculous hype of that era, that its success is a qualified one. To go from inadequately funded start-up to sales of $150 million in two years is impressive by any standards. To reach revenues of $15 billion in another ten is extraordinary.

On the success factors for market creation, Amazon scores almost top marks. Its strategic vision was radical and crystal-clear, and it adapted quickly and boldly to new circumstances. Its competitive advantage was built solidly on a set of highly distinctive capabilities, which it has taken care to renew and enhance. It was exceptionally entrepreneurial and ambitious but intensely disciplined from the beginning. It understood from the outset the importance of a compelling value proposition and was consistent in articulating and justifying it.

Jeff Bezos, unlike the vast majority of Internet entrepreneurs, has proved an exceptionally smart strategist. He is also one of the very few business founders to have led his company all the way from start-up to multibillion-dollar success.

Webvan

The vision for Webvan was certainly radical – to 'reshape the retail landscape' with an entirely new distribution system that would compete head-on with conventional supermarkets and eventually with Wal-Mart and Amazon.

Its author, Louis Borders, had led one revolution in bookselling, a decade before Amazon, partly by defying conventional wisdom about size. The typical bookstore to be found in most American malls carried 40,000 titles, whereas a Borders one had 150,000–200,000 and therefore offered customers much greater choice. Louis, a talented mathematician, had also developed an artificial intelligence computer system to manage inventory, adding more books on topics that were selling and eliminating those that were not.

In 1992 he had sold the chain of bookstores that still bears his name to K-Mart. Borders' success had a big impact on how Webvan was conceived, viewed and launched: the entrepreneur's track record gave him considerable credibility with investors and the confidence to think big. In 1997 he declared that Webvan would be either a $10 billion company or zero. Two years later he decided that $100 billion would be closer to the mark. Zero turned out to be the right number.

Technology was at the heart of his vision for Webvan. Louis was fascinated above all with the process of handling inventory in a completely different way. He believed that home delivery would restore the personal touch to retailing that had been lost with the coming of malls, but the key to success would be automated distribution, built for scalability. What would count, he believed, would be who could be first to scale, not first to market. 'Expert systems and other artificial intelligence systems enable the customization of inventory to precisely fit a neighbourhood store or an Internet customer... the basic idea behind Webvan is that inventory moves to the person.' As one of his colleagues put it, 'This is the first back-end re-engineering of an entire industry.' Louis calculated that with 8,000 orders a day, each Webvan centre could achieve profit margins of 12 per cent, compared to 4 per cent for normal supermarkets. The average lifetime value of a customer would be $40,000. This assumed, of course, that the customers would come to the Field of Dreams he was building, and stay.

Louis's original idea was to sell just about everything, from ready-to-serve meals to consumer electronics and clothing. He had envisaged 3 million stock-keeping units, or SKUs, plus neighbourhood stores with kiosks. He believed he was addressing the $2.3 trillion US retail market.

The full vision was a little too ambitious for Benchmark Capital, Louis's first venture capital partners, but they were excited by the fundamental concept. The idea of having stores was dropped, there would be far fewer distribution centres and the initial focus would be on groceries. Louis, however, intended to add soon video rental, film processing, computer installation, a pharmacy and dry cleaning. In 1997 Benchmark put up $3.5 million seed capital, as did Borders himself and Sequoia, another VC firm. This, however, was a mere drop in the ocean – they would need $33 million just to set up the first distribution centre in Oakland. But raising capital proved to be the easy bit. Customers were more problematical.

Hubris

Early in 1999, Webvan raised $150 million from a group of prestigious investors that included Yahoo, CBS, Knight-Ridder and LVMH. This enabled it to build the first warehouse in Oakland, which would serve the Bay Area.

It opened in June 1999 and looked like the set for a science fiction movie. It covered 330,000 square feet, divided into colour-coded segments, yellow for dry goods, green for chilled and blue for frozen. Thousands of 'totes', or shopping baskets, made their way along a 4.5 mile network of conveyor belts to 'pods', or assembly areas. The electronic order on the tote triggered lights that showed the packers precisely what needed to be added at each point. The packing workers never needed to move more than nineteen feet to fill any item in an order. In a hub-and-spoke system modelled on FedEx's, completed orders were taken by truck to one of twelve docking stations in the Bay Area, and then by van to the customer's home in a pre-designated thirty-minute time slot.

Webvan's market positioning was gourmet food at ordinary prices.

There were over 300 kinds of vegetable, 350 cheeses and 700 wines on offer. A culinary director was appointed to supervise the preparation of meals. Advertising was aimed primarily at the wealthy, and was as much about the merchandise as the service. The first ad was a double-page spread of a lobster standing in a living room. '7.15 pm. Fresh seafood is delivered free to a couple in Oakland... Whether it's cake mix, caviar or live lobster, the selection and quality is amazing.' Orders trickled in, but never reached more than a few hundred a day in the first months – the centre was designed to handle 8,000. Half of the customers who tried the service never came back.

At this point, however, given the general mood of excitement about the Internet, such minor details passed virtually unnoticed. The main focus of attention was the ever-rising stock market. The opportunity to raise unprecedented amounts of money, and use this to expand faster than Wal-Mart had done, was irresistible. In July 1999 Webvan closed another $275 million of financing, in exchange for a mere 6.5 per cent of its latest valuation of $4 billion. It also signed a $1 billion order with Bechtel to build the next twenty-six warehouses, telling them that this would be as important to them as the Hoover Dam had been. In the same month, *Business Week* declared that Webvan 'may be the most innovative e-commerce venture to date'. In August, with sales for the half-year of only $395,000 and losses of $35 million, it filed papers for its Initial Public Offering.

In September, just before the IPO, Webvan announced that George Shaheen, the managing partner of Andersen Consulting (now Accenture), would be its new CEO. This was a major coup: Sheehan had made Andersen the top technology consulting and outsourcing company in the world, taking it from revenues of $1.1 billion in 1989 to $8.1 billion in 1998. When Sheehan solemnly told investors that Webvan was creating an intimate personal relationship with its customers through delivery people who actually stepped inside their homes, they swallowed it. He also got away with assuring them that conventional grocery stores could never compete effectively: 'They're adding cost to existing cost; the physical store becomes an albatross.'

Sheehan's credibility and the rise of NASDAQ to 3,000 on 2 November ensured another stunning IPO two days later. Despite forecasts of losses rising from $78 million in 1999 to $302 million by 2001, Goldman Sachs issued 25 million shares at $15 each. By the end of the day their value had risen to $25: Webvan, a fledgling business with scarcely any customers and an unproven business model, was now valued at $8 billion.

Nemesis

The rapid roll-out of distribution centres continued throughout most of 2000: Chicago, Los Angeles, Orange County, Portland, San Diego, Seattle. From January, however, the writing was on the wall, as the stock market started what became a dramatic slide. In March, Webvan's stock was down to $11 and by June it was $9, a third of its valuation at the time of the IPO. Sales were picking up, but painfully slowly. The Oakland centre peaked at 2,380 orders a day, only 30 per cent of capacity, and never reached profitability. The regular customers were happy with the service and the average order size reached $115, but there were not enough of them. Only 2 per cent of Web users were ever buying food online.

By the time it acquired its closest rival, Home Grocer, in September for $1.2 billion in stock, confidence was seeping away. Burning $125 million in cash every quarter, it became clear that the company would soon run out of money – and there was no prospect of raising any more. By November, the stock had fallen below $1 and never recovered. In a desperate attempt to staunch the negative cash flow, the emphasis shifted to closing centres rather than opening new ones. In April 2001 Sheehan was forced to resign. In July the company filed for bankruptcy, and $1.2 billion of investors' money disappeared.

It is easy to mock Webvan as a great folly, but it is sobering to realize that most of those involved were immensely talented people with serious records of business achievement. The idea of a distribution network organized for home delivery was not in itself ridiculous. What was colossally arrogant was to imagine that a model which was largely theoretical and untested could deconstruct the whole retail industry, starting with one of its most efficient sectors. Only a super-abundance of capital and the feverish speculative atmosphere could account for so many bad business judgements.

The whole edifice depended on high volume and a rapid take-up by customers. Yet no serious research was done on just how dissatisfied consumers were with existing supermarkets and what value they would place on home delivery. This was surely one of the (rare) cases where focus group discussions with different demographic groups could have been highly revealing. The venture would only fly if large numbers of people were so frustrated by tedious shopping trips that they would be strongly inclined to opt for, and probably be prepared to pay for, Webvan's service. The common problem with such research is that respondents cannot easily comprehend a new proposition and tend to reject it because of its unfamiliarity. In this case, they would have been comparing a very simple proposition with a very familiar experience.

Instead, Borders and his buddies at Benchmark, all highly intelligent people, relied to a considerable extent on hunches and reasoning based on what they and their families would do. If the target market consisted entirely of busy multimillionaires living in the Bay Area, this might have worked, but only as a niche business. Louis Borders, though, was aiming to be bigger than Wal-Mart. He believed that his model would be one of the most disruptive technologies of all time. What he failed to understand was that competition is only disruptive if it results in a value proposition that is irresistibly compelling to customers and that incumbents cannot easily match. Neither was true in this case.

The proposition basically boiled down to convenience, and this on its own was simply not enough, certainly not in the short term. Customers were not seeing the benefit of the cost advantages of the new model, because these depended on volumes that were never achieved. The quality of the merchandise was generally excellent – but it was no better than that offered by many other retailers, who also offered customers the familiarity and reassurance of seeing what they were choosing. And offering greater choice assumed that that was what consumers wanted, when most of them were frequently confused by too many brands. (In this, as in several other respects, the market for books was very different.)

The other major misjudgement concerned the competition. This was not racing with Amazon for 'the last mile', as Louis sometimes seemed to think, but bricks-and-mortar retailers. Webvan seriously underestimated the extent to which organizations like Wal-Mart had turned the management of their businesses into something like a science. The reason why average margins were 4 per cent (though Wal-Mart's were higher) was that the industry was highly competitive and efficient. The leaders really understood the logistics and economics of distribution from experience that it would have taken Webvan years to acquire. They also enjoyed considerable customer loyalty, of which Webvan took virtually no account.

Sheehan's comment about old-fashioned retailers needing to add cost to existing cost was a classic consultant's sound-bite – seemingly profound, but basically baloney. Tesco in the UK was already showing how it could offer home delivery more economically than a stand-alone business, simply by taking the goods from supermarket shelves and putting them in vans. It was foolish to imagine, as so many did during these boom years, that online retailing was something that could only be done by a new caste of digerati and that old-economy dinosaurs could never catch up. Webvan did not face much direct competition from the latter, but that was mainly because its own service posed so little real threat to them. If, however, it had enjoyed more success, it would probably have been destroyed by their counter-attacks.

Webvan's failure therefore was as much about distinctive capabilities as marketing. It did not understand those of the supermarket chains and overestimated its own. Envisaging and establishing the new infrastructure was a remarkable achievement, but it was rather like reinventing the wheel. As with so many business concepts of this era, like e-learning, innovative technology and scale were seen as ends in themselves which would magically transform a market.

Nothing like enough thought was given to how these would translate into value for customers. The real albatrosses turned out to be not physical stores but Webvan's magnificent distribution centres.

4

PROPOSITIONS AND DISCIPLINE

The test of a first-rate intelligence is the ability to hold two opposed ideas in the mind at the same time, and still retain the ability to function.

F. Scott Fitzgerald, *The Crack-Up*

In Chapter 2 we looked at two of the most challenging success factors for market creators, distinctive capabilities and radical vision. Here we consider the other two: a proposition that customers find irresistible and an organization that achieves the elusive quality of disciplined entrepreneurialism.

An offer they can't refuse

The fundamental test for any new business is the story it has to tell to customers. If it does not provide a compelling reason why they should buy, all the capabilities in the world will be of little avail. Yet this very basic idea is something that many businesses and commentators ignore. Perhaps it is too simple – or too close to the vulgar nitty-gritty of selling.

Webvan, described in the previous chapter, is a classic case – it paid almost no attention to why customers should use it, and who exactly they would be. Surprisingly for a mathematician, Louis Borders gave very little thought to the probability of them coming back, in what proportions, and how often. Most of his attention, and that of his equally clever colleagues, went on the brilliant system he had devised for revolutionizing retail distribution. This was a business in pursuit not so much of customers as of an abstraction.

The economics of Webvan's business model meant that it had to appeal to a mass market quickly: in order to finance its immensely expensive and unproven distribution and delivery network, it had to persuade enormous numbers of consumers to abandon established grocery outlets and to order their groceries online, in most cases paying a delivery charge. However, so far as customers were concerned, what was on offer – the convenience of home delivery – was nice rather than need-to-have. Despite all the investment in infrastructure, sales stayed pitifully low.

The big problem was demand: only for a narrow market segment of well-heeled, but time-poor, consumers was the convenience of avoiding a trip to the supermarket seen as a significant benefit. Most consumers preferred to choose fresh produce for themselves and to seek the best prices at the store. They may not have been passionately devoted to their supermarkets but they were familiar with them. For Webvan's plan to succeed, a major change was required in the lifelong habits of hundreds of thousands, eventually millions, of consumers. Such changes can only be effected gradually – and the benefits to the consumer need to be substantial and crystal-clear. Convenience often forms part of a winning proposition but is rarely sufficient to trigger a major shift in behaviour.

Tesco, the largest supermarket chain in the UK, made a considerable success of home grocery deliveries, reaching online revenues of £1.2 billion in 2006, by running them as an adjunct to its normal operations. In this way it benefited from existing economies of scale and scope and from building on long-established customer loyalty. It was able to grow slowly initially and could cover its incremental costs at much lower levels of sales volume than Webvan. Albertson in the US also succeeded early with a simple model.

The comparison between Webvan and Amazon is striking. Jeff Bezos's starting point was his insight that an online bookstore could offer customers greater choice and convenience and lower prices than conventional bookstores. Technology and logistics were used as means to this end, rather than ends in themselves. Scalability only became important in enabling Amazon to continue to provide good service at high volumes. Its *raison d'être* was a powerful proposition to customers.

Starbucks and Naxos also built their businesses around making customers an irresistible offer. For customers who had never previously tasted espresso, cappuccino or caffe latte, Starbucks offered a new and sometimes addictive experience. Classical music lovers hesitant about spending £15 on a CD would happily take a punt on Naxos at a third of the price.

All the outstanding successes in creating new markets or transforming existing ones have been based on uniquely compelling and easily understandable propositions for customers. Google offered millions of Internet users a way of finding information on just about any subject, instantaneously and free, and advertisers an easy, affordable way of reaching customers who already had an interest in their products and services. The Apple II was the first personal computer that people without previous technical experience could use confidently. The Macintosh took user-friendliness to a new level.

In all cases, the new proposition is based on a set of capabilities or assets that are to some extent unique to the provider – it cannot be generated without them. In the case of lowest-cost producers like Wal-Mart and Dell, the proposition derives mainly from their unequalled efficiency. BSkyB offered British television viewers packages of live football coverage and recent feature films that were unobtainable elsewhere and proved irresistible to millions.

The proposition must be exceptionally strong to persuade customers, particularly the first ones, to change long-established habits, to desert trusted suppliers or to try something new. In many cases they are taking a chance with a business they do not know and therefore may not trust greatly. Every new business has to pass this hurdle of winning new customers and keeping most of them.

The Internet boom, when entrepreneurs' focus was more on rises in share prices than on business fundamentals, encouraged a reckless disregard of the need to satisfy real customers. The reason most dot.com ventures were flaky, apart from the similarity of so many of their strategies, was that their propositions were feeble. The business plans of many, which had raised millions in venture capital, paid scarcely any attention to the issue of why customers should use their services.

Value propositions define a business more sharply than mission statements ever can. The most fundamental questions any business needs to ask of itself are: what exactly do we have to offer to customers and why should they buy from us rather than someone else?

The creative destroyer

For many engineers Jim Clark was the iconic entrepreneur of the Information Age. After a tough childhood, he was thrown out of school at sixteen for insubordination and joined the Navy. Discovering exceptional abilities in maths and physics, he started studying at night school, eventually obtained a PhD in computer science and became an associate professor at Stanford. He still had insubordinate tendencies, but also a gift for attracting talented engineers. With a group of graduate students he invented something he called the Geometry Machine and was later to be known as the workstation.

This was the first microprocessor-based device that enabled realistic, graphical simulation of complex objects in three dimensions. It eventually transformed engineering design, making graphical representation affordable for thousands of designers and engineers. With six graduate students from Stanford, in 1981 Clark formed Silicon Graphics, Inc., which became a leader in the new industry sector for high-powered computer workstations and attracted some of the best engineers in Silicon Valley. It became *the* place to work. George Lucas and Steven Spielberg used SGI workstations to create state-of-the-art special effects in movies like *Jurassic Park* and *Terminator II*.

However, most venture capitalists, and companies like IBM, HP and DEC, could not see the Geometry Machine's potential. Clark eventually raised finance from Glen Mueller of the Mayfield Fund, but on terms he later felt exploited his business naivety. As the company needed further infusions of capital his stake was further whittled down. He himself had little interest in running the business, but was infuriated when Mueller made the conservative Ed McCracken CEO. They came to dislike each other intensely. Clark felt that 'fucking Ed McCracken', as he invariably called him, had no feel for technology and not much grasp of where the industry was heading; McCracken regarded him as a wild maverick. Both had a point.

Silicon Graphics took off in the mid-1980s. Its IPO in 1986 was a stunning success and it was hailed as 'the new Apple'. Clark was a hero to his engineers but, although nominally chairman, he felt increasingly alienated from the board. They may have been more reasonable, but Clark was more far-sighted. He realized, probably before Bill Gates, that the remorseless advances in power and fall in price of PCs meant that workstations would eventually be submerged, and that Microsoft was likely to come to dominate the high end of computing along with so much else. He wanted Silicon Graphics to make a low-cost product to compete with the PC, effectively to cannibalize itself. Predictably this strategy did not appeal to McCracken and the board.

They were equally sceptical about his other big idea, the telecomputer, the television as a computer. Clark led SGI's participation in the Orlando interactive television project, into which Time Warner and others were pouring hundreds of millions. When McCracken wrested control of the project from Clark, it was the last straw. He resigned in January 1994 and announced that he would be forming a new business. He did not yet know what it would be, but he was determined never again to let VCs and professional managers call the shots. In his next venture, the person who found the next new,

new thing, i.e. him, would be the main winner. The engineers who helped him to make it happen would come next. And the suits would be a distant third. We shall see how well that worked in the Netscape story in Chapter 7.

The paradox was that Clark did not really want to run anything. He simply wanted to be a one-man force for creative destruction, endlessly starting new things and becoming enormously rich. Who then was actually going to build these businesses, if not the despised professional managers? This was the dilemma at Silicon Graphics, decided by the pyrrhic victory of the pros, and while they produced good financial results for a while, they had little idea where to lead the business after Clark's departure. It was to be a major fault-line at Netscape, where the technologists held sway, and Clark only gave management a fraction of his easily diverted attention.

Disciplined entrepreneurialism

Silicon Graphics encapsulates the tension in entrepreneurial companies between the creative, sometimes anarchic, elements and the forces of managerial order. Achieving the elusive balance of disciplined entrepreneurialism – maintaining both creative energy *and* professionalism – is a trick that very few organizations pull off.

Many creative people and organizations are hopeless at practical tasks, and practical people and organizations frequently lack a creative spark. There is an inherent tension between originality and efficiency, radical thinking and disciplined activity. Too much brilliance with no discipline means products don't get delivered on time and costs run wild. Too much emphasis on efficiency and control crushes originality and inhibits experimentation. Outstanding entrepreneurial organizations achieve the elusive golden mean.

Most executives and management pundits, whether they acknowledge it or not, tend to fall into either the conservative or the radical school of thought, and to be dismissive of the other. They have spent so much time battling the other's excesses that their views are polarized. What marks out innovative but efficient organizations is what appears to be a fusion of right-brain and left-brain thinking, symbolized by some apparently oddly assorted partnerships between people with very different but complementary skills. The brilliant Stanford PhD drop-outs Sergey Brin and Larry Page recognized that the battle-hardened Eric Schmidt had much to contribute to Google. Pierre Omidyar, eBay's idealistic founder, worked closely with the professional CEO, Meg Whitman, when they made the crucial transition from folksy community to global business.

The most remarkable of these partnerships was that formed by the lifetime NTT engineer Keichi Enoki, charged with launching DoCoMo's i-mode service – a mobile telephone version of the Internet for Japanese consumers. Enoki knew that traditional telephone company culture would stifle the kind of innovative approach required to get a radically new service off the ground. It would need a different kind of organization with a different culture from the parent, and new blood. He recruited two executives with no previous involvement in telecoms or technology but brilliant marketing and strategic skills, Mari Matsenuga and Takeshi Natsuno.

Their strategy was based first and foremost on satisfying consumer needs – everything else, including technical excellence, was subsidiary. To the engineers who ran telecom companies like NTT, notorious for delaying product launches until every technical detail had been perfected, this was heresy. So was the decision to make i-mode not a vertically integrated operation but a collaborative venture with hundreds of independent service providers. DoCoMo adopted the technical standards most likely to encourage these companies to participate, rather than those favoured by most handset manufacturers and network operators. The i-mode service was launched in February 1999 and by the end of the year had 3 million customers. A year later it had 14 million customers and 44 million by 2005, far more than any other mobile data service.

Enoki's courage and vision in appointing Matsenuga and Natsuno, and giving them their head, was a decisive step. Without them and their ideas, DoCoMo would never have developed the enlightened strategy it pursued, the capabilities to orchestrate such a collaborative undertaking and propositions powerful enough to attract millions of consumers. Also, i-mode benefited from the reluctance of most NTT employees to join the new organization – those who did were willing and able to embrace new approaches.

Some of these partnerships have been harmonious. Others, like Tony Ryan's and Michael O'Leary's at Ryanair, and Sam Chisholm's with several of his colleagues at BSkyB, stormy. Dynamic tension is probably the norm and can be productive, but if it comes to a pitched battle and one school of thought ever achieves complete victory, the result is usually disastrous. The lunatics almost took over the asylum at several points in Apple's history, and similar failures of self-discipline have been the downfall of many ego-driven start-ups, notably during the dot.com era.

In most cases of corporate venturing, the typical outcome is the pyrrhic victory of the professional managers – rules and corporate criteria outweigh the needs of the market, decision-making becomes hierarchical and 'sensible', energy and entrepreneurialism seep away. This happened even at Apple, during the disastrous reign of John Sculley, where traditional marketing and pricing criteria were applied to a market that was still evolving. Likewise IBM bureaucracy suffocated the PC business that the accidental entrepreneur Don Estridge had built. Something similar happens at many start-ups when venture capitalists seek to ease the original entrepreneur out of executive authority and replace him with a professional manager.

Conflicts between the forces of order and subversion are almost inevitable when a company is growing fast: the eBay pioneers resented the growing influence of the corporate types, and there was an open rebellion over the decision to buy off arch-enemy AOL with $750,000 – and at Cisco, the hard-nosed businessmen John Morgach and Don Valentine ruthlessly disposed of the services of Sandra Lerner and Len Bosack, the idealistic founders of the company.

Commentators like to view these battles as good guys (creatives) versus corporate bullies. Almost invariably this over-simplifies and is often simply untrue. The brilliant Jim Clark might have achieved more at Silicon Graphics if he had been capable of working with rather than against Ed McCracken. The pragmatic alliance between eBay and AOL made strategic sense; it was a necessary step towards becoming a grown-up business, as well as a community: if the ruthless AOL, with its enormous customer base, had competed directly with eBay, it might have destroyed it; and Cisco needed to become a professional marketing operation if it was to penetrate corporate markets and maintain its lead over rivals.

The history of Apple highlights the perils of unbridled creativity and ego. Steve Jobs in the 1980s inflicted enormous damage on the business he had co-founded (and that he now leads so successfully). In the interim Apple went to the other extreme and allowed a 'professional' CEO, John Sculley, to come close to squeezing the life out of it. Miraculously, Apple remained an immensely creative company that produced a succession of stunning products.

This balance is therefore a very rare accomplishment. The trick is to become professional up to a point while retaining something of the maverick spirit that gave birth to the business. In the long run, as the business gets bigger and requires more and more predictability and therefore more discipline and convergent thinking, the anarchic tendency tends to be squeezed out, and with it often the creativity.

Entrepreneurialism is popularly associated with the idea of the individual business creator, but strategic visionaries need other people with complementary abilities to realize their visions. In most of the cases we have considered, one individual is the creative force behind the business, but it takes a team with complementary skills to make it a reality.

In a few cases (Michael Dell at Dell, Jeff Bezos at Amazon), that individual takes the lead in building the team and leading the business in the long term, but these are the exceptions and even these leaders turn to others to complement their own abilities. For the most part entrepreneurs like Howard Schultz at Starbucks and the founders of Google have concentrated on strategic or technical issues and delegated decision-making on detailed operations to people with more management expertise or interest. Sensible entrepreneurs choose managers different from and less naturally entrepreneurial than themselves.

Good teamwork does not necessarily mean constant harmony or that working life is comfortable – it rarely is at rapidly growing start-ups. However, the success stories have all been good at getting large numbers of people with different skills and backgrounds to work effectively together.

Market creators tend to be innovative in several ways, but the critical innovations are those that have a bearing on business success, rather than personal satisfaction: the power of the proposition, differentiation from competitors, effective organizational structure and supply chains, and binding relationships with customers. The market creators profiled here rarely pursued innovation for its own sake, but constantly sought ways of making products more attractive to customers, of differentiating themselves more sharply from competitors, of developing stronger relationships with suppliers and customers.

Amazon, Ryanair, Starbucks, Dell and Cisco are classic examples of constant incremental, customer-biased improvements. Amazon concentrated consistently on changes that would make it easier to buy books, from reviews and suggestions on other purchases, to 1-click ordering. Almost every change introduced at Ryanair, from online reservations to airport selection and charging staff for their coffee, has been aimed at cutting costs and lowering fares.

Michael Dell's philosophy was always: 'I'd rather design the winning system to sell and support customers than design an incredibly technically proficient microprocessor that nobody wants to buy... There are lots of technologists who will create wonderful things and then go try to find people to buy them. We started with the customer, and then worked our way back.'

Xerox PARC on the other hand epitomizes the pursuit of innovation and intellectual perfection as an end in itself, with no concern for getting a product out of the door. In this it was not unlike the R&D arms of many a corporation which produces stunning product ideas that the main organization has no idea how to market.

Can mature organizations be entrepreneurial?

Some mature companies are entrepreneurial but they are very rare. Only a handful of our successful cases – Nokia, DoCoMo, BSkyB, Sony, and the resurgent Apple – were not entirely new start-ups and they were highly unusual organizations. Nokia was a completely reinvented company, one of the tiny number of cases of self-administered creative destruction that created more than it destroyed. DoCoMo's i-mode operation was deliberately established as autonomous and separate from its parent. Sony was for most of its life an outsider in the Japanese business establishment and prepared on numerous occasions to bet the company on a new idea. BSkyB likewise drew strength from seeing itself as an outsider, even when it had become a powerful incumbent. Apple, of course, has never really behaved like a mature company.

Pundits like Richard Foster argue that mature companies are only really good at convergent thinking, which focuses on clear problems, and produces well-structured solutions quickly. They can be effective at handling small, incremental changes but hopeless at transformational ones. The more optimized organizations are for handling well-structured operational problems, the less likely they are to be able to tackle creative challenges. The skills required for divergent thinking – conversation, observation and reflection – are stifled by conventional corporate control systems.

The key difference is that entrepreneurial companies are oriented towards exploration and discovery. They make progress through trial and error, through pursuing many options simultaneously, through making lots of little mistakes and learning from them – what Eric Beinhocker calls deductive tinkering. Mature successful businesses are optimized for exploitation and execution. They want to measure progress, eliminate redundancy and avoid mistakes. They are focused on markets, technologies and processes that they know and understand. Radical ideas threaten the status quo, are risky and seem to offer at best uncertain returns. When radical ideas are generated in large companies, those who develop them often cannot get them accepted and leave to pursue them elsewhere.

This is particularly common in cases of technological innovation: Intel was created by a group of engineers who left Fairchild, itself in its day no mean entrepreneur in electronics. AMD was another breakout from Fairchild, and HP, the grandfather of Silicon Valley, served as a training ground for countless executives who went on to found their own companies. HP was outstandingly good at what Clayton Christensen calls sustaining innovation but was only occasionally willing to pursue radically new ideas. IBM in the 1980s was even slower, and many of its alumni went on to found outstandingly successful new companies like EDS, SAP and, astonishingly, Craigslist.

For most of its history, though, IBM has been better than most at reinventing itself. Other exceptions to the general rule include 3M in adhesives, Johnson and Johnson in healthcare, L'Oréal in cosmetics and GE across a wide front. Generally as companies get larger and more successful, they tend more and more to value discipline and efficiency, which inevitably tends to squeeze out the entrepreneurial spirit.

The four attributes of market creators considered so far – distinctive capabilities, radical vision, compelling proposition and disciplined entrepreneurialism – are the most difficult to attain and their combination is extremely rare. Market creators also require four other qualities, which we shall consider in Chapter 6. But first we look at two businesses which fell seriously short.

5

SHOOTING STARS

AOL and Netscape played enormous parts in popularizing the Internet and seemed set for superstardom. Netscape created the market for easy-to-use Internet browsers and paved the way for the mass usage of the Web, but never established a sustainable business model. AOL was stunningly successful at cultivating and marketing the new concept of online community and for a brief period became the most highly valued and powerful of all online companies, but its model too was not sustainable.

Netscape

The founders

Netscape was the creation of two unusually innovative men, Marc Andreessen, a 22-year-old recent computer science graduate and Jim Clark, the 52-year-old founder of Silicon Graphics (see Chapter 4).

Andreessen had developed Mosaic, the first Web browser suitable for a mass market, while working part-time at the National Center for Supercomputing Applications (NCSA) and still a student at the University of Illinois. Previously, the Web had only been a research tool for scientists like its inventor, Tim Berners-Lee. Using it had been hard work, as the browser Berners-Lee had released with his www protocols was text-based and only ran on expensive NeXT workstations.

Mosaic was a brilliant piece of software, easy to use and install, which displayed Web pages inside a familiar window, with scroll bars, buttons and pull-down menus. It allowed graphic images as well as words and made the Web colourful and fun. Users could see hyperlinks easily and move from page to page just by clicking on links. Within weeks of Andreessen posting a Unix version of Mosaic in January 1993, thousands of people downloaded a free copy. When in October they added versions for Windows and Macintosh, downloads accelerated. Over the next two years, the Web grew at 25 per cent per month and became a mass medium, with millions of users.

Andreessen was the ideas man in the team, while his main partner, Eric Bina, a full-time programmer at NCSA, wrote most of the 9,000 lines of code. Bina was initially against the idea of incorporating images in the browser as this would greatly increase the size of files and possibly choke up Internet traffic. This concern was conventional wisdom for several years, but Andreessen insisted that it would add enormously to the attractiveness of the Web for users, and that transmission capacity could easily be increased. He proved to be right on both counts.

The idea of making a business out of Mosaic did not occur to him. Andreessen was not yet in Silicon Valley but rural Illinois, where there was minimal infrastructure for new business creation. However, he was annoyed that the credit for his creation was being appropriated by the NCSA, and in December 1993 he took a job with a software company in California. There he received an email that changed the course of history:

Marc. You may not know me, but I'm the founder and former chairman of Silicon Graphics. As you may have read in the press lately, I'm leaving SGI. I plan to form a new company. I would like to discuss the possibility of your joining me. Jim Clark.

Clark had already founded one successful company and was to make billions floating two more before the decade was over. Yet he was not exactly a businessman: indeed, he rather despised conventional business people, and always remained temperamentally an outsider. He was first and foremost an engineer, and his main passion during much of this period was building the biggest, most technically sophisticated yacht the world had ever seen.

Start-up

Clark's interest in Andreessen was as the possible core of a new business, rather than in the browser itself. When they met, Andreessen told Clark that he was finished with Mosaic. Over the next few weeks they talked about interactive TV and a possible online Nintendo network. Then Clark made one of the frequent changes of direction that so alarmed his conventionally minded colleagues. All this talk of building an information superhighway through interactive television was irrelevant: it was the Internet that was becoming the superhighway.

> All of a sudden it was clear to me when I looked at the Internet that I was looking at the personal computer in 1985. It was this slow, clunky technology, but people were using it. And it would get faster. I realized that this was the thing I'd been groping for… How could anybody make money on the Internet? I didn't have a specific answer yet, but I figured that with the Web-and-Mosaic-enabled Internet already growing exponentially, you couldn't help but make money.

He told Andreessen, 'If you can hire the entire Mosaic team to do this, I'll invest in it. Screw the business plan and the conventional investors.'

Clark's words encapsulate both his brilliant flashes of insight on the big picture and the reckless distaste for rigorous business thinking that was to dog Netscape throughout its short life. In its meteoric rise and fall, it never quite developed a coherent business model, let alone a strategy. This did not prevent it from scaling some dizzy heights.

Clark went with Andreessen to Illinois, where they recruited seven former NCSA colleagues, including Bina. Clark wrote out their contracts on his laptop on the spot, giving them all generous allocations of stock. He put in $4 million of seed capital and in April 1994 Mosaic Communications was incorporated. (The name was changed to Netscape Communications in November, following a lawsuit by NCSA, which cost the fledgling company $2.3 million.) The first business plan targeted the business market and aimed to 'develop, deploy and widely license a next-generation, commercial-grade Mosaic client, server and authoring suite'. Quite what this meant in terms of products and customers was not clear.

Clark hired Rosanne Siino from Silicon Graphics as director of public relations, and she immediately persuaded the press to run excited stories about Andreessen. *Fortune* named Mosaic as one of its '25 Cool Companies' in June, before it even had a product, let alone a customer. This undoubtedly helped in interesting investors, but may have alerted Microsoft earlier than necessary. Many of Netscape's subsequent problems would arise from too much hype and uncontrolled early growth.

Because business press coverage generated so much free publicity and because it soon became clear that the company would be a big hit on the stock market, Netscape's priorities became distorted. Potential investors were if anything a more important audience than customers, and the measure of success became market capitalization rather than sales or profits. In this Netscape set a trend followed by many subsequent Internet companies. Clark left the product development and leadership of the technical teams to the talented but inexperienced Andreessen, while he focused on raising more funding.

Andreessen's main task was to completely rebuild and improve on the Mosaic software with new code. The core of his team were his fellow alumnni from NCSA. Former Silicon Graphics people were knocking Netscape's door down, but their experience was in workstations, not PC software, though this was not seen as a problem at the time. Hundreds of programmers were hired very quickly, many of them from the student hacker school of software development. There was a great deal of horseplay, shouting and late-night pizza deliveries. Not much thought was given to architecture, building support among outside developers, or the threat from Microsoft. Andreessen was obsessed with speed and 'Internet time'. He wanted to roll out products faster than any software company had ever done.

The company operated without a full-time CEO until January 1995, when Jim Barksdale, who had been a senior executive at large companies like IBM, Federal Express and McCaw, arrived. He was not a technologist and was even alleged to get his secretary to handle his email. Andreessen, who had no management experience at all, continued to have a free hand as Chief Technology Officer, and was frequently Netscape's public face, rather as the young Steve Jobs had been at Apple.

Within the company there was a long debate, never entirely resolved, on business model, pricing and product. Mark Homer, the new head of marketing, wanted to charge $99 for the browser, $5,000 for the basic server and $25,000 for the commerce server that incorporated encryption. Andreessen wanted to give away the browser to create a mass market. 'You can get paid by the product that you are ubiquitous on, but you can also get paid on products that benefit as a result. One of the fundamental lessons is that market share now equals revenue later, and if you don't have market share now, you are not going to have revenue later.' His strategy was undoubtedly successful in securing rapid diffusion and market penetration but did not address the issue of how revenues would be generated long-term.

Netscape eventually decided to ask most users to pay $49 for the browser after a ninety-day free trial. Corporate customers largely paid up; consumers largely did not. The prices for the servers were set at $1,500 and $5,000, but browsers still accounted for two thirds of revenues for the first year.

The beta version of the browser was posted in October 1994, and within days hundreds of thousands of copies were downloaded. In December, the alpha version of Netscape Navigator 1.0 was released, and over 3 million copies were taken in the next three months. Demand for server software was buoyant too, as thousands of firms rushed to build websites. By March 2005 quarterly revenues had reached $7 million, and Netscape had 75 per cent of the browser market. Revenues hit $12 million in the second quarter, but without yet making profits.

In the feverish climate that was then developing, this did not seem a serious barrier to going to the stock market. Indeed, Netscape's flotation was the most celebrated one of the century.

IPO

Clark started out with a big grudge against venture capitalists, who he felt had robbed him of most of the shares in Silicon Graphics. He was determined not to allow them to do that again – he expected first-round investors in Netscape to pay three times as much for their shares as he had. Most baulked at this, but John Doerr of Kleiner Perkins was receptive. After burning his fingers on pen computers and an unsuccessful games venture, Doerr needed a hit and was ready to be convinced by the argument that 'This thing is so big, the only thing to do is dive right in.' He invested $5 million on Clark's terms.

This, and the fabulous success of the subsequent IPO, represented a major shift in the balance of power. For the next few years at least, venture capitalists, who had been accustomed to dictating terms to entrepreneurs, would have to meet their demands.

In 1995 Clark started pressing for an early IPO. According to Lewis, he knew he was going to need a lot more money to finance his boat, and he had a hunch how popular the issue could be. Barksdale was initially opposed to going public before the company had established a track record. Clark, however, received predictably enthusiastic support from Frank Quattrone of Morgan Stanley. In February, Quattrone organized a private placement: a group of media companies bought an 11 per cent stake for $17 million, valuing Netscape at

$150 million. In June, Spyglass, Netscape's main competitor, which was supplying its browser software to Microsoft, had a modestly successful IPO. Clark persuaded his board that they should go in August, just before Microsoft launched its long-awaited Windows 95.

The prospectus announced that 3.5 million shares would be sold at a price of $12–14, implying a valuation of $500 million. During the subsequent investors' roadshow, interest was so strong that they decided to issue 5 million shares at $28. In fact, when trading started on 9 August, the stock opened at $71 and closed the day at just over $58, valuing the company at $2.2 billion. Clark's shares were worth $565 million, Kleiner Perkins's $256 million, Andreessen's $58 million. Each of the other former NCSA team members was now worth over $20 million.

Clark was jubilant:

> Netscape obviously didn't create the Internet. But if Netscape had not forced the issue on the Internet, it would have just burbled in the background. It would have remained this counter-intuitive kind of thing. The criticism of it was that it was anarchy. What the IPO did was give anarchy credibility.

Unfortunately, this was all too true. It also encouraged the people running Netscape to believe that they could walk on water and slay the Beast from Redmond. The reverse was to prove the case.

Competition

Clark understood the strategic realities better than anyone. In December 1994, without telling any of his collleagues in Netscape, he had made Dan Rosen of Microsoft a remarkable approach:

> We want to make this company a success, but not at Microsoft's expense. We'd like to work with you. Working together could be in your self-interest as well as ours. Depending on your interest level, you might take an equity position in Netscape with the ability to expand that position later... Given the worry that exists

regarding Microsoft's dominance of practically everything, we might be a good indirect way to get into the Internet business.

This offer was ignored, but in June 1995 Rosen, without making reference to Clark's email, told Mike Homer that unless Netscape gave Microsoft a seat on its board and equity, they would put it out of business. Clark got his lawyer to inform the US Department of Justice (DOJ), because, as he told Lewis, he 'didn't think Barksdale was going to do anything about it'. He himself told the DOJ in 1995 that the threat was not Microsoft's Internet service business, MSN, which had attracted much public anxiety, but what Microsoft could do in the browser market. It is questionable whether at this stage Microsoft saw the situation as clearly as he did. It was not until the end of the year that Gates made the Internet central to his strategy. The DOJ took no action for a further two years, and Clark left it to his lawyers to press them.

After the summer of 1995, Clark's main attention was on his yacht and on his plan to make his next billion dollars. This was nothing less than a rather hazy scheme to revolutionize the $1.5 trillion American healthcare industry, which culminated in another stunning IPO in 1999. He remained chairman of Netscape and occasionally told Barksdale that Microsoft was going to take over the browser business as it was taking over workstations. Barksdale appeared, according to Charles Ferguson, to believe that he knew better. He told Ferguson in September that Microsoft would never catch up with Netscape. The Internet, he said, rewarded openness and non-proprietary standards. He also took little notice of Clark's, characteristically shrewd, suggestion that the real action was now in portals and that Netscape should be developing its business there. It did for a while have the most frequently visited home page, as the browser made Netscape's the default setting.

The conventional wisdom in 1995 was that Netscape had established an unassailable lead in browsers. Microsoft's first effort, Internet Explorer 1.0, had been a reassuringly feeble imitation. Netscape Navigator 2.0, which came out in beta in September 1995, seemed to clinch its lead. It extended its market share to over 80 per cent of the browser market and revenues in its second year reached $346 million. This version also incorporated Java software, which Netscape joined Sun Microsystems, Microsoft's loudest critic, in hyping noisily. Java offered the possibility of a threat to Microsoft's dominance of the desktop, as it might in the future enable third-party applications to be written directly for the browser, with no involvement of the PC operating system.

If this was provocative, Andreessen's subsequent comment that Windows might soon be reduced to a 'poorly debugged set of device drivers' was reckless. At this stage the Java threat was entirely hypothetical – it never did materialize – and Netscape did not remotely have the resources to attack the most formidable competitor in the industry. Posing as the champion of the software community, most of whom loathed Microsoft, won applause in Silicon Valley, but guaranteed a ferocious response. In December 1995 Bill Gates announced that Microsoft was now 'hard core about the Internet'. Internet Explorer would be improved and distributed free, as would Web server software. Microsoft would do everything it could to take market share from Netscape, including licensing Java. Netscape's share price, which had reached $170, fell $30 on this news.

Gates was as good as his word. Microsoft mobilized an army of programmers, many of them previously occupied with interactive television, and flexed its competitive muscles. According to evidence presented to the subsequent anti-trust trial, it deliberately integrated Internet Explorer with Windows. Gates asked AOL executives, 'How much do we have to pay you to screw Netscape?' (A few months earlier, Gates had been threatening to bury AOL, then the largest online service provider and an obstacle to his ambitions for MSN.) He now made AOL an offer it couldn't refuse – an AOL icon on every Windows desktop. In March 1996 AOL announced that Internet Explorer would be its default browser. CompuServe, then an important ISP, announced a similar deal a few days later. Microsoft subsequently told Compaq, then the leading supplier of PCs, that unless it reversed its decision to replace IE with Navigator as the default browser, it would lose its licence to Windows 95. It quickly complied and a few months later dropped Navigator entirely. Over the next two years, Microsoft persuaded most ISPs and even Apple to make IE their default browser.

In August 1996 Microsoft announced IE 3.0, a product as good as Netscape's. Version 4.0 the following year was generally acknowledged to be superior. Now Netscape was paying the price for its lack of solid architecture. It took longer to develop new versions of its browser because it had not designed interfaces for later additions. It tried a new strategy, bundling mail and news readers into the browser together with a Web page composition tool, but this made the software program larger, slower and less reliable. Barksdale then decided to rename Navigator 'Netscape Communicator' because it was a 'general-purpose client application'. This, however, confused the market. In a last desperate move, in 1998, it made Communicator open source and codenamed it Mozilla. By 1998, when the DOJ's anti-trust case opened against Microsoft, Netscape's share of the market had fallen to 40 per cent and Microsoft's had reached 50 per cent. At this stage everyone assumed that Microsoft would dominate the market, as indeed it very soon did.

In 1999 AOL acquired Netscape for $4.2 billion. This was not a bad exit for stockholders, considering the scale of the defeat it had suffered, and what happened to share prices the following year. AOL would continue to develop further versions of Mozilla, but Netscape had ceased to be a significant force in the Internet industry.

Strategic audit
Netscape fell short on most of the success factors for market creation and leadership:

Clear, radical strategic vision Netscape's strategic vision was never entirely clear, and it was based more on its technology than on serious insights about customers. Clark and Andreessen knew that the Web was going to be big and that browsers and server software would be critical to its take-off, but that was about it. Clark was sure he could make money out of it, and he was certainly right about that, but the amounts made were largely due to share transactions and the Gold Rush mentality of the era rather than earnings. By any objective investment criteria, Netscape was overvalued both at the time of its IPO and when it was acquired by AOL, let alone the stratospheric valuations it reached in the interim.

The founders can be forgiven for vagueness initially, since scarcely anyone had a clear idea in 1994 of how the Web would evolve. They started with great insight and a brilliant innovation, which made possible an information and communications revolution, but it was not a fully-fledged product. They had only the vaguest idea of who their customers would be, other than just about everyone, or how much they would pay – in most cases, nothing. It would have been difficult for anyone in Netscape's position to define a satisfactory business model. This illustrates the difficulty most technology pioneers face in capturing returns on their innovations.

Distinctive capabilities Netscape had at least three distinctive capabilities, which took it a long way: a brilliant, and initially unique, piece of software, the ability to move fast to develop it further, and a better understanding than most of the likely evolution of the Web and of the need to capture market share and mind share quickly.

However, it also had some serious weaknesses simply as a software developer. None of its top three people, Clark, Andreessen and Barksdale, had any previous experience of managing the professional development of software or marketing it. Most of those it recruited had no experience of developing software for PCs and, according to one of its severest critics, Charles Ferguson, who viewed it at close quarters, Netscape was not just strategically naive but simply unprofessional.

Ferguson's main criticism is that Netscape's initial products and approach to production lacked proper software architecture: they did not have a carefully designed overall structure with clearly defined interfaces that would enable the construction and smooth development of successive generations of industrial-scale products. This made Netscape's products difficult to integrate with those of other developers. The lack of interfaces, or APIs, meant that third-party developers could not add functions to the browser or embed it in other software products. Steve Case claimed that this was one of the factors in AOL's decision in 1996 to choose Internet Explorer as its default browser.

Microsoft may have lacked Netscape's flair but it could never be accused of a lack of professionalism. This, as well as ruthless deployment of its power over ISPs and PC vendors, was a significant factor in its crushing of Netscape.

Compelling customer proposition From 1994 to 1996, Netscape's value proposition to consumers, particularly those who obtained the browser free, was virtually irresistible, as reflected in its 80 per cent market share. However, once Microsoft offered Internet Explorer and server software free, and the quality matched Netscape's, the power of Netscape's proposition plummeted.

The proposition to businesses on servers, the crucial one for long-term revenues, was always rather less compelling. The market for server software was more crowded, less homogeneous and more conservative.

Entrepreneurial but disciplined organization Netscape was unquestionably entrepreneurial but woefully short on discipline. It did not form well-balanced teams and in particular it did not achieve the elusive combination of creativity and practicality. If the brilliant Andreessen had been working with an experienced mentor, progress might have been slightly slower but steadier. And he would not have been allowed to make declarations of war on the most powerful competitor in the business.

Clear consistent leadership Clark was right not to be a hands-on chairman, but he should have appointed a CEO he really trusted, rather than one who would appeal to investors, and he should have insisted that the company take the competitive threats seriously.

Coherent strategy and business model Netscape never really had a solid business model and its strategy was essentially based on the theory of first mover advantage and the delusion that it would always be better and faster on the draw than its competitors.

First mover advantage only works when barriers to entry are high and/or the company is able to form such strong relationships with customers that they cannot be prised away (see Chapter 8). Neither applied strongly in the case of the browser, particularly as Microsoft offered its own completely free. It was also able to capitalize on other advantages to squeeze Navigator hard – its greater financial resources, its brand, its leverage over intermediaries like ISPs its complementarity with the operating system. It showed that it was perfectly capable of catching up with Netscape in terms of product features, not least because these were comparatively easy to replicate.

If Netscape had made its software the industry standard, it would have been in a much stronger position, but it is questionable whether this was really possible. Once Netscape had appeared to pose a threat to Microsoft's domination of the desktop, and once Gates had realized the importance of the Internet, he was bound to crush such a rival. As Eric Schmidt remarked to Ferguson, 'there is nothing harder to do than to set up a positive returns monopoly in the presence of an existing monopoly. It is the hardest business challenge in the world.' The problem, as Ferguson retorted, was that Barksdale failed to understand the challenge he faced.

Strong customer relationships and brand Netscape certainly built a high profile quickly, though it is questionable whether this constituted a strong brand to which large numbers of customers were loyal, one that represented a guarantee of quality. One notable brand attribute prized by its most vocal supporters was that it was not Microsoft, that it appeared to be its first serious challenger. Internet users were highly aware of Netscape, not least because, by a trick of the browser settings worthy of its rival, its website became the home page of most Web users. Nevertheless, there is no evidence that many of them, beyond a fanatical hard core, formed a strong emotional attachment to it.

Acquire capabilities to operate efficiently at scale Netscape was slow to bring in senior management of any description until Barksdale arrived in January 1995. Andreessen had been leading most of the company, rather well considering his total lack of experience, and Clark was far too hands-off. Even if Barksdale was not the corporate numbskull that Ferguson portrays, he clearly was not the man to go into battle with Bill Gates. Clark's actions suggest that he viewed him rather as he viewed most professional managers, at best a corporate front-man who would keep investors reassured. What Netscape needed was a CEO Clark could genuinely respect, with some of his own strategic insight and deep understanding of the strategic potential of technology.

Collaboration with other organizations Unlike the other successful technology leaders in the networked economy – Cisco, Microsoft, Apple, Intel, Dell – Netscape failed to build a community of developers, an ecosystem of suppliers and complementary businesses. It is questionable whether it had the capabilities to be a platform leader, but it did not appear to recognize the need.

Spot change in the competitive environment Clark had the best grasp, but his attention was elsewhere. Andreessen was undoubtedly right about the need to move fast and capture the imagination but ridiculously arrogant regarding Microsoft.

Netscape's best strategy would probably have been that implied in Clark's infamous email of December 1994 – to try to coexist with Microsoft. At that point Gates was preoccupied with the launches of MSN and Windows 95. It was Netscape more than anyone else who in 1995 alerted him to the importance of the Internet and the possible threat it could pose. If Netscape had plainly signalled that it simply wanted to develop the browser and Web server space, rather than challenging Microsoft on the desktop, it might have survived. After all, Microsoft only won the browser 'market' by giving away the software, after spending a fortune on developing a clone of Navigator. It was an expensive, mainly defensive operation. And it only did this after Netscape had committed the unforgivable blunder of 'threatening the platform'. Eventually it might have felt compelled to seize control of such a key aspect of its domain, but certainly much later.

However, Netscape should not be written off as a failure – or only in terms of its delusions of grandeur in 1995–6. It did not succeed in establishing long-term market leadership, but it is questionable whether this was an achievable goal. It created some fine products, contributed enormously to the development of the Web and e-commerce, and made a lot of people rich. Most of those who had shares in Netscape did well. The price AOL paid in 1998, $4.2 billion, was almost twice the value at the time of the IPO, and that was inflated in view of its modest earnings. For a venture capitalist like Kleiner Perkins, it was a superb investment. Clark, Andreessen and Barksdale are also considerably wealthier and, one hopes, somewhat wiser.

AOL

AOL was the all-conquering chameleon of the Internet boom years, adapting itself and its business model endlessly, metamorphosing from online gaming to software, bulletin boards to direct marketing, online service provider to advertising and media colossus.

Many online services started up in the decade before the Internet found a mass market, but few would have picked the company that was to become AOL as a future giant of the Internet. Jim Kimsey, its CEO from 1985 to 1991, admitted that 'I wouldn't have given us a one in a thousand chance of surviving.' Making billions was not even a dream.

Control Video Corporation started life in Virginia in 1981 as an online games business, the brainchild of Bill von Maister, a hyperactive entrepreneur. Maister had a talent for dreaming up business ideas and persuading venture capitalists to invest in them, but not much for running them. In two years Control Video Corporation burned its way through $20 million of venture capital and ran up massive debts, while earning scarcely any revenues. One of its investors, Frank Caufield of Kleiner Perkins Caufield & Byers, called on an old friend, Jim Kimsey, to lend a hand, and he reluctantly became chief executive. Another investor, Dan Case of Hambrecht and Quest, suggested that the company might like to use the services of his 24-year-old brother, Steve.

Jim Kimsey and Steve Case, neither of whom had a technical background, were to become the main creators of a company which was to reinvent itself several times over the next fifteen years: its name during Kimsey's tenure was Quantum Computing Services. They were an odd couple: Kimsey was a party-loving former army officer and bar owner; Case, who had been a junior marketing executive at Procter & Gamble and Pizza Hut, was so introverted and inexpressive that his colleagues called him 'The Wall'.

Kimsey's chief contribution to Quantum was simply enabling it to survive. His aim was to turn it round and then get out, but this took seven years. The company struggled to turn a profit throughout his tenure, and he sometimes had to threaten creditors with declaring bankruptcy to get them off his back. Survival sometimes involved questionable accounting practices, a habit the company found it hard to break. However, Kimsey gave it decisive leadership, tried out a number of business models and groomed Case as his heir apparent.

Kimsey told Alex Klein he thought of Case as a Cyborg. 'I mean that in a nice way. He ate, lived and breathed this shit... It was his life.' While travelling the country for Pizza Hut, trying out new toppings, Case had become hooked on online services himself. 'I thought there was something magic in sitting in a hotel room and connecting to all of this.' His voyeurism was to give him an insight into the tastes of ordinary customers that few of his rivals had.

There were many attempts in the 1980s and early 1990s to offer online information and communications services, before the World Wide Web made the Internet a mass market medium. The most successful in terms of numbers of users was Minitel, a videotext service run by France Telecom. Thanks to a massive subvention from the French government, France Telecom was able to give away millions of terminals to consumers and small businesses and to create an instant market. This precursor to the Net enabled hundreds of home shopping and banking services to flourish and eventually reached 25 million users. But to nearly everybody's surprise, the most popular activity on Minitel was what became known as *Messageries Roses*, early chat rooms where the main topic of conversation was sex.

The largest online information service in the US was the stodgily respectable Prodigy, a joint venture between IBM and Sears Roebuck, a top-down, hierarchical service which assumed that consumers would use their PCs essentially as dumb terminals. Despite their sinking over a billion dollars into it, it never came close to showing a profit. When Prodigy eventually offered subscribers email, it limited them to sending no more than thirty messages a month, and strictly forbad chat of a sexual nature. Slightly smaller, but profitable, was CompuServe, used mainly by the technically literate. Its most popular services were email and bulletin boards, but it also allowed access to premium content at a price.

Until the mid-1990s proprietary, commercial services like these were entirely separate from the Internet, which remained the commerce-free preserve of academics, anarchists and techies. Scarcely anyone imagined it would become a mass medium until the arrival of the Web browser. However, public interest in cyberspace in general took off in a big way from about 1990 onwards, though most consumers had only a vague idea of what they could expect to find there. By the mid-nineties public interest was reaching epidemic proportions.

What the early online services did show was that few consumers were willing to pay explicitly for content, though some did buy financial information and pornography. Most dipped into the free information on offer, but what really excited them was the ability to communicate with others in new and generally spontaneous ways. Howard Rheingold coined the term 'virtual community' to describe this curious new phenomenon, but nobody had seriously tried to package and market it. This was to be AOL's contribution to civilization.

Virtual life

Kimsey's initial hope, when he reluctantly became CEO, was to sell the struggling company to a computer company like Apple. When that did not happen, he took the business into developing software that would enable PC users to communicate online, initially in partnership with Apple and Commodore, and subsequently in Quantum's own right. A critical step was acquiring much more user-friendly software than anyone else had, but winning the first early adopter subscribers was painfully slow, and Quantum took time to work out exactly what it was it had to offer them. Money was tight for years, and Kimsey had to conduct frequent lay-offs simply to stay in business.

For a while Quantum was running three separate online services – for users of Commodores, Macs and PCs. In 1989 it combined these into one larger overall community, which was still much smaller than Prodigy or CompuServe. Case gave it the shrewd name of America Online, which paid off. It described what was on offer in a non-technical way and for consumers it sounded reassuringly mainstream. Two years later, America Online, or AOL, was adopted as the name of the company itself.

One of AOL's key features was that it enabled its users to create 'virtual rooms'. Sometimes these areas were defined by information, like sport and news, but more often by the interests and tastes of different groups of individual users. What AOL and similar services offered was an alternative cyber world, where a variety of communities could converse, fantasize, play games, find friends and lovers, and generally interact in previously unimagined ways.

Some users became so addicted that they spent hours every day online, and this became their social life. Many of the chat rooms were about sex, and these generated at least half of the traffic and revenue. AOL, with chat rooms like 'Married But Flirting' and

'Crossdressers2', quietly encouraged this but did not itself provide sexual content. One of its many paradoxes was the contrast between this crucial aspect of its business model and its bland, middle-of-the-road public image.

AOL built up teams of people to facilitate the new and evolving services. Some were moderators of discussion groups, others developers or packagers of various kinds of content, some combining both roles. Many were unpaid, enthusiastic users, prompting *Wired* to call AOL 'the cyber sweatshop'. As time went on, the amount of content grew, so that it started to resemble what we would now call a portal, but the mix was always rather fluid. AOL did not charge for content and looked for cheap, unbranded sources – it made its money from keeping customers online. Interactive services like games and role-playing proved particularly popular, but the biggest channel by far was always 'Personal Connection', where subscribers provided their own content.

Case spent many hours online, mostly observing. His great insight was that the key to opening up a large market was getting beyond the technology. His mantra was 'make it easy and make it fun'. Most of those running and using online services in the early days were themselves technologists and took for granted the computer jargon and complications that baffled and frustrated most ordinary folk. Case realized that to reach beyond that narrow audience of nerds, AOL would need to simplify every step for customers.

There were several barriers to be overcome before the technically inexperienced customer could venture online: first becoming confident about using a PC, then acquiring a modem and attaching it to the PC and the telephone, signing up with a service provider and getting the communications software to work. Each of these was a point where the faint-hearted could easily give up. A major breakthrough at AOL was to put the start-up software on a disc and give it away free – the real revenue came from the monthly subscription of $9.95 and the $6 an hour connection charges. Even a simple initiative like putting a toll-free helpline number on the start-up discs overcame one of the barriers to getting started. Other innovations included borrowing Apple's point-and-click interface, searching on keywords, dividing the service into 'channels' on the lines of television, and having a voice tell people that 'you've got mail'.

Mass market

When Case became president in 1991, AOL had 150,000 subscribers and annual revenues of $20 million. Online services were beginning to look interesting to investors and the company went to the stock market in March 1992. The IPO valued the company at $66 million – modest compared to the enormous Internet valuations that started three years later. However, AOL was being valued as a technology company, still in start-up mode. It rarely made a profit, but its share price proceeded to rise astronomically. For the first time the company had the funds to invest in expansion. Case became CEO, stepped up recruitment and started to spend serious money on marketing.

What shot AOL into hyper-growth was a decision to distribute large numbers of start-up discs to potential customers speculatively. Case was initially sceptical but changed his mind when the direct mail campaign achieved not the normal 1 per cent conversion rate but 10 per cent. AOL then rolled out discs by the million, in every way that it could – in magazines, video stores, even in packets of frozen steaks. Eventually hundreds of millions were given away. Anyone who had a computer and didn't quite know how to get started online could not fail to have a couple of AOL CDs to hand. It was very much easier to do this than sign up with almost any other service provider.

This was the single most successful promotional campaign of the Internet decade. From 250,000 subscribers in July 1993, AOL grew to a million by August 1994, to 2 million the following February and 4 million by the end of 1995. It was during these two years that it shot past Prodigy and CompuServe and made itself the clear market leader. The success of this campaign caused one of several crises of capacity at AOL, when it was flooded with traffic and many users could not get through, but the howls of anguish, when the company was dubbed 'America on Hold', did not seem to put off new subscribers. Throughout the 1990s, demand was growing at ever-accelerating rates, as more and more consumers decided to try the new medium that everyone was talking about. Although many dropped out, there were always more anxious to sign up.

What differentiated AOL from its competitors was marketing: its comparative customer friendliness, its reassuring brand and aggressive promotion. Unlike Prodigy, it also offered a limited form of access to the Internet and the Web. AOL was able to position itself both as an Internet service provider and as a safe, easy-to-use managed service. For most of the 1990s, 75 per cent of AOL subscribers scarcely ever ventured outside AOL's walled garden into the wild Web. Many were unsure of how to do so – the main thing they all wanted to do was use email and just explore other things within the garden. It was a halfway house, which suited the many unsophisticated consumers who were unsure of how to proceed in a strange new environment and wanted a helping hand.

The digerati of Silicon Valley looked on AOL, in the backwaters of Virginia, with something close to contempt. It was cheesy and clunky, the K-mart of Cyberspace. Why would any self-respecting surfer stick with that when Netscape had opened up the whole Web to them? Forrester Research and *Wired* magazine frequently proclaimed the imminent demise of what they called a dinosaur. The critics had a point but assumed that a mass market would behave in the same way as knowledgeable, enthusiastic early adopters of the Net. It was to take several years and the arrival of broadband and reliable search to make most consumers comfortable and confident surfing the Web, but in the meantime AOL was becoming the most powerful player in the online world.

One influential commentator who could see why AOL was succeeding was Mary Meeker. The Queen of the Internet gave it her endorsement in the *Internet Report* in 1996: 'Nobody does a better job of making general information available for a mass market than AOL.' Not entirely coincidentally, her bank, Morgan Stanley, had recently joined with Goldman Sachs and Merrill Lynch to help AOL raise another $100 million.

AOL and Netscape were the first Net companies to carry out

Meeker's favoured strategy of getting big fast. This was not just a question of pushing up the share price. At the back of everybody's mind in both companies was the looming threat from the most formidable competitor of them all. The difference between AOL and Netscape was that AOL did not assume that it would easily defeat Microsoft.

Beating Bill

AOL attracted the early attention first of Bill Gates's former partner, Paul Allen, then of Microsoft itself. Allen bought stock at the IPO in 1992 and went on to build up a holding of nearly 25 per cent, prompting the board to develop a poison pill share structure to repel him. In 1993 Gates invited Case and Kimsey to Redmond to talk. According to popular legend, he mused on the possibilities in a rather chilling way: 'I can buy 20 per cent of you or I can buy all of you. Or I can go into this business myself and bury you.' Kimsey was not entirely opposed to being bought, if the price were right. Case was deeply suspicious and told Gates that the company was not for sale.

When Microsoft announced that it would develop its own online service, MSN, many commentators jumped to the conclusion that this was AOL's death sentence, but in 1993 online was a long way from being Microsoft's top priority: that was the small matter of the launch of Windows 95, the new operating system that was to clinch its domination of the PC desktop. However, Microsoft's declared intention spurred everyone in AOL to expand the business as fast as possible. A new executive, Ted Leonsis, whipped them up into a frenzy: 'Some day your children will ask you what you did in the war. How we stopped Bill Gates from taking over interactive services.'

Case lobbied anti-trust regulators about Microsoft's plan to bundle MSN with Windows, and when Netscape emerged as a major challenger to Microsoft, he wrote to its CEO, James Barksdale, suggesting that they join forces in the 'fundamental imperative of attacking the common enemy', comparing them to Roosevelt and Churchill allying themselves with Stalin against Hitler.

What changed things fundamentally was not these manoeuvres

or anything that AOL did. Although Gates announced in December 1995 that he was now 'hardcore about the Internet', MSN's rivalry with AOL was still only of secondary importance to him: beating Netscape was paramount and AOL, with its millions of users, could make a decisive difference in the browser war. The price agreed 'to screw Netscape' was that, in exchange for making Internet Explorer its default browser, AOL would have parity with MSN on the Windows desktop.

This was an enormous prize that justified a change of alliances. Every time a user switched on the PC the AOL icon would appear on the screen. Case hesitated only momentarily. The deal gave yet another impetus to the already enormous momentum behind AOL's growth. AOL's consumer franchise was overwhelmingly its greatest asset; this endorsement only fortified it. Microsoft went on to marginalize Netscape, but AOL built up an unassailable lead in service provision. Gates later acknowledged that this agreement had destroyed MSN's chances of winning leadership in that market.

What exactly was AOL?

This was a question that puzzled many people, including some within the company. Although AOL was often perceived as a technology company it was not really – it was not even all that effective at using the network technology it needed to handle its millions of users. According to some people it was little more than an electronic bulletin board, though much bigger than any other and more user-friendly. It was gradually evolving into something that was beginning to look more like a media business, and developing an ego to match. In the early nineties AOL had seen itself as more akin to a news stand than a content provider in its own right – it did not so much add value to the content it bought as rework it. Mostly it dealt with little-known providers like Hoovers and The Motley Fool, but in 1994 it negotiated its first deal with a prestigious name, *Time*: it got free advertising in the magazine in exchange for a small share of its revenues from connection charges. This turned out to be more beneficial to AOL than to *Time*, but *Time* did not then have other online distribution channels available to it.

Case greatly admired John Malone's TCI, the cable TV giant, and liked the idea of AOL as a sort of cable channel for the computer. One of the attractions of Cable TV as a business is the tight hold operators have on their customers. Case deliberately recruited two executives with experience of the media world and larger personalities than his own.

The first and most colourful was Ted Leonsis, who was both a brilliant rabble-rouser and a visionary. Given that online numbers dropped off when *Seinfeld* was showing, he argued that *Seinfeld*, rather than CompuServe, was who AOL was now competing against. AOL should turn itself into a major consumer brand. He told the massed ranks of AOL troops, 'We are the number one company in the number one growing industry in the world.' He turned sections like news and sport into 'channels', started to develop the advertising side of the business, and stepped up AOL's own advertising.

In 1996 he was replaced as the main front man by Bob Pittman, a glamorous, charismatic figure who had been involved in many media ventures, including some at Time Warner. He had played a big part in creating MTV, one of the most successful cable channels, of which he had said, 'We don't just shoot for 14 year-olds, we own them.' He now told his delighted new colleagues at AOL that they too owned 'consumer eyeballs'. AOL would not be paying content owners any more, but charging them for access to 'their' consumers. 'We're not a technology company, we're a media company.'

Pittman accelerated the emphasis on selling advertising and signed big deals with emerging e-commerce players like Amazon. This became imperative after AOL moved to fixed-price charging for users and suddenly needed fresh sources of revenue. However, Pittman realized that AOL was becoming the most valuable property on the Net. Its success at keeping those eyeballs within AOL's walled garden, for the present, meant that it had a massive audience to offer to advertisers, and virtually all new Web ventures were desperate to drive traffic to their sites. This was why the cyber cockroach, as some called it, was so indestructible.

What AOL was becoming most of all was a very aggressive marketing company. Although everybody repeated the cliché that content was king, in the jungle of the Web it was customers who were crucial, and AOL had more of them than anyone else. Their real value was their effect on the share price, rather than the revenue they generated. This became the main metric by which the company's success was measured and stock options the main motivator for employees.

Close to the wind

In 1996 AOL was riding on the crest of a wave. In three years it had gone from being number three in an obscure, unprofitable industry to superstardom in the networked economy that everyone wanted to get a piece of. It had just shaken off a potentially lethal threat from Microsoft, and its market capitalization had risen to $6.5 billion.

Yet cracks were starting to appear in the facade. Despite continued growth, or partly because of it, an alarming 6 per cent of customers were leaving every month, due to appalling congestion on AOL's network and prices that were much higher than those of other ISPs. Case and Pittman might talk about AOL's content and compare themselves with Viacom and Time Warner, but what increasing numbers of customers wanted most was reliable access to the Internet and not to be charged through the nose for the time they spent online. Many ISPs were making no time-based charges, while AOL customers were paying $6 a hour.

Case's dilemma was that AOL's shaky profitability depended on those time-based charges. If he moved over to an all-you-can-eat tariff structure, margins would melt away – and congestion would get worse. So he delayed as long as he could. In October his hand was forced – MSN announced that it would be charging $19.95 a month for unlimited usage. AOL promptly followed suit.

This led to another surge in new customers, and to existing customers staying online longer. The congestion problem was becoming a ghastly vicious circle, as some users stayed online indefinitely in order to avoid not being able to get back on later. The protests and complaints from furious customers were deafening and widely reported. AOL's reputation was suffering badly and its business model turned upside down.

This crisis coincided with another. There had long been complaints within the financial community about AOL's accounting practices, in particular its treating marketing expenditures not as an expense to be charged against current income but as capital investment. A struggling start-up might get away with this, but not a large public company. And the costs of distributing hundreds of millions of CDs were not trivial. In October 1996 it wrote off $354 million, more than it had ever made in profits, as 'deferred customer acquisition costs' and ran up a big loss for the year. The Motley Fool, ironically a content partner, put it succinctly: 'The Emperor hath no pants, AOL hath no profits.'

It should have been clear that AOL's finances and business model were far from solid and that its competitive position was nothing like as impregnable as it seemed. MSN might not have been the big threat, but there would soon be others. The financial markets had been treating it as an Internet start-up whose early losses were inevitable, but AOL was not a start-up, and there were big question marks over how much value it really delivered to customers and how long their loyalty could be relied on.

In normal times these crises would have precipitated a wave of selling and quite possibly a change of management or a takeover, but these were not normal times. The share price did indeed drop to a low of $25, but it soon started to bounce back up. A massive investment in network infrastructure greatly improved customer service and new ones kept arriving. By September 1997 there were 9 million subscribers and the share price was up to $70. This enabled it to pull off the delightful coup of acquiring its formerly larger rival, CompuServe, which had not adapted well to a new competitive environment. This brought the total number of subscribers to 12 million. In a complex three-way deal, it also handed over ownership of its troublesome physical network to WorldCom, another buccaneering opportunist. It seemed as if its troubles were far behind.

They were certainly put to the back of most people's minds, as share prices in Internet companies rose from the merely inflated to the simply ludicrous, and on to the completely insane. AOL's own rose sixfold in 1998 alone and had plenty more to go. More than any other Net business, AOL knew how to make the most of the short-term opportunities presented by this speculative frenzy. Thousands of businesses, old and new, were desperate not to be left behind in the race for cyber-stardom and instant riches. AOL had something that nearly all of them wanted badly – not just naive start-ups, but some of the biggest stars in the business world. It showed a remarkable talent for extracting the best possible price.

Gouging clients

The head of AOL's advertising sales operation was Myer Berlow, a man who on at least one occasion took a large knife to a meeting with a client, placed it on a table and threatened to use it on a colleague. Berlow's aggression, however, was nothing compared to that of David Colburn, who headed the Business Affairs division. After Berlow's team had won the customer, tough guys from Business Affairs would negotiate a new and generally much higher price. They were AOL's muscle, and Colburn was the Luca Brazzi of the family, personally fearless and able to strike fear into the breasts of most of those he came up against. He had won his spurs as the chief negotiator with Microsoft during the browser wars, and took the credit within the company for having got AOL's icon on to the Windows 95 desktop.

At Microsoft he had been dealing with a fellow heavyweight. Faced with most dot.com customers, the contest was distinctly unequal. Colburn was able to bully advertisers into paying sometimes twice what they had originally agreed because of all those eyeballs. Nobody else on the Net had a fraction of AOL's consumer franchise; nobody else could deliver remotely as much traffic to advertisers' sites. And most of these clients were desperate. They had virtually no revenues and only a finite amount of venture capital cash. If that ran out before they could manage a successful IPO, they would simply disappear. Advertising on AOL could transform their fortunes. The mere fact that they were doing so made their prospectuses more credible to investors. Colburn had an offer that scarcely anyone in the Net world could refuse.

Colburn's goal became to extract as much as possible of each target's available cash, typically 50 per cent of the venture capital it had raised. The salesperson's first objective was to find out how much money might be in the bank. If an IPO was imminent, so much the better – the price should be that much higher and shares were also possible. These guys were going to become millionaires thanks to AOL – it should get a piece of the action.

Prospective customers were frequently presented with sudden deadlines – if they did not agree to the terms demanded, AOL would sign with a rival. Another popular wheeze was 'accidentally' to insert into a sales presentation a slide with the name of a competitor to the prospect on it. The salesman would feign embarrassment and apologize for the mistake, but the message had been delivered: if the target did not bite, a competitor probably would, with consequences that could be lethal.

One tactic AOL sometimes tried in order to maximize the advertising take was to declare that it would only have one client in any one sector. This was an effective way of concentrating the minds of all the others and raising the price. Hard-nosed businesses like Enron understood this logic and were happy to pay $60 million to be the main energy company advertising on AOL. In other cases paying astronomical sums for advertising pushed companies into failure a few months later.

This was not so much high-pressure salesmanship as extortion. The mood within the company was an extreme version of that of many investors at the time – near-hysterical levels of greed and fear. Most people in AOL were obsessed with the value of their options, from Colburn, already worth hundreds of millions of dollars, to the hungry new soldiers. At meetings obscenities and screaming became everyday forms of communication. No target should escape unfleeced.

In this feverish atmosphere, the bar on what the company could hope to achieve was being raised all the time. In 1998 it acquired

Netscape for $4.2 billion in stock, but Netscape was now small pota-toes. A giant like AOL could set its sights much higher. There was no limit to what it could conquer.

The Greater Fool

The only rational justification for continuing to trade in a market where prices bear no relation to real value is the theory of the Greater Fool. So long as you could find someone who was foolish enough to pay even more for the stock than you had, it made sense to stay in the game. If you could find a fool with an undervalued asset who was prepared to swap it for your highly inflated one, that was a particu-larly good way of cashing your chips. Without being quite that cynical about it, that in effect was the deal that Steve Case pulled off. His fool was Jerry Levin, the CEO of Time Warner.

Time Warner was the world's largest media conglomerate, a loose federation of proudly independent movie, publishing, television and music businesses, ruled by feudal barons, who squabbled endlessly – an old-fashioned organization for a company that had become conscious during the Internet boom of being stuck in old media. Like News Corp, Viacom and others, the stock market was punishing it for the crime of being insufficiently digital. Its share price in 1999 was a supposedly feeble $60 something, whereas AOL's reached a mighty $94.

New technology and especially the Internet had seemingly passed Time Warner by. Before the Internet era dawned the company had sunk hundreds of millions of dollars into a gigantic interactive tele-vision experiment in Orlando. In January 1993 Levin declared, 'Our new electronic superhighway will change the way people use televi-sion... the Full Service Network will render irrelevant the idea of sequential channels on a TV set.' The Full Service Network was as top-down and technology-driven in its approach to predicting what might interest consumers as IBM's Prodigy and equally unsuccess-ful. By mid-1995 fewer than fifty homes were connected, and in 1997 it closed down.

In 1994 Levin announced Pathfinder, a Net-based attempt to make

Time Warner the leader in interactive media. For a few months Pathfinder was the most popular site on the Web, but it was soon deserted for the hundreds of new ones. It cost $20 million a year to run and even at its peak only generated revenues of $4 million. Like most traditional media companies, Time Warner had no idea of how to construct a viable business model for the Web that did not cannibalize its print products and Pathfinder's efforts were not helped by endless internal feuding. The independent businesses would not cooperate with what they called the 'Nav Bar Nazis' organizing Pathfinder.

Time Warner did, however, have some real achievements in new media in the nineties: acquiring Ted Turner's CNN and investing in upgrading its cable television networks. It was in fact much better prepared for the broadband era than AOL. Indeed, those cable networks were a major part of the merger's attraction for Case. Levin, however, believed that Time Warner needed what he called a 'digital construction'. The phrase alone reveals the woolliness of his thinking. He was frustrated by the obstinacy of the independent fiefdoms and thought that a new unit might be the way to impose a 'digital override' on them. He admired AOL, which reminded him of the HBO he had played a big part in creating. He was also a strong believer in 'the deal' as the defining, transforming moment in business. His career had hinged on his role in Time's (unhappy) merger with Warner Bros. in 1989 and on the acquisition of Turner Broadcasting in 1996. So when Steve Case called him in October 1999 and suggested a merger, he was open to the idea. This could be the deal to crown his career. It was to bury it.

The essence of the eventual agreement was that the 61-year-old Levin would be CEO of the new group and the 42-year-old Case chairman; AOL shareholders would have 55 per cent of the shares in the new company, and Time Warner shareholders, 45 per cent. Judged by their respective share prices in January 2000, this was generous to Time Warner. Judged by any other standard, it was a steal.

Time Warner was the larger, more profitable and, by any measure, the more substantial company. At the time that the merger went through a year later, it had sales of $27.3 billion, four times AOL's $6.9 billion, and net income of $1.9 billion against AOL's temporarily bloated $1.3 billion. It had 12.6 million cable customers and 130 million readers of its magazines, not to mention all the viewers of its movies; AOL had 20 million low-margin users and some rather fleeting advertisers. Time Warner's competitive advantage was based on highly distinctive capabilities, valuable brands and strong customer relationships. AOL's, as was soon to be apparent, was much more evanescent. As Rupert Murdoch, delighted to see a media rival blunder, remarked, it was a brilliant piece of financial engineering by Case and his fellow shareholders. 'He jumped in and bought something with $6 billion in cash flow.' But for many Time Warner employees, whose retirement plans were linked to company stock, it was a tragedy.

Strategically it made little sense for Time Warner. AOL had no digital magic to sprinkle on its supposedly tired old businesses. Its management skills, with the arguable exception of people like Pittman, were largely irrelevant to Time Warner's mix of businesses and the cross-promotional opportunities were nothing like as extensive as Case and Levin believed. Levin's crowning deal was a monumental folly.

Case was well aware that the deal overvalued AOL. He explained to his people, anxious that the value of their shares was being diluted, that the underlying rationale was 'capital preservation'. He knew that a correction in Internet share values was inevitable – it started in March 2000, just two months after the merger was announced. He also knew, though he took care not to tell Levin, that AOL's sales pipeline was not looking too healthy. The timing was perfect.

Unravelling

The initial reaction to the public announcement of the merger in January 2000 was one of awe. For Peter Huber in the *Wall Street Journal* it marked 'the beginning of the end of the old mass media and the end of all serious debate about the triumph of the new... Gerald Levin has grasped what many... just can't believe: For the old media now, it's go digital or die.' *The Economist* declared, 'For once the superlatives and the hype seem justified.' It was soon to change its tune.

The first snag to hit the merger was a lengthy investigation by the Federal Trade Commission, which meant that it was not consummated for another year. Meanwhile the share prices of both companies fell on the announcement, and that of AOL, sharing in the fate of all Internet stocks from March onwards, fell much further.

The NASDAQ crash turned the sales warnings of November into a major crisis for AOL. Dot.coms teetering on the edge of bankruptcy pleaded to be allowed to renegotiate contracts. There was no way that many of them would ever be able to pay up. If the full extent of the crisis were known, the merger might well not go through.

Fortunately for AOL, Levin refused to consider that his judgement might have been faulty and the irrepressible Colburn had some more tricks up his sleeve. Enormous ingenuity, and pressure, was applied to making the numbers. Creative accounting, a long-standing AOL talent, was applied to booking as advertising revenues some very unusual barter and other deals. The details of most of these were only revealed in 2002, when eventually hundreds of millions of dollars of revenues were acknowledged to be bad transactions and written off. However, in his earnings report for the final quarter of 2000, Case felt able to say that 'AOL's advertising growth is right on target... The current advertising environment benefits us because it will drive a flight to quality.' Relieved analysts, desparate for a ray of hope, bought the line that at least AOL was resisting the dot.com meltdown. AOL's market capitalization in December 2000 was only down a third, at $112 billion, and Time Warner's was $88 billion. The merger was effected in January 2001.

The atmosphere within the new company, however, was poisonous. The cultural differences were enormous – AOL-ers treated Time Warner people as stuffed shirts and were cordially loathed in return. Case and Levin were on bad terms before the merger went through, and Case was soon involved in manoeuvres to oust his new CEO. Levin found that everyone in his old company was blaming him for their loss of wealth, and in December 2001 he announced his retirement. However, he ensured that the presumed heir apparent, Bob Pittman, did not get his job. That went to an old Time Warner hand, Dick Parsons. Pittman's star was fading because AOL's poor results were now plain for all to see. Advertising revenues were falling precipitously and dragging down the share price of the whole group. Some old Time Warner businesses, particularly television and movies, with the success of *The Sopranos*, *Sex and the City* and *Lord of the Rings*, were booming, but AOL was now the problem child of the group. And that was before the accounting fiddles were revealed. Analysts speculated that the company would probably be worth more if AOL were not part of it.

Pittman and Colburn were pushed out in 2002, and in January 2003 Case bowed to the inevitable and resigned as chairman. Parsons abandoned the search for synergy between the businesses, and they went back to simply seeking to be the best in their respective markets. Later in the year, the hated word AOL was removed from the company name. It was now merely the Internet division of the group, and something of an embarrassment.

That AOL should have fallen from the precipitous heights it reached in 1999–2000 is scarcely surprising. What is remarkable is how it came to be quite as powerful as it did. It benefited from massive momentum, market confusion and powerful, though temporary, network effects. It became America's default ISP at a time when just about every American wanted to get online, and had more communities and bigger ones than anyone else. Its brand was the only one many people knew and its size, its share price and its enormous customer franchise made it irresistible to advertisers. It enjoyed a virtual monopoly at a time when billions of dollars were being pumped into the biggest speculative bubble of the century.

AOL's marketing was initially very effective. Scarcely anyone else in the Net economy, other than Jeff Bezos, was thinking about customers in the mid-1990s. Steve Case had been doing so for years and had a better take on ordinary consumers online than the brilliant software engineers of Netscape and Microsoft or pundits like

Forrester and *Wired*. It was also very much more adaptable, and opportunistic, than most.

All these things became weaknesses when the tide turned. The structure of the Internet world changed fundamentally, and not just because of the stock market crash. Internet access, especially dial-up, became a commodity business, with many providers offering it free. Using the Web became not just easy but normal – consumers had much less need of the comfort of a familiar provider. Broadband access and search played a major part in the normalization process. AOL had failed to invest in broadband capacity. By 1997–8, it was clear that this was transforming both the Internet access industry and the nature of the consumer experience of being online. Advertising, especially that relating to the Net, went into a long slump, and AOL lost its monopoly. Banner advertising was about to be outshone by search-based methods.

AOL grossly overestimated, to itself as well as to others, its capabilities as a media player, suffered from delusions of grandeur and confusion about what it really had to offer customers, and its core capabilities were not that distinctive. Many suppliers were able to provide Internet access, often better and cheaper than AOL, and managing community ceased to be a mystery.

When AOL's primary concern became boosting its share price it got cavalier about customers. It failed to build lasting relationships with them based on loyalty and trust, relying instead on the illusion that it 'owned' them. The quality of service to users was frequently appalling and it treated its advertising clients with callous cynicism. Large numbers of both groups came positively to loathe AOL and the brand acquired negative connotations. Its quicksilver adaptability became a substitute for serious strategy and when it had pulled off the biggest deal of all, largely through smoke and mirrors, it had nowhere to go but down. Its business had come to depend on continued rises in share values, both its own and those of the companies it was fleecing, and on unsustainable levels of advertising revenues.

AOL neglected to renew and extend its capabilities and attracted too many slick salesmen and shysters, rather than the kind of human capital it needed to compete effectively with, say, Yahoo. Its concept of synergy, both before the merger and during it, was vague and mostly illusory.

The company remains, however, the largest ISP in the US, though its customer numbers have shrunk steadily each year, from 28 million in 2000 to 9.3 million at the end of 2007, and revenues even more.

6

WHAT IT TAKES

There is nothing more difficult to carry out, nor more doubtful of success, nor more dangerous to handle, than to institute a new order of things.

Niccolò Machiavelli, *The Prince*

If I'd asked people what they wanted, they would have asked for a better horse.

Henry Ford

In this chapter we examine four other important qualities shared by all successful market creators – radical business models, genuine concern for customers, clear, consistent direction and sharp focus on the market.

Radical business models

The fatal flaw of both Webvan and Netscape was their business model: neither had a realistic plan for attracting enough paying customers.

A business model describes how a business works, in particular how it creates value for customers and makes money for itself. The term came into vogue during the dot.com boom, which gave it rather a bad name, since most of these businesses failed to create value either for themselves or for customers. Their models were also remarkably similar – attract lots of people to the website, try to keep them there as long as possible, and hope to 'monetize the eyeballs' by selling advertising. In the early days this model only really worked for one company, AOL, and that not for long.

Innovative business models are not just a phenomenon of the new economy. It was in 1891 that J. C. Fargo invented the traveller's cheque that made American Express famous. The beauty of this model was that travellers paid for their cheques up-front and American Express earned interest on the money before they were cashed. Fargo had discovered the delights of 'float', until then known only to banks. It is a model still employed today by discount telephone companies selling prepaid cards.

The 'razors and blades' model invented by King C. Gillette in 1901 is also still going strong: sell the basic product at a low cost and make better margins on the blades (or ink cartridges for printers, or air time on mobile phone networks). Microsoft's Hotmail is a variation on this model – the basic product is free, but users pay for additional storage.

The key to the success of no-frills airlines was the radicalism of their business models. Because they used their planes more intensively than conventional airlines, their costs were much lower. They could attract a different kind of customer, as well as take business away from established airlines.

All of our market creators had original business models and strategies: Nokia's intense focus on mobile communications in the 1990s and being the first phone maker to develop a close relationship with consumers made it a very different kind of company from competitors like Motorola and Ericsson; Dell's big idea was its model of selling PCs directly to customers and building to order, which gave it economies of scale and scope that others could not match; one of the keys to Sky's success was concentrating on subscription revenues rather than advertising and using its customer service operation to sell upgrades to premium programming.

Genuine concern for customers

Caring about customers has become a cliché of management-speak, second only to 'our people are our greatest asset'. Even if they are mostly more honoured in the breach than the observance, both express important principles and were taken seriously by all our winners. Concern for customers, though, was almost an obsession for our market creators. They recruited staff with this in mind and reinforced it both by training and by constant example. The companies' motives may vary, but organizations like Dell, eBay, Amazon, the Open University and Starbucks demonstrated time after time that they really did care.

eBay and the OU started from something like a moral conviction, and an emotional bond with their communities. Avoiding the term 'customer' only underlined how much they valued them. Members of eBay's community were passionately devoted to it and some treated other members almost as family. Pierre Omidyar himself, with his lack of corporate values and his laid-back, hippyish style, seemed part of that community.

The OU also built a community and its most important members, the *raison d'être* for the whole organization, were the students. This was the most revolutionary aspect of the OU, a university where learners came first, an organization built around supporting students, rather than run for the benefit of scholars. The most visible sign of this was course materials expressly designed to help learners who would be studying mainly on their own. Equally important was the network of part-time but dedicated Associate Lecturers, who advised and supported students locally. As with eBay, this community of students and teachers and their families developed passionate loyalty and a life of its own.

Amazon's *raison d'être* was systematically making life easier for customers and becoming the 'Earth's most customer-centric company'. Bezos was obsessed with the customer experience and ensured that everyone in the company shared his passion. The main reason Amazon survived the dot.com meltdown, when its share price dropped to $6, was that it consistently provided the best customer service on the Web.

Dell's model was likewise based from the outset on personal interaction with customers, which gave it an enormous edge over other low-cost suppliers, who were simply competing on price. Michael Dell was partly inspired by the observation that most retail suppliers offered very poor value to customers. He knew that a marketplace he described as the 'marriage of the unknowing buyer and the unknowledgeable seller' would not last. As Dell developed ever-closer links with its customers and knowledge about their configurations and needs, with teams of Dell people working on the sites of some of its larger business customers, it often knew more about a customer's IT operations than they did themselves. Like Wal-Mart, it has virtually made a science of customer knowledge and its own speed of response.

Obsessive concern for customers is so important for new companies because in the early days the relationship is potentially a fleeting one. It is too early for loyalty to have been earned or for a deep relationship to have been formed. The first few contacts or purchases are moments of truth, like the beginning of a love affair. If they are delightful, a bond is formed. If the business behaves poorly, or casually, even once, the spell is broken.

There is one notable apparent exception to this rule, Ryanair, whose customer service has sometimes seemed breathtakingly minimal – and whose CEO has occasionally berated customers with foul language. Companies with exceptionally strong propositions and comparatively weak competition can get away with treating customers cavalierly in the short term. However, it is important to distinguish between Michael O'Leary's words and his actions. His public irritation has been directed at customers with unrealistic expectations – Ryanair's low fares do not allow for refunds when flights are delayed or for providing hotel accommodation. The proposition is about very low fares for very basic transport – no frills means precisely that. But O'Leary cares passionately about what he considers the crucial elements in the customer experience – low fares and punctual flights. He is ferocious in his pursuit of anyone in Ryanair who lets customers down on these basics – not very often. Ryanair may define the customer experience more narrowly than most, but it is actually very good at delivering what has been promised.

In genuinely caring about the customer experience, market creators are no different from most successful new businesses. Any new business that is mechanical or complacent in this respect, as so many large organizations become, is taking big risks. Businesses whose main concern is controlling costs at the expense of customer satisfaction have forgotten the basic principle outlined by Peter Drucker: 'It is the customer who determines what a business is. It is the customer alone whose willingness to pay for a good or a service converts economic resources into wealth, things into goods.'

Clear, consistent direction

Like customers, leadership is another much-ploughed furrow in business writing but a rather hazy concept. The way in which our outstanding successes were led varied greatly, but they all achieved one important outcome: everybody in the organization was clear about what it was trying to achieve and were mostly happy to pull in that direction.

That is a much more unusual achievement than one might think. According to research by Kaplan and Norton, 95 per cent of employees do not understand their companies' strategies. Organizations implementing radically different strategies cannot afford to muddle along – they need everybody to be crystal-clear about what they are doing. What all our winners had in common, whether they dictated strategy from on high or developed it collaboratively, was that they communicated it clearly, ensured that people understood it and believed in it, and that it was implemented consistently.

However, their styles of leadership varied enormously. Pierre Omidyar and Jeff Skoll at eBay and Sergey Brin and Larry Page at Google were idealists, prepared to share leadership, both with each other and with professional managers. Michael Dell and Jeff Bezos were hard-driving autocrats, who focused their own efforts and those of their companies strictly on business results. Akio Morita and

Masara Ibuka sought to make Sony both a family and a place where engineers in particular could do their best work. Walter Perry at the OU and Keichi Enoki at DoCoMo were outstanding team-builders who from the outset relied on the contributions of many specialists. Steve Case was, in the words of his predecessor, a Cyborg, an introverted analyst, who left it to Ted Leonsis and Bob Pittman to rouse the troops.

Not many were self-consciously heroic leaders in the style of Jack Welch at GE or Lou Gerstner at IBM, battling against great odds to change the direction of an enormous organization or save it from destruction. Jorma Ollila, who really did save Nokia from disaster, was deliberately modest and anti-heroic in his style. In most cases, it was the vision that made the leader, rather than the other way round.

Two notable exceptions were Sam Chisholm at BSkyB and Steve Jobs at Apple. When Chisholm took command of Sky it was on the brink of bankruptcy. By a combination of bullying, wheedling and determination he kept it alive, and went on to win the rights to Premier League football and make Sky fabulously profitable.

Jobs's leadership of Apple in his second reign has been a triumphant success, after the calamities of the first. It was rarely clear who was in charge at Apple in its early days: it had too many leaders. Mike Markkula was a negligently hands-off chairman who allowed Jobs too much licence. Respect for authority was destroyed by the sacrifice of the first CEO, Mike Scott, who had ensured, against heavy odds, the successful launch of the Apple II. Jobs undermined the authority of several other key people, including John Sculley, the CEO he had foolishly recruited, which led to his own ejection. Sculley was consistent but out of his depth in the rapidly changing personal computer industry.

Netscape also suffered from divided leadership for much of its brief life. In Jim Clark's memorable words, 'Screw the business plan and the conventional investors.' Clark left the effective leadership of the company in its early days to the 22-year-old Marc Andreessen, whose only strategic imperative was speed of movement. Even after the appointment of a professional CEO, the brilliant but inexperienced

Andreessen was allowed to commit follies like publicly threatening to marginalize Microsoft. The danger, as Clark knew, was precisely the reverse, but he did not even tell Barksdale of his own secret approach to Microsoft.

Clear direction springs as much from the clarity of the original strategic vision as from strong leadership. It is also closely related to focus.

Sharp market focus

Sharp focus on a particular market is hardly an exclusive property of market creators but is a striking feature of all our successful cases. Its absence is a weakness in most cases of corporate venturing and inadequate focus was one of the fatal flaws of AOL and Netscape and of Apple in the mid-1980s.

Netscape never satisfactorily defined what business it was in – it made most of its money from browsers, but this became a market where the product was generally given away free. In the market for server software it was up against serious competition where its supposed first mover advantage counted for much less and it made the fatal mistake of thinking it could challenge Microsoft's dominance of the desktop. Apple likewise took a long time to acknowledge that it could not be a major player in corporate markets for computers. It wanted to have it all. So, of course, did AOL, and for a while it seemed as though it had pulled it off.

Virtually all of our long-term winners started with a very narrow focus, which they only broadened when they had opened up their markets and established their leadership. Amazon stuck strictly to online bookselling until 1997, when it moved into music. By then its processes and capabilities were well honed and worked equally well for CDs. It was only a few months before it was the leading online supplier of music too. It briefly over-extended itself during Bezos's Napoleonic phase in the late 1990s but subsequently recovered, enabling its customers to 'find anything they might want to buy online'. Its core business, where its capabilities are well honed, is confined to media – third-party sellers supply other products.

The organizations that have the greatest difficulty with market focus are ventures by established corporations. They are inevitably, and sometimes rightly, constrained by their existing interests. In particular they are reluctant to compete directly with other parts of the company or with their own customers. These problems proved almost fatal in the tragedies of Encyclopædia Britannica and IBM, recounted in Chapter 9.

One way around this conflict is to establish a separate, autonomous organization to pursue the new opportunity, with its own strategy, management and culture. IBM did this, initially very successfully, with its original PC business before it sucked it back into the fatal corporate embrace. Organizational and cultural autonomy was likewise critical to DoCoMo's success with i-mode. It deliberately recruited people with different skills, unencumbered with telecom ways of thinking, and focused solely on the success of the new venture.

Necessary and sufficient?

The eight essential attributes of market creators are summarized below. It seems clear that with one possible exception, all eight are necessary conditions for market creators. The exception is customer concern, but only where the supplier has a truly irresistible proposition and insignificant competition.

1. A clear strategic vision based on a radically different way of meeting a large, previously unsatisfied customer need.
2. A set of highly distinctive capabilities – technological, marketing and logistical – tailored to the needs of the market.
3. Value propositions that are so compelling that they change customer behaviour and shift loyalties.
4. An entrepreneurial but disciplined organization that balances creativity with practicality, is innovative but pragmatic, and creates effective teams.
5. A robust, radical business model that is not easily imitable.

6. Genuine concern for the quality and consistency of the customer experience.
7. Leadership that ensures the clear communication of strategic direction and consistent implementation.
8. Sharp focus on the chosen market.

A case can be made for other attributes, but most candidates are common to all successful organizations. The strongest contender is attracting and motivating exceptionally talented people, essential to the development of distinctive capabilities and disciplined entrepreneurialism. I have bundled this with nurturing human capital, one of the factors essential for long-term competitive success.

The eight attributes are not in themselves sufficient conditions for take-off, which depends crucially on the external business environment. Many ventures in the 1980s attempted something similar to what AOL and Amazon did so successfully in the 1990s, in France as well as the US, but the technological infrastructure was less sophisticated and funding not so easily available.

By far the most important precondition for the development of a new market is latent demand, the previously unsatisfied market need that the new enterprise uncovers. These markets took off in the way that they did because they struck a major chord with customers. In all the successful cases, demand greatly exceeded the expectations of the founders, let alone those of sceptics. Where the companies can take credit is in spotting the need before almost anybody else and developing capabilities and value propositions that answered it.

The availability of venture capital, the absence of regulatory obstacles and a climate favourable to new enterprise also played a crucial part in the history of most of these companies. These factors explain why so many recent market creators have been American, and so few European. The stories also show the big part that luck and timing play.

What is remarkable about these organizations is that they managed to get all of these things right during a crucial period. The first five qualities – radical strategic vision, distinctive capabilities, compelling value proposition, disciplined entrepreneurialism and radical business model – are particularly unusual, and their combination extremely rare. As we shall see in Chapter 13, an almost completely different set of attributes is required for enduring industry leadership.

We now turn to two of the most notable creators of the Internet as we now know it, who had all of these qualities in spades.

7

NETWORK MODELS

eBay and Google became the largest and most profitable businesses on the Internet, yet largely serendipitously. Each was made up of several networks, both physical and virtual: computers, server farms, users, advertisers, affiliates and sellers. These networks gradually became their most valuable assets and fuelled their extra-ordinary growth.

eBay

There are several myths about the birth of eBay, most of them partly true. One which has no foundation is that Pierre Omidyar started it to help his girlfriend complete her collection of Pez sweet dispensers. This story was fabricated some years later for PR purposes and, despite Omidyar's denials, is still part of the official narrative.

The story about it becoming a business by accident, however, is entirely true. Omidyar had never seen himself as a businessman and created AuctionWeb in 1995 with no thought of making money from it. It was one of a family of websites he developed in his spare time under the collective banner of eBay, and not the most important. It was 'more of an intellectual pursuit... It was just an idea I had, and

I started it as an experiment, as a side hobby basically, while I had my day job.'

Four years earlier he had been shocked by the IPO of a business he had helped to create, when a few favoured investors had been able to buy stock at much lower prices than others. This gave him the idea of creating a 'perfect marketplace', where buyers and sellers could meet on equal terms and arrive at a fair price. He made up the auction categories himself, publicized the service on newsgroups, and ran it on a shoestring. Traffic grew almost entirely by word of mouth.

It grew considerably faster than he had expected, and in February 1996 his Internet service provider decided that AuctionWeb must go on to a business server at a cost of $250 a month. This prompted Omidyar to ask sellers to pay him a percentage of the value of each transaction. Cheques started arriving at his home so fast that, unlike almost all other Internet start-ups, the fledgling business covered its costs from the outset. When revenues reached $10,000 a month in June, Omidyar gave up his day job and devoted himself full-time to writing new software and answering users' emails. He also brought in a partner, Jeff Skoll.

Still in the process of graduating from Stanford business school, Skoll was more business-oriented than the idealistic Omidyar and became his 'analytic powerhouse'. Like everyone else, he was initially sceptical that complete strangers would buy and sell from each other in any volume. He focused on developing the business and writing the business plan, while Omidyar nurtured the growing AuctionWeb community. They both expected major players like AOL to move into auctions and felt that the real business opportunity would be selling the software to them, but sales kept on growing by 30 per cent every month.

The most powerful eBay myth was that it was more of a community than a business: buyers and sellers were not described as customers. The more obsessive members of 'the community' spent ten hours online every day, mostly chatting to each other on AuctionWeb's bulletin board, like AOL's devotees, and Omidyar participated actively in their discussions. The community, or at least its more vociferous members, seemed to have a collective will and protested loudly if changes were introduced without consultation. They were eBay's fervently evangelical, unpaid sales force and customer service arm, helping others to get more out of the service, and the company's revenues to soar.

Jim Griffiths was an archetypal early member of the community, an eccentric, depressive part-time trader in computer components who gave advice to other users adopting the persona of Uncle Geoff, a cross-dressing farmer. When a bout of depression led to him disappearing from the bulletin board, a dismayed Skoll persuaded him to become AuctionWeb's second part-time employee. For $100 a month, he answered enquiries by email and proved adept at defusing the often vociferous disputes between traders. Subsequently he and other 'remotes' became full-time employees.

According to Omidyar, what they were doing was 'building a place where people can come together. They just happen to be coming together around trading... the whole idea was just to help people to do business with one another on the Internet.' Omidyar sincerely believed this, but it soon became clear that this was becoming a large and very profitable business. The first users and employees certainly felt they were part of a community with similar values. Omidyar's laid-back lifestyle, rusty old car and manifest idealism made him seem like one of them – and an effective, trusted leader.

Trust was a much-proclaimed value. Initially Omidyar trusted sellers to send him commission – in the early days the company had no means of knowing if a sale had been made – and Omidyar strongly encouraged users to treat each other as they would wish to be treated themselves. 'I founded the company on the notion that people were basically good, and that if you give them the benefit of the doubt, you're rarely disappointed. And statistics have borne that out.'

To encourage good behaviour, and drive out the few dishonest people, he developed Feedback Forum. Users gave each other ratings and AuctionWeb tabulated the scores. If someone's total score got as low as minus four, they were banned, but soon no-one would buy from a vendor who did not have a high positive score. Regular vendors were able to build a solid reputation which in normal circumstances would take years to develop.

The feedback system was a brilliantly simple, subsequently much imitated innovation. When every buyer and every seller was given a score for every transaction, and totals publicly displayed, there was little incentive to short-change others. With such transparency, trust as such ceased to be so important – it was in everyone's interest to behave well. This meant that total strangers could buy and sell second-hand items from each other in confidence.

AuctionWeb's first community was a geeky one of computer enthusiasts like Omidyar, but they were quickly outnumbered by a rash of new ones, nearly all compulsive collectors and traders, mainly of antiques and other second-hand goods.

The category that brought AuctionWeb to the public eye was Beanie Babies, a family of kitschy soft toys, which hundreds of thousands of children and many adults started collecting in the 1990s. The makers of these had deliberately created a shortage of certain characters to stimulate more interest in them. Collectors became desperate to complete their sets, but the only people who possessed them were other collectors. The markets for obtaining missing characters were local, and therefore very limited until the Internet took off. AuctionWeb opened up the whole of the United States to them, and became the place to track down missing Babies. By April 1997, when Beanie Babies were classified on eBay as a separate category, there were 2,500 of them listed for auction. In May sales reached $500,000, each toy selling for an average of $33, considerably more than the original retail price. By December, sales hit $2 million.

Collectors of, and dealers in, stamps, cards and coins were also swarming to the site, which proved a much more efficient marketplace for them than traditional local flea markets. From the end of 1996 onwards, AuctionWeb was adding new categories like antique dolls and chintz every week. By the end of 1997, computer items, once the core, only represented 14 per cent of listings, while collectibles were almost 80 per cent. Many professional dealers found that they could sell goods far quicker and at better prices online than through their shops or other channels, because they could reach so many more people. Others found that they could no longer pick up goods at knock-down prices – collectors were bypassing them and selling directly on eBay. Thousands of individuals turned their hobby into a business.

In 1996 there were 250,000 auctions and AuctionWeb's own revenues reached $350,000. January 1997 was the tipping point. There were 200,000 auctions in that month alone, and revenues for the year reached $41 million. Growth was becoming almost unmanageable.

AuctionWeb's cobbled-together computer systems could not cope even with the 1996 volumes. Mike Wilson, an old friend of Omidyar, had been hired to develop a new architecture but had to devote much of his time to dealing with crises, as the system crashed over and over again. Omidyar and Skoll did their best to hold back growth, putting limits on the number of listings and introducing credit approval procedures for new users – to vociferous protests from the community. Traffic, however, mounted inexorably, the number of listings growing by 40 per cent a month, and by the summer, crashes occurred almost daily.

Wilson was a brilliant but eccentric individualist, wholly in tune with community values. They loved his abuse of 'Microsloth' for crashes and his alter ego, 'Snoopy', who made irreverent comments and fixed problems for individual users. After endless delays and much anguish, the new, scalable site was finally launched in September 1997 and was now called simply eBay. AuctionWeb was buried. For the first time in its brief life, the company was ready to start actively encouraging traffic.

The summer of 1997 was a turning point in several respects.

Community to corporation

Early in 1997 Skoll had produced a business plan which, with the benefit of hindsight, appears ridiculously modest – in stark contrast with most business plans being written in Silicon Valley at the time. It assumed that AuctionWeb would be the company's 'crown jewel', but that the real revenue earner would be SmartMarket Technology, which would license auction systems to other businesses. Auction-Web could surely not continue its meteoric rise for much longer. Selling new players the technology seemed like a prudent strategy, though it was never energetically pursued.

Skoll's plan, however, was spot-on regarding the business's main assets – the number of buyers and sellers and the sense of community: 'A few services have attempted to create a "community-like" feeling by offering bulletin boards or asking users for their input on site improvements. These bulletin boards, however, are edited by the online auction companies, unlike the free-flowing conversation afforded AuctionWeb users.' What he did not spot was the self-reinforcing nature of its growth.

Omidyar had learned from his earlier experience of start-ups the value of an association with a prestigious venture capital firm, even if the company had no immediate need of cash. He knew that he needed strategic advice, contacts, a solid reputation and above all good people. He and Skoll had long recognized that they would sooner or later have to find a 'world-class CEO'. In fact they had often had difficulty recruiting even programmers in Silicon Valley, as to most people they did not look like a proper business.

They had similar problems with investors. They were approached by some Times Mirror executives, and might well have sold if the price had been right, but the top brass of Times Mirror could not understand how a business with no assets could be worth millions. It was a similar story with many VCs. The partner at Mayfield Fund described in the Introduction was typical. Nobody 'got' eBay.

Bruce Dunlevie and Bob Kagle of Benchmark Capital made the killing they did because they got eBay before anyone else did. When they explored the site, they were amazed by the range of goods on offer, and were even more impressed by the numbers: the business had negligible costs, carried no inventory and enjoyed gross margins of 85 per cent, compared with Amazon's 22 per cent. Plus revenues were growing at over 30 per cent every month, without any promotion at all.

In June 1997 Benchmark agreed to buy 21.5 per cent of the equity for $5 million, which valued the company at $23 million. The $5 million went straight into the bank and stayed there, as eBay had no immediate need of cash, and no intention of changing its frugal ways. Desks were still self-assembly and salaries modest. Benchmark's real value came from Kagle's strategic advice – one of his first recommendations was to clarify the brand, which was still confused between eBay and AuctionWeb. He introduced them to designers who cleaned up the messy-looking site and gave the brand a clear, strong identity, using the primary colours and white background still employed today. Benchmark's involvement also helped eBay to recruit its first professional managers.

In August, Steve Westly joined as Vice-President of Marketing and Business Development. Skoll had tried to recruit him in 1996, but without venture backing the business had then looked too flaky. Westly brought with him several other smartly dressed MBAs and there was an inevitable culture clash with the early eBayers, notably Mike Wilson. The newcomers did not quite get community and proceeded to cut deals with Netscape and others to generate traffic to the site. When Westly agreed to pay AOL $750,000 for six months' worth of leads, there was uproar.

Wilson and many others saw AOL as the antithesis of eBay and its values. The unique cyber community they all loved, and that had grown so excitingly organically, was degenerating into something merely commercial. The amount of money involved shocked most eBayers, after the thriftiness they were used to: even Skoll felt that the price might be too high. In the end the decisive argument, which swung Omidyar, was that this would guarantee that AOL, with its enormous customer base and understanding of community, did not compete with eBay. He and Skoll had always seen ISPs as a danger and AOL as the most formidable threat. The deal was signed in December and after some adjustments did indeed deliver significant traffic.

Serious-looking competition also emerged in 1997 and had been one of the factors in making Omidyar and Skoll consider a deal with

Times Mirror. OnSale had also started up in 1995 and was well-funded and well-organized. Its approach was different: it auctioned goods, mainly remaindered, that it had bought and was selling directly to customers. This appeared to many VCs to be a better business model, but it meant that customers did not see OnSale as an honest broker.

In 1997 OnSale launched its own person-to-person auction service and used software 'bots' to capture the email addresses of customers on eBay's site and approach them directly. Westly publicly denounced OnSale's underhand tactics; OnSale responded aggressively: 'We're creating a whole new way to sell goods that appeals to male hunting instincts, to male gamesmanship, competition and skill.' This was hardly a brilliant ploy for winning the trust of the eBay community, many of whom were women. OnSale's methods also favoured sellers over buyers, unlike eBay's 'level playing field'.

Auction Universe was another well-designed system, with many good features that eBay's site lacked. When it was acquired by Times Mirror it looked even more formidable. Omidyar and Skoll had always feared a major media player as a competitor.

In fact what saw off both these rivals, and more than a hundred other auction sites, was the simple fact that eBay had many more buyers and sellers in almost all categories than any of the others. Experienced online traders who tried the alternatives were invariably disappointed and soon came back. With 300,000 users in October 1997, eBay was benefiting from what were already enormously powerful network effects, which greatly outweighed any other advantages – it offered sellers more buyers, and buyers more sellers. Omidyar subsequently expressed it graphically: 'We had a big magnet, which was eBay, and all these little magnets came along and tried to pull people away. But eBay's magnet was so strong that it was hard for them to get started.'

eBay's victory has been represented as proof that first mover advantage always applies in online markets – the idea that the first business to stake out a new market invariably enjoys an overwhelming competitive advantage. This is a dangerous oversimplification.

First mover advantage applied in this case, more than any other considered here, because of network effects and positive feedback loops. But eBay's position was, and remains, a highly unusual one – its community of buyers and sellers is itself a network. Even the participants did not appreciate in 1997 just how strong the magnet had become, with 80 per cent of auction listings.

Pragmatic professionalism

By mid-1997, with an IPO in prospect, eBay had clearly reached the stage where it needed a professional CEO. This was where Benchmark made perhaps their biggest single contribution to the business. David Beirne, a former headhunter and now partner at Benchmark, engaged his old firm to find candidates. Everyone agreed that the right person must have a strong record in consumer marketing but must also be 'eBaysian'. S/he had to understand and respect the organization's values and be able to fit in with the existing team. This ruled out several otherwise highly qualified candidates.

Meg Whitman, despite twelve years in corporate roles – consultant at Bain, and executive at Procter & Gamble, Disney and Hambro – seemed a great fit but did not want to move from Boston, and knew nothing about the Internet. Beirne worked on her and she became intrigued. 'This was actually... something that couldn't be done offline... The second thing that struck me was the emotional link between eBay users and the site.'

Whitman saw the importance of going with the grain of the culture of community, and working closely with Omidyar. The mutual respect between them was one of the keys to a smooth transition. Omidyar became chairman and Skoll VP of strategic planning, and both remained full-time until 2000.

Whitman's immediate task when she joined in February 1998 was to prepare the company for an IPO in September. Many dot.coms in this feverish period treated the IPO as their main business objective and forgot about customers and the long-term future of the business.

The main reason for eBay's flotation was strategic – it needed to put itself on the map. Despite its phenomenal success, it was still not being taken seriously by the media, who like nearly everyone else did not really understand it. A successful IPO would enhance credibility with potential employees, associates and the wider community, including new customers. It would also help finance international expansion before others sewed up the best markets, and fight off threats posed by rivals like Yahoo and Amazon. It needed to make acquisitions, both defensive and offensive.

The IPO in September 1998 exceeded all expectations. Shares closed the day at $47, valuing the company at $1.7 billion. By November market capitalization had risen to $4.6 billion, and would go on rising astronomically through the boom of 1998–9. All the staff had been given stock options and seventy-five of them were now worth millions of dollars each, but they tried to keep some things the same. Omidyar and Skoll, despite both being well on the way to becoming billionaires, continued to drive to work in their battered old cars, and insisted that people must focus on the job in hand, not the stock price.

The received wisdom about Whitman is that she was eBay's heroine who transformed it into a world-class business. This is certainly true, but discipline and efficiency came at a price. Her initial priority was to professionalize marketing and customer service but the financial targets became paramount and much of the spirit of the old eBay was lost.

Whitman took a systematic approach to eBay's numerous niche markets – the first 'category specialist' appointed was herself a collector of Elvis memorabilia. Equivalent specialists were recruited to nurture and develop eventually hundreds of market segments. There was an influx of MBAs and former consultants who soon outnumbered the scruffy hippy pioneers.

Customer service remained problematical, in view of the enormous increases in scale. eBay could not continue to deal with its community in the personal way that Omidyar, Skoll and Uncle Geoff had done. Meg's answer was to automate it. Instead of live support boards customers were handled by a combination of email support and explicit, easy-to-follow instructions. Many old members disliked the transition to mass marketing but were consoled by the constant growth in the numbers of buyers and sellers.

The area which posed the greatest single threat to the business was one that Whitman initially thought she could safely leave to the experts – computer technology. Wilson and Omidyar were accustomed to occasional crashes, and this was not a field where Whitman had any expertise. The continued growth in the number of users and transactions was unprecedented and put enormous strain on eBay's systems. This culminated in a prolonged outage in June 1999 that shook everyone's confidence and battered the share price.

It was clear that the overall system was nothing like robust enough. Given that eBay's whole business depended on constant availability, it seemed reckless not to invest in large-scale spare capacity and back-up. Whitman also decided that the time had come to replace Wilson as head of IT: eBay now needed not so much brilliant improvisation as solid reliability – and scalability.

During the preparation for the flotation, eBay had approached both Amazon and Yahoo about possible collaboration. Bezos bluntly rebuffed it, telling Whitman and Kagle that Amazon, then very much the bigger company, would soon offer auctions and crush eBay. Yahoo proposed to absorb it. These remained the biggest rivals, but AOL, with 16 million customers, looked potentially more dangerous and Westly concluded an even bigger deal with it, whereby eBay would pay $12 million dollars over four years. In 1999 the price was raised to an extraordinary $75 million. In the light of what we now know about AOL, this looks suspiciously like the work of David Colburn and his heavies. It is hard to believe that this expenditure had as much to with generating new business as with buying off a bully.

In 1999 Yahoo and Amazon attacked aggressively with their own auction services. Yahoo dramatically increased its listings, by offering them free. Amazon offered free fraud insurance and localized listings – its customer service was already very much better. This prompted eBay to make some improvements, but it was overwhelming network effects that saw off these rivals. Amazon only managed 55,000 listings against eBay's 2.3 million, and Yahoo's sellers never achieved high conversion rates.

Whitman's strategy was to continue to maximize growth, to reinforce eBay's advantage and ensure its lasting domination of online trading. She expanded the range of categories beyond collectibles and aimed to raise the average price of items sold, redefining the brand as one that enabled people 'to buy or sell practically anything'. A major priority was to establish eBay as the leader internationally, but she ruled out acquisitions that would take it too far from what eBay did best, unlike the strategies then being pursued by Amazon and Yahoo.

The acquisition of Butterfield's, a traditional auction house, seemed to fit this strategy but turned out to be a mistake. In October 1999 eBay lauched Great Collections, a site for premium items. In the same year, Sotheby's launched Sothebys.com, in alliance with Amazon. Neither venture was successful – it became clear that traditional auctions of high-value items required the reassurance and the experience of the right kind of staff and premises – ambience was a crucial part of Sotheby's value proposition. Its customers had no interest in the humbler mass markets that eBay had opened up.

The most remarkable apparent diversification was into cars. In the early days, nobody had believed that they would sell online – the transaction seemed simply too valuable, too uncertain to buy unseen. In fact the number of vehicles listed and sold (under the Miscellany heading, where all embryonic categories started) crept up without any stimulation. In 1999 eBay acquired Kruse International, a large automotive auction company. Steve Hoffman, the executive charged with integrating it into eBay, successfully argued for a major commitment to automotive and to similar kinds of what eBay called 'practicals'. These expanded to include clothes, records, consumer durables and computers. Where eBay worked best, Hoffman pointed out, was in markets where buyers were passionate about the product and where existing selling methods were flawed. That turned out to apply to car sales as much as to collectibles. Market research revealed that most consumers loved their cars but distrusted dealers and disliked buying used vehicles. The new category attracted thousands of items and many dealers – who had to learn to operate in a transparent environment where their performance on every sale was scrutinized and fed back to other customers.

eBay Motors was the first category to merit a separate site, launched in 2000, with its own look and feel, together with specialist services like search by make and inspection and financing facilities. It was a huge success: by 2001, sales of automotive products reached $1 billion in value. The margins on this were deliberately set initially at a modest fixed $25 per sale and a rather higher $25 per listing, but with only five dedicated staff it was immensely profitable. eBay Motors is now the largest single category, with sales of $16 billion in 2006. Collectibles, at a mere $2.7 billion, were dwarfed by sales in Consumer Electronics; Clothing & Accessories; Computers; Books, Music and Movies; and Home & Garden.

This success was followed by moves into other 'practicals' and more and more professional merchants working within the eBay marketplace. There were soon hundreds of thousands of businesses, operating solely or primarily via eBay. Many of the original part-timers drifted away, often bewailing the new corporate culture. Within a few years, 95 per cent of sales were being made by professional sellers and 1.6 million people made their living primarily through eBay.

eBay has become the hub of a complex business network – and not just of traders. More than 15,000 software developers produce tools for eBay businesses.

Global marketplace

In 1999 eBay started serious operations overseas. It already had users in over ninety countries using the US site, but they had to communicate in (American) English and trade in dollars, which limited the appeal. The scale of the business opportunity, however, was now clearly understood by Yahoo and Amazon, to name but two, and eBay needed to move fast, to forestall both them and local competitors.

Two enterprising young Germans, having closely studied eBay's model, had launched Alando.de in March. In two months it gained 50,000 users and clearly could quickly become uncatchable in Germany. Whitman agreed to pay $42 million for Alando in June 1999, subsequently changing its name to eBay.de.

In October eBay.co.uk was launched. It faced significant competition but by 2000 had established a clear lead. In 1999 eBay also opened in Australia and in 2001 it acquired iBazar, the leading auction site in France, Italy, Belgium, the Netherlands, Spain and Portugal, with 2.4 million users. By the end of 2001 eBay was operating in seventeen countries.

The only one of these where eBay was not the leading auction site was Japan. It had declined the opportunity to partner with Yahoo and Softbank for fear of Yahoo learning too much about its business. Whitman subsequently judged this to be one of their biggest strategic mistakes. It launched its own Japanese site in February 2000, but has not yet caught up with Yahoo.

International operations played a big part in overall growth from 2000 onwards, after an extraordinary surge in the US after 1997. Total sales rose eightfold in 1998, to $745 million; tripled in 1999, to $2.8 billion; and doubled in 2000, to $5.4 billion. In 2000, eBay's own revenues were $431 million, 8 per cent of this. By 2006, as it steadily increased its fees, it took 11 per cent of a gross $52.5 billion – $6 billion.

Capturing competitors

In January 2000 a new competitive threat appeared with the launch of Half.com. It was the brainchild of Josh Kopelman, who had spotted a gap between the markets of eBay and Amazon – fixed-price trading. There are many kinds of goods for which the auction model is not the most efficient: it works best where the value of the item is uncertain. Also, as eBay knew from its own research, many consumers would never participate in an auction, disliking the uncertainty over whether they would obtain the item.

Kopelman's idea was to create a network of buyers and sellers very similar to eBay's but for standardized goods, like books and CDs. These would be easier to list than on eBay and as simple to purchase as goods from Amazon. Half.com would actually take payment from the buyer and pay the seller, thus making that part of the transaction more painless too. This enabled Half.com to take a much larger commission than eBay's, 15 per cent, and to become the fastest-growing e-commerce site.

Rather than develop eBay's own fixed-price capability, Whitman decided to buy Half.com. There was a risk of it developing strong network effects and of Amazon acquiring it, and so posing an even greater threat, and Whitman decided that this justified paying $340 million (in stock). In 2001 Half.com was integrated into the main eBay operation, where the 'buy-it-now' capability was introduced in many areas. Soon 30 per cent of eBay listings were offering the possibility of immediate purchase.

A pattern now emerged of eBay relying on its muscle and its money to eliminate or crush competition, rather than developing new organizational capabilities. Eric Jackson, an executive at the next disruptive challenger, PayPal, has given a revealing account of battling with eBay. Yet PayPal was less of a threat than a useful complement to eBay's operations.

One of the problems with completing purchases on eBay before 2000 was making payments. Most sellers were too small to be able to accept credit cards, so paying for goods at the end of auctions was a painfully time-consuming process for both parties, involving posting and clearing cheques before the items could be delivered. When sellers discovered a free service that solved this problem, they signed up in droves and encouraged others to do so too.

PayPal was an online person-to-person payments business, which found to its surprise that most of its users were eBay traders. Its founder, Peter Thiel, understood network effects and knew that he had to attract a critical mass of new members quickly if he was to win the inevitable race – several other players were soon in the field. To make PayPal as attractive as possible, he offered an account with $10 already in it to anybody who signed up, and another $10 for each new member introduced. When he realized that his largest and fastest-growing group of customers were eBay traders, who were loudly singing PayPal's praises on their auction listings, he quickly adjusted his strategy to concentrate on them. PayPal introduced a stream of new features to make it as easy as possible to use the service and for sellers to put the PayPal logo on their listings. By April 2000, it had a million subscribers, and by the end of the year 5 million, half of all regular auction users.

Although eBay had been thinking about payments for some time, it had hesitated before getting directly involved. In April 1999 eBay acquired a credit card payments specialist, Billpoint, for $86 million, but this turned out to be, in Whitman's words, 'a very nascent company'. In other words, eBay had paid $86 million for what was not much more than a business plan. At the end of the year eBay launched its own Billpoint service, charging 4.75 per cent on each transaction, in addition to its normal fees. Not surprisingly it attracted few takers.

eBay's reaction to PayPal's success was to try to impose Billpoint on traders. In language reminiscent of Microsoft, it introduced a series of measures to 'integrate Billpoint more tightly' with eBay. In order 'to keep its site clean', it announced that 'third party' logos could not be larger that 33 x 88 pixels, a quarter of the size of PayPal's – any auction violating this rule could be closed down without warning. At the conclusion of buy-it-now transactions, users were directed to a Billpoint form, with the clear implication that this was the only way to pay for it. When it launched eBay Stores for professional sellers, it announced that only Billpoint or credit cards would be acceptable forms of payment. Checkout, 'an improved and consistent buyer experience', came close to obliging buyers to use Billpoint if they had an account.

eBay had moved a long way from Omidyar's dream of a level playing field. Fortunately for PayPal, eBay users were vociferous in their protests. By an overwhelming majority they preferred the simpler, cheaper system to the one eBay tried to force on them. eBay was obliged to withdraw Checkout after three weeks, and PayPal continued to increase its market share. In 2001 it moved into profit and the following year had a successful IPO. Whitman now bowed to the inevitable and, in July 2002, acquired PayPal for $1.5 billion. Billpoint was closed down and PayPal became the preferred payment method on eBay.

Unfortunately, most of the entrepreneurial people who had built PayPal left within the next few months, finding the ponderous decision-making and emphasis on business process uncongenial. Jackson is amusing on the cultural differences between the nimble start-up and the now lumbering bureaucracy that had acquired it. According to him, decisions were only taken after lengthy meetings packed with khaki-clad executives making PowerPoint presentations to each other. His is, of course, a partial view, but it has the ring of truth.

Both businesses grew even faster after the merger, when they could concentrate on ways of cooperating with each other, rather than fighting. PayPal has become the effective standard for Internet payments generally. By 2007 it had more than 100 million registered users and accounted for 25 per cent of eBay's overall revenues. It has also become a monopoly itself, not least because eBay made it difficult for sellers to use other payment methods, like Google's.

In 2002 eBay obtained a stake in Craigslist, a classified advertising service which really was more of a community than a business (see Chapter 10). This gave it a seat on Craigslist's board, initially amicably occupied by Omidyar. But when eBay used the insights it gained to build its own rival business, Kijiji, relations became much less cordial.

In 2005, as growth in traffic was slowing, eBay made another major acquisition, this time in Internet telephony. Skype had been started in 2003 by Niklas Zennström and Janus Friis, working with a team of young programmers in Estonia. Its peer-to-peer system was the first application of Voice Over Internet Protocol (VOIP) software to appeal to a mass market. VOIP allows telephone calls to be made free – speech is digitized and sent in packets over the Internet, just like emails and Web pages. Within two years, Skype had attracted

50 million subscribers and caused panic among traditional telephone companies.

Skype had a business model even more radical than eBay's. According to Zennström, 'We want to make as little money as possible per user because we don't have any cost per user, but we want a lot of them.' It was able to add 150,000 users a day, at zero marginal cost. If only a few of them signed up for its paid-for services like SkypeOut and SkypeIn (for calls to and from conventional phones), it would make money. Skype was not actually making a profit at the time of the acquisition, but was generating revenues of $60 million.

eBay paid an astounding $2.6 billion to acquire Skype. Its rationale was that enabling buyers and sellers to speak to each other easily would 'reduce friction' and there would also be advertising opportunities. Even more curiously, Rajiv Dutta, eBay's CFO, argued that it was a good deal as it only represented 4.8 per cent of eBay's market capitalization, whereas it had paid 8 per cent for PayPal. PayPal, however, had been profitable. All this had distinct echoes of the ridiculous dot.com valuations in 1998–9. From subsequent remarks by Whitman, it appears that the main consideration was pre-empting rivals.

Skype is certainly a very disruptive business for telephone companies and enjoys significant network effects, but two years after the acquisition it had failed to find a way to make money out of all those customers. In 2007 eBay took a write-off of $1.43 billion and dispensed with Zennström's services. Whitman was looking rather less than world-class and later stepped down as CEO.

In 2006–7, eBay's attractiveness to buyers started to fade – they could find bargains in other places. Many found the site difficult to use and customer service poor. In 2007, for the first time, there was a slight decline in listings for auction. Many sellers, annoyed at frequent hikes in fees, were going to other sites like Amazon Marketplace, or using Google AdWords as a way to drive traffic to their own sites.

Perfect market to monopoly

There were three main reasons why eBay's revenues grew a hundred-fold between 1997 and 2005, and its profits even more: it uncovered and made more efficient a multitude of imperfect markets; it was carried forward on wave after wave of positive feedback loops, as more categories and countries joined the party; and it enjoyed immensely powerful network effects: its critical mass of buyers and sellers made it irresistible to both – no matter what its competitors did, they could never come close to the power of eBay's magnet. This was the case in 1997, when it had 300,000 users, and was clearly even more so in 2005, with 83 million worldwide.

This is not to belittle the achievements of its leaders. The system Omidyar designed managed to cope, just about, with the flood of demand, and the many-to-many business model exploited, better than any other, one of the most distinctive features of the Internet – the way it connects millions of users to each other: eBay is more virtual than any other large-scale business. It is in fact the community that does virtually all the work – eBay simply collects the rents.

What appeared to be a disadvantage in its early days – that no-one else took it seriously or really understood it – turned out to be an enormous advantage. It faced no serious competition until it was close to being invincible. As Omidyar put it, 'Our system didn't scale, so we didn't grow big enough to attract competition. Everyone thought we were flying beneath the radar screen on purpose.' Behind the ponytail and the counter-cultural lifestyle was a shrewd strategist. He and Skoll may have underestimated eBay's potential scale at the outset – but then just about everybody else dismissed it entirely.

Whitman deserves considerable credit for handling the transition. Turning a community of hippies into a global giant, and taking revenues from $225 million when she joined to $6 billion eight years later, is a feat that dwarfs even that of John Chambers who only managed tenfold growth at Cisco. However, the scale of her achievement was largely due to the strength of the immensely powerful currents she sailed so skilfully. Her frequent assertion that 'we are building one of the great companies of our generation' was pure hyperbole. Great companies create and build – eBay's most important innovations were made before Meg joined.

Whitman deployed eBay's formidable strategic assets single-mindedly to dominate the global markets for online trading and payments, and turned it into a gigantic money-making machine. But her instinct was to buy rather than to build, and sometimes to bully. The money she threw at Skype looks like the kind of hubristic error made by those who confuse size with greatness. It is ironic that the idealistic Omidyar's 'perfect market' should have mutated into a massive monopoly, with an iron grip on its sellers.

The only business whose growth surpassed eBay's was also founded by idealists seeking to solve problems no-one else ever had. Their modest mission was 'to organize the world's information and make it universally accessible and useful'.

Google

Like Apple, Google was born out of a friendship between two bright young men with a shared passion. In the case of Larry Page and Sergey Brin, the friendship developed into a highly effective partnership. The passion was for mathematical algorithms, which were almost in their genes. Brin's father was a maths professor and his mother worked for NASA; Page's parents were both professors of computer science. They had each grown up surrounded by computers, prodigies in virtuoso households, and had attended Montessori schools, where they learned always to question authority. Mike Moritz remarked, 'If Larry and Sergey were given instructions by a divine presence, they would still have questions.'

Page had always wanted to be an inventor – at school he had built an ink-jet printer mainly out of LEGO. However, he was determined not to be the sort who won no real recognition or reward – Page wanted to change the world. Brin was also highly ambitious, and not quite sure what he wanted to do with his life. His family had emigrated to the US from Russia in 1979, when he was six. A brilliant mathematician, he had taken his undergraduate degree at the age of nineteen, and gone straight to Stanford. There he dabbled in many topics, the latest of which was data mining.

Page considered the Web as a topic for his PhD dissertation, and became intrigued by something nobody else had taken much notice of – the links *to* Web pages from other sites. A good way to analyse these, he thought, would be to crawl the entire Web, download all the links and create a gigantic graph of them. Everyone warned him that this would be a Herculean task, even when the Web had a mere 10 million documents. It did indeed take him much longer than the week he originally envisaged, but Page believed in 'having a healthy disregard for the impossible'. He devoted every available minute to the project he then called BackRub. Brin's interests clearly overlapped with this and the two friends decided to join forces.

Page had not set out to create a search engine, but he had an idea that the number of links pointing to a site was a good indicator of its popularity and importance, analogous to the number of citations a scholarly paper received. And clearly some links mattered more than others. A link from a site which itself has lots of links pointing to it would be worth more than a link from most others. They started to develop a ranking system based on the number of links, and the links to links. The Internet, they discovered, had an inherent mathematical structure. Defining it proved to be a daunting task with over 500 million variables, but together they produced the complex (and still secret) algorithm they called PageRank.

When they compared its results to those produced by established search engines like AltaVista and Excite, they were astonished at how many irrelevant results the latter were giving. These were ranking results solely according to how many times the search term appeared in the text on a website and took no account of links at all. PageRank was proving to be a much better indicator of subject relevance and the foundation for a vastly superior search tool. It was also clear that as the Web got bigger, their model would get even better. They saw that it could revolutionize how people obtained information and decided to give it a new name, Google. This was a misspelling of Googol, the term for the enormous number 1 followed by 100 zeroes.

Googol was already registered as a domain, so Google, registered in September 1997, stuck. (A stroke of fortune, since it is unlikely that Googol would have evolved into a popular verb.)

Google's mission at this stage was modest: 'to make it easier to find high quality information on the Web.' It was made available to the Stanford community and rapidly became popular with a wider audience. 'Pretty soon we had 10,000 searches a day at Stanford. And we sat around the room and looked at the machines and said, "This is about how many searches we can do and we need more computers." Our whole history has been like that. We always need more computers.'

So they begged and borrowed a hard drive here and a CPU there, and ran up $15,000 of credit card debt, to create the LEGO-like network of PCs that remains the hardware foundation of Google's computer system. This distributed architecture turned out to work more reliably, deliver search results faster and cost less than one based on large servers. But feeding the monster in the early days was almost a full-time job – and always needed more money.

Becoming a business

The obvious solution was to license their technology to an established search business. They demonstrated it to all of them, but nobody was very interested in improving search. The leading guides to the Web, AltaVista and Yahoo, wanted to keep users within their portals as long as possible to maximize advertising revenues. A tool that encouraged them to go to other websites had little appeal. It did not occur to anyone at this stage that better search could be a means of selling a different kind of advertising. Page and Brin's own view was that 'advertising funded search engines will be inherently biased towards the advertisers and away from the needs of the consumers.'

The partners kept working on refining PageRank. They knew they were on to something big, but by the middle of 1998 the computer system itself was getting so large that they would soon run out of resources. It was also clear that growth both of the system and of usage were accelerating. They were introduced to Andy

Bechtolsheim, an early-stage investor who had helped to found Sun Microsystems. He could see at once that this search engine was so good there simply had to be a business there. Without any prompting he wrote a cheque to Google Inc. for $100,000.

At this stage Google Inc. did not yet exist, and the cheque stayed in Page's desk for a few weeks until 7 September 1998, when they incorporated the company and opened a bank account. They took on their first employee and dropped out of the PhD programme, from which they are still technically on leave of absence. The patent for PageRank remains with Stanford.

With Bechtolsheim's endorsement under their belt, and in the feverish investment climate of Silicon Valley in 1998, it was not difficult to raise a million dollars from more angel investors, including Jeff Bezos, over the next few months. They were also getting noticed outside Stanford, with *PC Magazine* putting Google in its list of the top 100 websites for 1998.

Working day and night, and soon numbering eight people, they were running out of space. Early in 1999 they moved into offices in Palo Alto and recruited their first salesman, Omid Kordestani, formerly of Netscape. He was first subjected, like many candidates, to a four-hour interview: the founders only wanted the best. Fortunately, he had started life as an engineer, an almost mandatory requirement.

Recruiting fast and buying their computers in 21-packs meant that Google would soon run out of money. It was going to take a long time to generate much licensing revenue, so they had to obtain serious funding. Aiming high as always, and aware of the non-financial benefits from association with the best, they targeted two of the top venture capital firms in the Valley, Sequoia Capital and Kleiner Perkins. Sequoia they hoped could be a route into Yahoo, another Stanford spin-off, and Kleiner Perkins to AOL, both important prospective partners.

This was a very good time to raise capital. Michael Moritz of Sequoia and John Doerr of Kleiner Perkins quickly decided they wanted to invest, and while they were relaxed about the lack of a business model, they insisted that Google appoint a CEO. Their other sticking point was that neither wanted to share the investment opportunity with a rival.

Brin and Page decided that they really wanted both VCs in, and when the negotiating got tough let them know that they would be prepared to go elsewhere if they did not get their way. Eventually Moritz and Doerr conceded and in June each put up $12.5 million for 10 per cent of the business. Page and Brin were equally determined not to be edged out, as had happened to so many technology entrepreneurs. They half-heartedly promised that they would recruit a CEO within a year but had no intention of surrendering management control. Imposing their own terms on two of the biggest stars in venture capital gave them enormous cachet in the Valley, and technologists flocked to join them.

With $25 million in the bank they embarked on a massive investment programme, constructing their own supercomputer from racks of PCs, and building a dedicated operating system based on the open source Linux. They went from 300 PCs to 2,000 in one month, and doubled this the following year, by now spread between three data centres. It was an inherently scalable system, and the bigger it got, the more robust it became. Despite constant growth in traffic, response times stayed remarkably fast.

Their paramount priority was making the search engine the best it could be – their goal now was to 'organize the world's information and make it universally accessible.' Commercial success would 'just be a great side effect'. Brin hired a marketing manager in 1999, but neither of them was entirely clear what her role should be. A plan for boosting the brand through advertising did not win the board's approval and this was subsequently acknowledged to be sensible. Google could not have competed with AltaVista, spending $120 million on promotion, but the brand gained from appearing to be above the commercial fray. Its ascent was a direct result of the search engine's popularity.

Marketing effort focused on getting the best possible press coverage. The strategy was 'to invest in the product and use PR as a tool for getting people to read and talk about Google. Once they'd tried it, they'd like it.' Praise was almost universal. Journalists were particularly enthusiastic users: the *New Yorker* described it as the 'search engine of the digital in-crowd' and *Time* said, 'Google is to its competitors as a laser is to a blunt stick.' Danny Sullivan of *Search Engine Watch* named it 'Outstanding Search Service'.

The number of daily searches rose from 3 million in August 1999 to 18 million by June 2000, when Google became the largest search engine on the Web, with a billion documents in its index. When it got the chance to bid for Yahoo's search business, it was determined to supplant the incumbent, Inktomi, and quoted a very low price. As part of the deal Yahoo took a $10 million equity stake.

If most of Google's money was going on hardware, most of the founders' time went on finding talent, and they systematically sought to recruit the brightest and the best they could. The NASDAQ crash that started in March 2000 enlarged their choices: suddenly thousands of talented people were looking for jobs and no-one else was hiring. For some time Page and Brin insisted on interviewing every candidate personally, nervous of the danger of betas feeling threatened by alphas and hiring more betas. For all its laid-back, friendly external demeanour, Google the company was going to be strictly for alphas.

They were also determined that the organization should retain other characteristics of the elite university world most Google people had come from. There were long hours of intense work activity, but Googlers, unlike Microserfs, were supposed to have fun and games too. When they moved into purpose-built premises, the Googleplex, early in 2000, they had even more to offer recruits than stock options and cool technology – it was a self-contained, cosseted world that Googlers need scarcely ever leave. Gourmet food, games, laundry, hair-stylists, masseuses and personal trainers were all on tap – and mostly free. Commuters from San Francisco were bussed in on company vehicles fitted with wi-fi access so that they could google uninterruptedly. In the course of 2000 the number of employees went from forty to 150.

Google was intent on getting big fast, but this was not so much a land grab as a race to build the best technology and the best possible way of finding information. Google's revenues were in fact to get bigger, faster, than any other business had ever achieved, but it had not yet worked out how to generate them.

Finding a business model

Google's business plan assumed that revenues would eventually come mainly from licensing sales, but Kordestani was finding selling hard going, with customers intent on driving the hardest possible deal for what they still saw as a commodity product. With monthly expenditure soon exceeding $500,000, it became clear that other approaches would have to be tried. Inevitably this meant some form of advertising. The founders loathed the idea of loud banner ads. People were getting to love the uncluttered simplicity of Google pages, belying the complexity of the technology beneath.

By late 1999 they decided that they could tolerate something less intrusive – text-only ads which would be displayed to users selecting certain keywords. These were sold, like most forms of advertising, on a CPM basis – the cost of an estimated thousand 'impressions' – rather than on the pay-per-click model pioneered by banner ads. However, these were only attractive to a few big companies, and it looked as if Google would be obliged to resort to the dreaded banner ads. However, that fallback option was suddenly removed by the slump in online advertising revenues in 2000, along with the near collapse of the digital economy. 'We always thought we could swim to the boat,' recalled Brin, 'but there was no boat.' They were forced to look hard at what an innovative rival was doing.

GoTo was the most successful creation of Bill Gross's IdeaLab. Gross had made two important discoveries about the way the Web was being used and abused. Keyword search was being ruined by spam – websites, frequently pornographic, craftily inserting popular but irrelevant keywords in their sites, and enticing unsuspecting users into their lairs. His other discovery was that most traffic 'delivered' to websites was 'undifferentiated' – most customers lured there had no previous interest. So even at 7–10 cents a click, the customer acquisition costs were too high. However, if advertising could deliver visitors who were actually looking for products or services, businesses would be happy to pay more.

The problem with search for users was that it was frequently rigged. A small industry emerged of experts in manipulating websites and their 'tags' to give them higher rankings on search results. A higher ranking could effectively be bought, so the results of most search engines were simply not reliable.

Gross's solution was a radically different proposition to advertisers. He realized that the term keyed into the search box is a valuable way of qualifying the interest of a consumer – and that advertisers would pay for it. He developed a search engine for GoTo which auctioned the responses to particular keywords – the higher the bid, the higher the listing, not unlike the Yellow Pages model, where the more you paid, the bigger the display ad you got and generally the more phone calls and sales leads. Yellow Pages' business model is essentially the same – users get the directory free and advertisers pay for ads that will reach more of them. The difference for advertisers on GoTo was that they paid not a fixed amount up-front but only for customers who responded to their ad.

GoTo launched in 1998 and by the middle of 1999 had 8,000 advertisers. It got even bigger by syndicating its service on other people's websites, notably AOL's. By the end of 2000, traffic from syndication was nearly twice that on its own site and it was still growing fast. The merits of GoTo's model were obvious, but it was one that was designed primarily for advertisers. There was no way that Brin and Page would be prepared to compromise the quality of their search results with advertising, but they could see a way to borrow the core of the idea and still preserve Google's purity.

In October 2000 they launched AdWords, a simple, self-service system that small businesses could use. This confined ads to unobtrusive text, under the subtly worded, barely perceptible heading 'Sponsored Links'. These were clearly separate from search results, initially above them, later on the right-hand side (now both). Initially ads were ranked by how much clients had bid for keywords. Later they introduced a crucial refinement: ranking ads according to a formula that took account not just of the size of the bid, but also of their popularity. The more users clicked through, the higher the ad rose – and advertisers were charged on the basis of the number of clicks. This harmonized the ads with Google's search logic, and made them more convenient for users. It also maximized Google's revenues from popular ads and spared clients whose ads had less appeal from wasting their money. Unlike most forms of advertising, effectiveness was immediately obvious, and Google's cost of selling it a fraction of that of conventional ads

Brin and Page were sufficiently anxious about the launch to sully their cherished home page with a promotional message: 'Have a credit card and 5 minutes? Get your ad on Google today.' It worked – within weeks, revenues shot up, and reached $19 million for the year as a whole, most of it earned in the last two months. In 2001 revenues rocketed to $86 million and made Google cash-positive. The business was not only saved but on an unstoppable upward trajectory.

AdWords was the turning point, but it was only the stunning success it was thanks to Google's large and growing consumer franchise. When it launched, Google was serving 60 million searches a day. Concentrating on giving users the best possible experience created the strongest brand in search and the largest possible audience for advertisers. The network effects and feedback loops were almost as powerful as eBay's. When Gross suggested a merger in 2001, Brin and Page had no hesitation in refusing. GoTo (renamed Overture that same year) might have been better for advertisers but it had little to offer Google. Apart from their convictions, they had good business reasons not to risk tainting Google's reputation. They did, however, have to settle a subsequent lawsuit from Overture, accusing them of breach of copyright.

Although Google has since developed a dazzling, not to say bewildering, array of mainly free products for consumers, virtually all its revenues come from two advertising products : AdWords is the core service for advertising on search results; advertisers select their own target keywords and pay on a per-click basis for traffic that arrives at their websites. These ads are placed both on normal Google searches and on those carried out on 'partner' sites. AdSense was introduced subsequently and is mainly aimed at publishing websites like newspapers – Google inserts ads relevant to the content, and the site owners, like partner sites in AdWords, share in the revenue generated.

For the small advertiser, and most of Google's clients are small businesses, the process is largely automated, inexpensive and transparent. They only pay for qualified sales leads, can set a modest daily budget and fine-tune their key words and bids. It has also proved cost-effective for larger advertisers like Amazon.

'Not a conventional company'

For a company that disdained conventional marketing and did no advertising of its own, Google scarcely put a foot wrong on branding. It started with enormous consumer goodwill from making available, at no cost, a tool that delighted millions of Internet users. Partly because it did not blow its own trumpet, it persuaded most users to accept its own image of itself: that it was not your normal ruthless, greedy corporation, that it was mainly devoted to making the world's knowledge freely available, and that it was simply very cool.

This was all pretty much true, though not the whole story. The founders did not hesitate to lay off some of their engineering managers in September 2001, when they decided that the organizational structure was hindering progress. Google's extraordinary levels of profitability were a well-guarded secret until the IPO in 2004. Its advertising was so low key that most users were not even aware of its existence. When it adopted 'Don't Be Evil' as a corporate mantra in 2001, scarcely anyone sniggered. Compared with nasty Microsoft, recently condemned in an anti-trust case for crushing Netscape, few people had any doubts as to who was now the good guy in the software industry.

This virtuous circle was a big boost to its ability to recruit the brightest engineers, an increasingly important area of competition. So were the big bucks they were earning from a business model as juicy as eBay's. Competitors who found themselves outbid or under-cut, as Overture was when Google ousted it from AOL in 2002, could be forgiven for recalling George Burns's dictum that 'sincerity is everything. If you can fake that, you've got it made.'

But Google is sincere. The first of the ten commandments in its philosophy, proclaimed in 2003, was 'Focus on the user and all else will follow.' This means, 'providing the best user experience possible. While many companies claim to put their customers first, few are able to resist the temptation to make small sacrifices to increase shareholder value.'

Google has steadfastly refused to make any change that does not offer a benefit to users. 'The interface is clear and simple. Pages load instantly. Placement in search results is never sold to anyone. Advertising on the site must offer relevant content and not be a distraction.' It has been consistent about this, though rather less so about its second commandment, 'It's best to do one thing really, really well.' That one thing was, of course, search. It has certainly not neglected it, endlessly exploring new fields to search, but an abundance of riches enabled it to do many more things too.

Within the top ranks of the company, it was clear from 2001, when sales revenues quadrupled to $86 million, and even more in 2002, when they rose fivefold to $439 million, that Google was going to be enormously profitable. The secret remained with that small group. Not only did most people in the Internet world still think of search as a sideshow, but scarcely anyone realized what a goldmine search-related advertising was going to be. Quietly pursuing maximum growth before others spotted the opportunity helped to build an unstoppable lead.

Yahoo took a while before it realized how important and profitable search could be. In 2003 it acquired Overture for $1.63 billion, and ceased using Google itself. Microsoft was also slow to recognize how much of a threat Google posed. It was becoming not just the single most important company on the Web, but a serious contender for leadership of the software industry. Some of its online products, like Google Docs, competed directly with Microsoft's Word and Excel and were free to consumers. With the Web becoming as important a computing platform as Windows, Microsoft was increasingly looking like the slower-moving incumbent. (In 2008, Google launched a browser, Chrome, which some described as an embryonic 'mini-operating system' for Web applications, and looked as though it might realize Marc Andreessen's dream of marginalizing Windows.)

Not the least of Google's unusual qualities was the way that it was led.

Shared leadership

Nominally Page had been CEO and Brin chairman since they incorporated the company. In fact they had taken all important decisions jointly. Moritz and Doerr kept trying to get them to hire a conventional CEO, but none of the candidates met the founders' exacting standards. They were particularly dismissive of those with marketing backgrounds, and pointed to the examples of Bill Gates, Michael Dell and Jeff Bezos as young founders who had managed without adult supervision – they did not really see why they needed a CEO.

Moritz at one point threatened to pull out Sequoia's investment, as technically they had breached their commitment to appoint a CEO within a year. Doerr tried a more oblique approach, getting the young men to meet leading business figures, only some of whom were possible candidates. Andy Grove and Jeff Bezos were among those who gave them the benefit of their wisdom and experience, but Bezos remarked to Doerr afterwards, 'Hey, some people just want to paddle across the Atlantic Ocean in a rubber craft. That's fine for them. The question is whether you want to put up with it.'

At the end of 2000 Eric Schmidt had been CEO of Novell, the struggling networking software company, for five years and was negotiating a merger. He would soon be looking for another job and had heard that he was on the list of possible candidates at Google, but had no interest in joining a search company. When Brin called him, ostensibly to discuss a former employee, he was not inclined to accept the invitation to come round for a chat. Doerr persuaded him that he should. 'This is a little jewel that needs help in scaling.'

When he turned up, he was disconcerted to find his résumé projected on to a wall in the partners' room. He had had a career most people would call stellar – PhD at Berkeley, research at Xerox PARC and Bell Labs, Chief Technology Officer at Sun Microsystems, and now CEO of Novell. At Sun he had pioneered Java, battled with Microsoft, and made the interesting prophecy that when the network became as fast as the processor, the hardware would 'hollow out and spread across the network'. Brin opened the conversation by telling him that Novell's strategy was stupid, and the two of them bombarded him with questions and criticisms. After ninety lively minutes, Schmidt was surprised to find that he liked these argumentative young men – and that the company they had created was a remarkable one, with a well-thought-out strategy and a strong culture.

The founders soon decided that Schmidt's experience of competing with Microsoft would be valuable and that he might be good enough to be, if not their boss, at least a member of a triumvirate. In March 2001 he became part-time chairman, joining full-time in June, when Novell's merger was complete.

His title was CEO, but some have argued that his position is really that of a COO, and that Brin and Page take all the big decisions. According to this view, Schmidt was appointed mainly to keep the VCs and Wall Street happy. This seems highly improbable: Schmidt had plenty of money and career options, had no need of a sinecure, but knew he had a contribution to make. He has spoken publicly and with good humour of his relationship with the dynamic duo and his respect for their achievements and brainpower. His job was 'to put a business and management structure around the vision and the gem that Larry and Sergey had created'.

The broad division of responsibilities has been that Schmidt has managed operations, finance and marketing. According to him, 'Sergey is the master deal-maker, Larry is the deep technologist, and

I make the trains run on time.' It seems clear that leadership is genuinely shared between the three of them – they confer virtually every day. As Page put it in his letter to shareholders in 2004, 'Differences are resolved through discussion and analysis and by reaching consensus. Eric, Sergey and I run the company without any significant internal conflict, but with healthy debate. This partnership among the three of us has worked very well and we expect it to continue.' They had their first healthy debate almost as soon as Schmidt joined. His priority was to introduce some discipline to a company that was growing like gangbusters but still had the culture of a fraternity house. A bigger, professional company needed a solid financial system. The founders were strongly opposed to spending millions on an Oracle database, when the company was doing fine using Intuit's off-the-shelf Quicken software, but Schmidt won that argument – eventually. Subsequently Doerr would introduce Intuit's chairman, Bill Campbell, who had experience of *enfants terribles* at Apple, as a part-time coach for the management trio.

There was another disagreement in 2002, and this time Schmidt was outvoted. Google was bidding for AOL's search business against Inktomi and Overture. As with most affiliate deals, Google would provide its search engine free, and share the resulting ad revenue with the site owner. AOL, strongly conscious of the value of its enormous customer base, was accustomed to extracting the best possible deal and demanded both large financial guarantees and stock options. Schmidt, conscious that Google had run dangerously short of cash the previous year, was alarmed by the extent of the financial exposure, which could destroy what was still a very young business. The founders, however, were adamant that Google must win this deal at any price. The subsequent acceleration in growth of revenue justified their judgement.

Schmidt was subsequently gracious about the decision: 'I understood that if you ran out of cash, you were done. They were more willing to take risk than me. They turned out to be right.' He, of course, was also right to put the case for caution. Brin later acknowledged that this move could have bankrupted them.

Two years later Brin and Page did something almost as bold. When they heard that Overture, now part of Yahoo, had outbid Google for AOL Europe's search business, they diverted their private jet to London and demanded a meeting. They made it plain that they were determined to win AOL's business at just about any price, and did. Their goal was to dominate the search business and Brin in particular had every intention of doing the deals necessary to achieve this himself.

Revenues from affiliate sites, which ranged from the *New York Times* to Wal-Mart, accounted for half of Google's revenues in 2004. The most surprising of these deals was with another search company, Ask Jeeves. In 2002, Ask Jeeves was struggling financially, but it had a fairly strong brand and had recently acquired good search technology of its own. Its CEO, Steve Berkowitz, approached Google proposing that it should carry Google ads. The two companies decided that they had everything to gain and little to lose from 'coopetition'. This enabled Ask Jeeves to become profitable within a few months and Google to gain substantial advertising revenues it would not otherwise have had.

Another important source of revenue came from overseas. It was Schmidt who pointed out soon after joining that 60 per cent of Google's searches were coming from outside the US, but only 5 per cent of advertising revenues – mostly from people using the do-it-yourself service. Kordestani was charged with setting up a European sales operation, which brought almost immediate returns. By 2005, non-US revenues had reached $2.4 billion, 39 per cent of the total.

Patronizing Wall Street

Google postponed going to the stock market as long as it could. It had no need of cash and no desire to reveal to its competitors how profitable it was. Press coverage which cast doubt on the sustainability of its business model helped to keep the outside world guessing. In 2003, when its sales revenues reached $1.5 billion, it had generated a cash surplus of $395 million, and was to amass close to a billion in 2004. Going public also raised the spectre of pressure to please financial markets obsessed with short-term results, and a dilution of its distinctive culture. However, the growing number of its shareholders and the size of its assets meant that it would soon be obliged to disclose its financial results.

To the consternation of Wall Street, the founders decided to do the IPO their way. Instead of the price of the shares being set in advance, and doled out to favoured investors, who would generally make immediate gains, they wanted to conduct a Dutch auction. They would sell $2,718,281,828 worth of shares – the mathematical constant e multiplied by a billion. Anyone who wanted to could bid for the stock before it started trading. This they felt would give small investors a better chance, reduce the power of underwriters to make a killing, and maximize the financial return to Google.

Page's letter to shareholders, filed with the SEC in April 2004, opened with the defiant words, 'Google is not a conventional company. It does not intend to become one.' He went on to spell out just how inconventional:

> Sergey and I founded Google because we believed we could provide an important service to the world – instantly delivering relevant information on virtually any topic. Serving our end users is at the heart of what we do and remains our number one priority.
>
> Our goal is to develop services that significantly improve the lives of as many people as possible. In pursuing this goal, we may do things that we believe have a positive impact on the world, even if the near-term financial returns are not obvious.

He made it plain that Google was not impressed with Wall Street conventions. 'Outside pressures too often tempt companies to sacrifice long-term opportunities to meet quarterly market expectations... A management team distracted by a series of short-term targets is as pointless as a dieter stepping on a scale every half-hour.' They would not predict earnings in advance, nor sacrifice long-term opportunities to meet quarterly market expectations. He quoted the celebrated investor Warren Buffett: 'We won't "smooth" quarterly or annual results: if earnings figures are lumpy when they reach headquarters, they will be lumpy when they reach you.'

In order to retain some of the freedom of a private company, Google would have a dual-share structure, like family-controlled media companies. Existing shareholders would have ten votes per share, while new shares would only carry one each. Page was explicit about his aim of retaining control and avoiding takeover:

> The main effect of this structure is likely to leave our team, especially Sergey and me, with increasingly significant control over the company's decisions and fate, as Google shares change hands. After the IPO, Sergey, Eric and I will control 37.6% of the voting power of Google, and the executive management team and directors as a group will control 61.4% of the voting power.

He also explained why they would continue to do what they rather than Wall Street thought was the best way to treat employees:

> Google is organized around the ability to attract and leverage the talent of exceptional technologists and business people. We have been lucky to recruit many creative, principled and hard-working stars. We hope to recruit many more in the future. We will reward and treat them well.
>
> We provide many unusual benefits for our employees, including meals free of charge, doctors and washing machines. We are careful to consider the long-term advantages to the company of these benefits. Expect us to add benefits rather than pare them down over time. We believe it is easy to be penny wise and pound foolish with respect to benefits that can save employees considerable time and improve their health and productivity.

He warned them, 'do not be surprised if we place small bets in areas that seem speculative or even strange', and proudly proclaimed the principle of 'Don't be evil':

We believe strongly that in the long term, we will be better served – as shareholders and in all other ways – by a company that does good things for the world even if we forgo some short-term gains. This is an important aspect of our culture and is broadly shared within the company.

Predictably this irritated some while reinforcing the admiration of fans. Critics were gleeful when the complex auction system ran into technical problems and the offer price had to be lowered. However, the flotation on 19 August 2004 was the most successful since the end of the dot.com boom, raising $1.67 billion. It might have raised more if Google had allowed its financial advisers to do it their way: the initial share price of $85 reached $108 the next day and $200 by November. Following the announcement of sales revenues of $3.2 billion for 2004, and net income of $399 million, it continued to shoot up, reaching $300 by the summer of 2005, and $500 in 2006. By 2007, Google was generating revenues of $16.6 billion and profits of $4.2 billion and was valued more highly than eBay and Amazon combined.

Product proliferation

A more serious criticism of Google than cocking a snook at Wall Street or its air of intellectual and moral superiority is that it might be in danger over-extending itself. It has moved far beyond search, introducing dozens of new products from email to news, word processing to payments, maps to satellite photographs. It has encouraged its employees to spend 20 per cent of their time working on new projects they think will most benefit Google. Both AdSense and Google News were prototyped in '20 per cent time'.

Danny Sullivan, once Google's most influential admirer, has been scathing about the endless proliferation of new products and 'betas'. 'Give me a break from Google going in yet another direction when there is so much stuff they haven't finished, gotten right or need to fix.' He reminded Google of its principle that 'it's best to do one thing really, really well', whereas it is now 'doing 100 different things rather than one thing really, really well'.

That, of course, is very much a search perspective, and search has only ever been a means to a greater end. Apart from AdSense, none of the frenetic activity has yet produced innovations as significant as the three that made Google's fortune – in search, computer architecture and contextual advertising.

Gary Hamel on the other hand lauds this 'brink of chaos' model as *The Future of Management*. According to him, Google is showing us how 'to build companies… fit for the twenty-first century'. It understands that 'entrepreneurial success is a Darwinian process' and it wants the company to be a microcosm of Silicon Valley, capable of evolving as fast as the Web itself. Product development is conducted by hundreds of tiny teams, often only three strong, each one a little start-up. A guiding principle is 'Just try it.' Google is happy to release products that are only 80 per cent right as betas. They also expect 80 per cent of new products to fail.

That is only realistic, and even modest success can help the main business. Anything that stimulates Web usage also stokes the growth of search and advertising, which partly explains initiatives like investing in free wi-fi access in the Bay Area.

Google offers something that few other companies can. 'Talented people are attracted to Google because we empower them to change the world.' 'We're doing things that make people better educated and smarter – that improve the world's intelligence.' Material rewards are also on offer. One team came up with a new advertising algorithm, 'Smart Ads', that predicted click-through rates. This enabled Google to weed out weak ads and improve click-through rates by 20 per cent. The team was rewarded with a bonus of $10 million. Not surprisingly, Google was rated number one in *Fortune*'s list of the best places to work in the US in 2007 and 2008.

All this hasn't stopped the public criticism. There were storms over the decision to cooperate with the Chinese authorities in 'filtering' search results and over G-mail, launched in 2004. This challenged Microsoft's and Yahoo's free online email services, offering each user a gigabyte of storage free – effectively unlimited – and 500 times more than Hotmail. Google proposed to make money from it by inserting 'contextually relevant' advertising. This outraged many commentators because of the implication that Google would be reading people's private correspondence. Brin and Page pointed out that nobody would actually be reading the content, other than their computer programs. However, it became clear that holding that amount of information about people, and the potential to link that to their searches, raised important questions about privacy.

Many see G-mail and Google Apps as steps towards users, both corporate and consumer, storing all their data and applications on the 'Google Grid' or the 'computer in the cloud', rather than on their PCs. Already some devotees are happy to run much of their lives on Google computers – photos, calendars, email, social networks and credit card information. Google Apps enables businesses to use not just its software but its now enormous network of computers for the modest cost of $50 per employee per year. For new businesses 'Software As a Service' is a viable alternative to investing in servers and software licences. Some see this as part of a more fundamental plan to organize the world's knowledge. Page has declared, 'Ultimately you want to have the entire world's knowledge to be connected directly to your mind.' Many people find this and Brin's statement that 'the perfect search engine would be like the mind of God' a little chilling. They are certainly hubristic.

It is not enough merely to intend to do no evil. Holding records of individuals' correspondence and of all their searches, not to mention fleets of camera cars taking pictures for Street View, is rather Orwellian, even though Google's present management has no evil intent. This alarms those who see few limits to how far Google can go. The worry is that this may apply to the founders. Schmidt has said it took him six months of talking to understand 'how broad Larry and Sergey's vision was... I remember sitting with Larry, saying, "Tell me again what our strategy is," and writing it down.'

For all the frenzy of innovation in the ideas factory, no new project since AdSense has generated significant revenues. The serious extensions of its business capabilities have come through acquisitions rather than home-grown initiatives. Several have been of media rather than technology companies: dMarc Broadcasting, which it acquired in 2006, had a way of connecting advertisers to radio stations and was part of a wider strategy to extend Google's model of targeted, measurable advertising to other media. It followed that up in March 2008 with Double Click, the largest provider of the once-despised online banner ads, paying $3.1 billion in cash.

Double Click was an established, profitable business, but Google's boldest acquisition, YouTube, was only a fledgling, with negligible revenues and an unproven business model. YouTube only started life in March 2005, founded by Chad Hurley and Steve Chen, two young former employees of PayPal, where they had learned the importance of building networks quickly. They had found an easy way for people to upload videos and made YouTube the most popular website for posting and looking at video clips, mostly user-generated. By November it was getting 3 million page views a day, and reached 100 million by April 2006, ten times as much as Google's own home-grown video service.

Even more than Google in its early days, this was a business whose spiralling growth required ever more resources, yet, in October 2006, Google decided to pay $1.65 billion for it. Google's now vast network infrastructure and economies of scale may make YouTube's enormous bandwidth and storage costs relatively affordable, but generating revenues may be a bigger challenge. It has yet to find non-obtrusive ways of placing advertising around videos. It has experimented with 'brand channels' for corporate customers like Warner Brothers Records to promote artists, and 'participatory video ads', which could be rated, shared and tagged like amateur clips, but neither looked like money spinners to compare with AdSense.

This may change if Google can apply its expertise in Web advertising to a business with some resemblances to itself in its early days. Another business YouTube resembles, however, is Napster, the music file-sharing service that was forced to close down in its original form because of legal action by the music industry. Although YouTube has done deals with several media players, some argue that it is illegally profiting from their copyright. At the time of writing, Viacom was suing it for $1 billion.

Furthermore, its apparently altruistic attempt to digitize most of the world's books was encountering serious opposition.

Organizing the world's knowledge

Google's founders, at one point at least, ranked the goal of organizing the world's knowledge more highly than business success. The Web, however, contains only a small fraction of the world's knowledge. The serious stuff is to be found in books, and very few of these are easily accessible. Virtually no sizeable library has the financial resources to put its collection online.

In 2002, Google started work on digitizing the University of Michigan's collection of 7 million books, at its own cost, provided that it could offer users limited access to the contents. It subsequently struck similar deals with the universities of Oxford, Stanford and Harvard and with the New York Public Library. In 2004, Google announced its Book Search service, which aspires eventually to include all 32 million books currently catalogued.

Several publishers welcomed the initiative as most readers have no means of knowing about the 175,000 books published each year. Laurence Kirschbaum of Time Warner believes that 'Google is now the gatekeeper. They are reaching an audience that we as publishers and authors are not reaching. It makes perfect sense to use the specificity of a search engine as a tool for selling books.'

However, in 2005 several publishers, including some who had signed up for Book Search, sued. They were happy for Google to help them sell new books but objected to its copying of others without consent. This dispute is still going on. Freedom of information advocates are concerned that if Google bought off these publishers, it could create barriers to competitors. Since these include Microsoft, which has enjoyed accusing its irritating rival of throwing its weight around, this does not seem too alarming. But Google has given an impression of arrogance, which could be its undoing.

Too big too fast?

In June 2004, Google had 2,292 staff; three years later the number had risen sixfold to 13,786, almost all of them unusually clever. Absorbing and managing this many talented people is a tall order, and anecdotal evidence from disillusioned Xooglers suggests considerable internal confusion.

Google could be straying far from its area of real competence and stretching its ability to manage quite so many projects. These may not necessarily damage its competitive position, but Google appears to be pursuing at least four enormous goals: organizing the world's knowledge, building the biggest network of computing capacity available on the Internet, dislodging Microsoft as the leader in software, and becoming a serious player in mainstream advertising.

Each of these looks heroically ambitious. The chances of one of them failing must surely be high, and Google would then be tested by serious adversity for the first time. The danger is that Google could be succumbing to the hubris that has afflicted nearly all businesses that have suddenly achieved enormous success: they start to believe that they can walk on water and to make serious mistakes. Its growing power and seemingly limitless ambition are making it more feared than loved, alarming not just other software and Internet companies, but advertising agencies, TV networks and movie studios. Nor can it take for granted its present position of dominance in search-based advertising.

At present, Google has one of the strongest brands in the world and enormous loyalty among users – who of course pay nothing for the service. This shows no sign of changing: Google's share of the search market has continued to increase, reaching 70 per cent in the US in 2007 and higher in other countries. Almost certainly that consumer franchise will be sufficient to retain the loyalty of advertisers.

It is more than likely, however, that other suppliers will emerge with new and different propositions for advertisers. Contextual advertising could be just the first wave of a new marketing communications industry that measures effectiveness infinitely better than the blunderbuss approaches of most display ads in both the physical and the virtual realms. It is currently the most important, but the rise of eBay, the iPod and i-mode, not to mention Google itself, show how quickly new media can establish themselves. And the fall of Encyclopædia Britannica, Netscape and AOL show how quickly a medium can be eclipsed.

Google is an immensely talented company and now has formidable strategic assets, but it is no more invincible or immortal than any other.

THE BIGGER PICTURE

8

NEW MARKETS AND NETWORKS

There is a tide in the affairs of men
Which, taken at the flood, leads on to fortune;
Omitted, all the voyage of their life
Is bound in shallows and in miseries.

William Shakespeare, *Julius Caesar*

For whosoever hath, to him shall be given, and he shall
have more abundance; but whosoever hath not, from him
shall be taken away even that he hath.

Matthew 13.11–12

eBay and Google were, together with Microsoft, the businesses that benefited most from powerful forces that until recently scarcely anyone had heard of: network effects and feedback loops. This chapter describes how these and some of the other distinctive features of the networked economy have helped a few lucky winners to take almost all the prizes.

First we look at the fundamental ways in which new markets have always been different from mature ones. Our economy may seem uniquely entrepreneurial, but the stories of Jeff Bezos and Bill

Gates have much in common with that of a great entrepreneur of a century ago.

Henry Ford

Henry Ford created a mass market for motor cars and played the greatest single part in the creation of the largest industry of the twentieth century. It started with a radical idea, in fact a string of them.

At the beginning of the last century, the automotive business was a craft industry. In America alone there were 275 manufacturers by 1911, most of them making tiny numbers of fancy cars for tiny numbers of rich people, the only ones who could afford them. Ford was initially no different, but Henry had an idea that alarmed his investors – to make a very simple, much cheaper car that ordinary folks like farmers could afford. He aimed to 'democratize the automobile... to build a motor car for the great multitude'.

The idea was partly about the division of labour, but even more about a revolution in costs. Instead of workshops where skilled engineers spent weeks lovingly producing beautiful, expensive vehicles, he organized teams of semi-skilled workers in a large factory, each performing a few simple tasks repeatedly, and between them producing a tin Lizzie every three hours. The first Model T appeared in 1908 and cost only $825, when most cars started at $2,000. In 1909, he sold 18,000 – and by 1912, 170,000. Ford was determined to bring the cost down further, building bigger and better factories. From 1914, conveyor belts took the car to the worker for assembly, reduced the time to make each one eventually to 93 minutes, and brought the price down to $360. Two years later, half of all cars made in America were Fords, and in 1920 it sold a million of them.

Ford's vision was about mass marketing as much as mass production. He developed what was perhaps the most universally compelling customer proposition of the century – convenient, affordable, flexible transport for everyman and his family. He also developed an entirely new business model, which was copied by other manufacturing industries, and set up a national network of franchised dealers to sell and service Ford cars.

His most shocking innovation, described by the *Wall Street Journal* as 'an economic crime', was to offer selected workers a wage of $5 a day for shifts of only eight hours. The going rate at the time was $2.34 for nine-hour shifts, but staff turnover was terrible. Paying more meant that he could attract the best workers – and enabled them to afford cars themselves.

The Ford Motor Company shared many qualities with recent market creators: it built formidable capabilities and economies of scale that for some years no other manufacturer could match. There was also paternalistic discipline for the workers, enforced by Ford's Sociological Department, but it did not extend to finances: Ford refused ever to have the company's books audited. He also refused until 1927 to bring out an alternative to the Model T, which famously could be any colour so long as it was black.

This inflexibility almost turned triumph into disaster. Others had managed to master the techniques of mass production, and one of them, General Motors under Alfred Sloan, developed capabilities that Ford lacked – in branding, styling, organizational structure and market segmentation – 'a car for every purse and purpose'. Sloan's better-managed company edged Ford out of industry leadership, but it recovered and went on to enjoy a profitable oligopoly with GM and Chrysler that lasted for decades.

Many innovations have transformed the automotive industry since the 1920s, but mostly incremental rather than disruptive, and capabilities and propositions evolved gradually. In the 1970s, however, the American industry was shaken by massive hikes in the price of oil and by competition from Japanese and German manufacturers with better production methods and quality control, and has never really recovered from these shocks. The eclipse of the distinctive capabilities of GM and Ford by Toyota, Honda, BMW and Volkswagen has eliminated their former competitive advantage.

Plus ça change

Henry Ford was a classic market creator who eventually lost leadership of his industry. He possessed most of the attributes of recent successes, and many of the weaknesses of imperious entrepreneurs. The Ford story also illustrates the fundamental features that make all new markets different from mature ones:

- Suppliers are meeting a significant customer need that was not previously being satisfied, or only poorly. The extent of this need – latent demand – is the main determinant of how large the market becomes. In a few cases, like low-cost air travel, the need had long been clear but was ignored by incumbent suppliers. In most radically new markets, the need had not been clearly articulated beforehand. Nobody knew in 1980 how much they 'needed' a personal computer, let alone a search engine, or how much difference a skinny cafe latte could make to their lives. Yet these products quickly came to be seen as essential by millions of customers because they offered them something of real value.

- New markets are conceived by a radical innovation or a series of them – in technology, product design, business process or model. The market for PCs was driven primarily by technological developments that later 'discovered' all kinds of customer needs. Amazon's innovations were Jeff Bezos's strategy and model for the business and the IT systems his colleagues developed. They also built on the innovations of those who created the Internet and the Web.

- Business models seem bizarre, and other features of the market are difficult to make out. At the outset it is unclear who will sell what to whom, if indeed any money ever does pass hands – viable business models can take years to emerge and frequently never do. What look like promising new markets often turn out to be damp squibs or tiny niches. Even in the most exuberant cases, it is impossible to predict how fast they will grow or how big they will become.

- To compete in the new market, suppliers must possess unusual capabilities. The lack of these is the most common barrier to other suppliers, including incumbents in markets invaded or subsumed by the new one. Typewriter manufacturers (other than Olivetti and IBM) had no means of competing with PC makers, nor corner stores with the prices and choice offered by giant supermarkets. Mail order companies, however, found that they already had many of the capabilities to be effective online retailers.
- These capabilities enable the first supplier(s) to make compelling new customer propositions that address the unsatisfied need.

In the new markets described here, the unmet needs, levels of innovation, business models, capabilities and customer propositions were all clearly distinct (with the benefit of hindsight) from those in any previous market. The reason sales rocketed was that the needs they identified or discovered came to seem vital to their new customers. They were thirsting, whether they knew it or not, for better coffee, and for more convenient ways to buy books and to trade with each other. The market creators were successful primarily because of their imagination in spotting these needs, their creativity in devising ways to meet them, the exceptional capabilities they developed and the compelling propositions they were able to make to customers.

Not many entrepreneurs set out to create a new market, nor do they do so literally in the way that a designer creates a new product or a composer a symphony. To what extent is it meaningful to speak of a company creating a market, and is it a sensible goal?

What all the market creators described here did was open up 'competitive white space' that other suppliers were ignoring. Their vision and the actions they took were critical to how and when each market developed and the shape it took: it was their capabilities that largely defined the new market. Consequently all of them initially enjoyed enormous competitive advantage – something close to a monopoly for a time. This is the prize that market creation offers.

There are essentially two strategies for holding on to it. One is to keep capabilities so distinctive and customer propositions so compelling that competitors cannot match them. The other is to cultivate strategic assets that act as barriers to competitors. The two are not of course mutually exclusive – the safest strategy combines both.

Strategic assets can take several forms – from the legal monopolies of post offices to the brands and distribution channels of Coca-Cola. When the idea of a new economy appeared, a new set of strategic assets and concepts captured the imagination of many entrepreneurs and investors – first mover advantage, scalability, feedback loops and network effects offered the hope of quickly dominating a large new market. Many of those who tried were more intent on pursuing this dream than on cultivating the essential attributes of market creators.

First mover advantage

The most hyped of the new concepts during the Internet boom was first mover advantage, when being the first to enter a market was seen as synonymous with winning leadership of it. Land grab was the main goal of thousands of businesses and investors – according to some analysts the first to stake out the new territory would automatically become the long-term winner. First mover advantage played a crucial role in eBay's sudden success, in Sky Television winning its battle with BSB, and in Jeff Bezos's get-big-quick strategy.

Making the first move, however, is not the same as creating a market. It may help in building a lead, but it is no guarantee of holding on to it and has proved to be a siren song that led many businesses astray. For every case where it appears to have been decisive, there are many more where the pioneer was quickly displaced. Netscape is a classic example – as was Sony in VCRs, Apple in personal computers, not to mention Adair, which produced the first primitive PC kit in 1975. Microsoft has scarcely ever been first into a new market. It made itself master of the universe to a considerable extent by painstakingly copying other people's ideas, and sometimes imposing on them, and occupying their markets. Google likewise benefited from not being first into search or contextual advertising.

First mover advantage then is far from being a universal rule, and is scarcely a rule at all. A good reason for being the early bird is when there is only one worm: a unique resource – a particular location for a retailer, an exclusive licence to a technology or the rights to uniquely attractive programming – can be critical to success. Sometimes a single supplier can tie up the lion's share of a new market, but generally through developing capabilities that competitors cannot match, forming binding relationships with customers or building other barriers to entry. One of these, or a combination of them, gave all of our subjects who became industry leaders much of their lasting competitive advantage. However, being first was nearly always a means of achieving them, not in itself the key to success, and sometimes yielded only fleeting fame. Netscape, emphasizing speed above almost everything else, got it the wrong way round. Microsoft caught up with it on capabilities and deployed formidable strategic assets of its own.

Barriers to entry in embryonic industries are often low initially – their absence spurs pioneers like Starbucks and Amazon to move as fast as they can, for fear of competitors overtaking them. Like most of our market creators, they made life difficult for competitors, primarily by honing their capabilities obsessively. They were better at doing what they did than anyone else and carried on improving. They also cultivated their brands and customer loyalty, but mainly by maintaining and enhancing their distinctive capabilities and customer propositions.

Some companies, however, get such a tight hold on customers that competitors cannot prise them loose. Two kinds of strategic assets have this effect – switching costs, which Cisco, Microsoft, BSkyB and Nokia enjoyed, and network effects, discovered by eBay, AOL and Google. Being early, if not necessarily first, played a big part, but it was these assets that were crucial. What tightened the grip in all cases were positive feedback loops that tipped the market towards one supplier.

Locking customers in

Switching costs are the penalty that customers face if they try to change supplier or technology. Switching would mean ditching their earlier investments in families of equipment, so they need a particularly strong inducement to do so.

When Sony launched the first CD player in 1982 nearly all consumers hesitated. The most likely buyers had large collections of vinyl LPs and expensive turntables to play them on. To buy a CD player and replace their LPs meant spending a lot of money, particularly as record companies decided to charge a premium for the same music in CD form. The appeal of the medium had to be very strong to overcome these switching costs. It was – but it took a few years.

In the case of business customers, investment in a particular kind of equipment, both its direct cost and the time spent learning how to use it, can run into millions, and the inhibitions are much stronger. Switching costs can be incumbents' greatest asset and are generally the biggest obstacle challengers have to overcome in technology markets. The battle is essentially between them and the challenger's proposition.

Some suppliers enjoy a combination of switching costs and network effects, when their equipment or format becomes the standard for the market. This situation, where it is virtually impossible to tempt customers away, is sometimes described as customer lock-in – they simply cannot escape. IBM enjoyed lock-in in mainframe computing for several decades, thanks to its customers' enormous investment in its proprietary technology and the comfort they took from their peers doing the same thing – 'You could never get fired for buying IBM.'

We shall see in the next chapter how dramatically that changed in the course of the 1980s. IBM had a brilliant initial success with its PC but soon found that customers in this new market had much lower switching costs. It was easy, and very much cheaper, for them to buy clone PCs from manufacturers like Compaq and Dell, who eventually became the leaders of the new industry. The real victor, though, was a PC standard based on Intel hardware and Microsoft's operating system, later dubbed Wintel. Virtually all large business users of PCs are now locked in to Wintel, rather than, say, Macintosh, though they do have a choice of PC supplier. To switch to Apple's more expensive machines would mean ditching all the company's investment in Wintel machines, software and learning.

An individual consumer thinking of making the same switch is only mildly inhibited – his or her investment in a single PC is probably only a few hundred dollars, given ever-lower hardware prices. Indeed, Apple is winning over many style-conscious young consumers to iMacs, in the wake of the success of the iPod and cool Apple stores. Switching costs work for Apple too, though its customers' investment tends to be more emotional than economic. Even in the darkest days wild horses would not have persuaded hardcore Mac users to desert, no matter how low Dell's prices.

Positive feedback and network effects

Positive feedback and network effects are close relations of the tipping point, made famous by Malcolm Gladwell. He called the rebirth of Hush Puppies and the sudden drop in crime rates in New York 'social epidemics', where a few 'infectious agents' (meaning influential people) triggered the snowball-like growth of a trend. Two more widely applicable metaphors are positive feedback loops and virtual networks. Feedback and networks, both physical and virtual, play an enormous part in the new economy, thanks to the proliferation of communications media and ever greater connectedness, and explain why social epidemics spread, and virtual networks grow, so quickly.

Positive feedback loops amplify small changes to produce a bandwagon effect: success breeds success, and failure, failure; the strong tend to get stronger and the weak weaker. Growth – or decline – accelerates. Positive feedback is critical to the development of networks, producing virtuous circles like the growth of eBay or vicious ones as when Netscape's market started slipping away from it.

Between 1975 and 2000 one international group after another –

air traffic controllers, scientists and, crucially, businesses – adopted English as their lingua franca. The more people who spoke it, the more others wanted to. The number of people speaking English as a foreign language rose from 100 million to 700 million, while the popularity of other languages spiralled downwards.

The Wall Street crash of 1929 triggered a cascade of bank failures and deflationary, protectionist measures by governments across the world. Massive unemployment and foreclosures on mortgages fuelled panic and despair. By 1933 world trade had shrunk by two thirds and Hitler was Chancellor of Germany. Another cascade of financial failures in 2008 threatened a new depression.

The concept of feedback loops has long been familiar to engineers and systems thinkers. They play a big part in all cases of rapid business growth, including those with no notable network effects. Henry Ford enjoyed positive feedback when he was building his economies of scale – the more he invested in automation and good-quality workers, the lower his unit costs and the higher his sales, which further increased his economies of scale. Things just kept getting better and better until the early 1920s, when there were no more big economies of scale to be achieved and competitors were copying his methods and innovating in new ways.

Markets with lots of positive feedback tend to tip towards a particular supplier or technology. At first the shift from LPs to CDs was slow, but once there was more music available on CDs, and consumers became aware of their quality and convenience, the market tipped decisively. Soon millions of people were converting and sales of vinyl dwindled.

A network generally becomes more valuable to its members the bigger it is, the more members, or nodes, there are. This is true whether it is a telephone system, the members of a stock exchange or the subscribers to a dating agency: they have more people to speak to, to trade with, to fall in love with. A virtual network can include the body of people who use the same equipment, listen to the same music or belong to the same 'community'. Once a network gets to a certain size, as we saw with eBay, it becomes extremely difficult to tempt members to leave it. Bob Metcalf, who developed Ethernet, the local area network used to connect personal computers, coined the 'law' that bears his name when he said that the value of a network is proportional to the square of the number of its members. Although not literally true, this is a useful rule of thumb.

In the early days of a network, it is so small that it is not its size that attracts other members – it generally needs to get to a critical mass before most people start to take notice. That is why growth follows an S-curve – very slow at first, accelerating sharply if and when positive feedback kicks in (the tipping point), and slowing down later as the network approaches maturity. Peter Thiel was one of the few entrepreneurs in the 1990s to understand this: his strategy of paying new PayPal customers $10 to join was designed to get to the tipping point as quickly as possible – an expensive strategy, which paid off in his case. Typically, it is enthusiastic early adopters who nudge a network towards critical mass.

Some networks have long gestation periods. Fax was technically possible early in the twentieth century but only really caught on in the early 1980s, when cheap machines came on to the market and there were robust, automated telephone networks in most Western countries. By the end of the eighties just about every business large and small was using fax every day.

Fax was soon eclipsed when email became contagious and even more ubiquitous. The original email systems were closed, proprietary ones, used mainly for internal communications by a few large organizations. When the Internet emerged as the common standard and big companies switched, there were suddenly hundreds of thousands of email users on the Net, helping this particular network on to the steep part of the S-curve. Very quickly, every business and most individuals could see a reason for joining the Internet. When nobody you knew used email, there was little point, but once most of your friends and colleagues did, it became almost essential.

This growth was fuelled by the positive feedback stoking the parallel growth of the Web, once most Internet users had Netscape browsers: more users encouraged more organizations to set up websites; more content and more interesting services encouraged more consumers to venture into cyberspace. It can be argued that email was the Internet's 'killer app' in the 1990s, the single most important factor driving demand. But the number, scale and intensity of the feedback loops mean that we cannot ascribe the Internet's explosive growth to a single cause. Complex systems invariably have many, and the Internet was a very complex system by 1995. In the case of cyberspace nobody 'owned' the customers, though AOL thought it did. The owners of proprietary email systems often lost theirs, and e-commerce merchants and media owners competed fiercely for the attention and custom of Web users.

All the companies described so far, apart from Webvan, found themselves benefiting from positive feedback loops, but these are easier to spot with the benefit of hindsight. Netscape, eBay and Google are all fairly obviously network businesses, but feedback and tipping also apply when virtual networks of customers choose between rival technologies or standards. Sony lost the VCR war, largely because JVC's strategy of licensing its technology to other suppliers meant that there were many more VHS machines available at lower prices. An initially slight numerical advantage was reinforced when there were more film titles available for rental on VHS. The market tipped decisively against Sony, and this accelerated when consumers began to conclude that Beta was going to be the loser. Sony became the victim of shifting expectations and sales of Beta machines dwindled almost to nothing. Finding oneself in a positive-feedback loop is one of the ways that luck can affect business success and failure. Sony was unlucky, as was Kodak faced with the digital revolution; eBay and AOL in the late nineties were particularly lucky.

Feedback and network effects played a big part in Apple's marginalization in the 1980s. There were soon many more IBM-compatible PCs in the marketplace than Apple machines, firstly because IBM sold so many to businesses, and then because clone manufacturers started churning out cheap copies by the million. First-time buyers, hesitating over Apple, mostly decided to go for the PC, not just because it was cheaper but because that was what everyone else was doing. There was already more software available for PCs, and it was easier for PCs to interconnect with each other than with Macs. The mass market, both business and consumer, tipped decisively against Apple.

The main beneficiary was a company that in 1980 had sales of only $7.5 million. Ten years later, Microsoft's revenues were $1.1 billion, and it had become the most powerful, most feared business in the IT industry.

A brief history of world domination

The story of how Microsoft extended its dominion to the most valuable segments of the desktop software market is a long and murky one. The extent to which it exploited its control of the operating system not just to 'embrace and extend' standards, but also to extinguish smaller software companies by adding proprietary features is hotly contested. Microsoft argues that incorporating in its operating system features like browsers that were previously distributed as individual programs mainly benefited consumers. There is no doubt, however, that it set out deliberately to wrest control of some markets from the companies who had created them, and withheld information on Windows from its rivals so that its own programs interacted with the operating system much more smoothly.

In the 1980s, Lotus 1-2-3 was the leading spreadsheet program for the new IBM PC. Lotus worked better on the PC than VisiCalc, which had done so much to feed demand for the Apple II. Lotus became the standard in spreadsheets, as did WordPerfect in word processing. The early versions of Microsoft's programs, Excel and Word, had only achieved market shares of 20 per cent by 1990. What was to make the difference was Windows, Microsoft's attempt at an operating system with a graphical user interface like the Mac's. However, early versions of Windows were not a great advance on DOS, with its forbidding lines of what looked like computer code, and not many customers bothered to upgrade.

Teams of Microserfs toiled away under Bill's eagle eye, and by 1990 not only did they have versions of Word and Excel that matched

WordPerfect and Lotus, and worked particularly well with Windows, but they also had a version of Windows that became a hit: Windows 3.0 sold 4 million copies in its first year and sealed Microsoft's dominance, both of the software industry and of the desktop.

WordPerfect and Lotus were slow to release Windows-compatible versions of their programs, partly because of their reluctance to see the Beast of Redmond become any more powerful. This hesitation was fatal – it enabled sales of Word and Excel-for-Windows to move into a slight lead by 1992. Positive feedback from the success of Windows 3.0 accelerated this and Microsoft pressed its advantage with its customary hard drive. Mike Maples spelled out the strategy: 'If someone thinks we're not after Lotus, and after WordPerfect, they're confused... My job is to get a fair share of the software applications market, and to me that's 100 per cent.'

What turned victory into the virtual annihilation of its rivals was bundling together Word and Excel in the 'suite' of programs labelled Microsoft Office. This was sold at a discount to their stand-alone prices, and threw in PowerPoint for nothing, a very compelling proposition. The new vogue for networking computers together, first in local area networks, then over the Internet, made it irresistible. Standardization of applications software suddenly became almost as important as it was in operating systems. Once people started exchanging files, it was infinitely easier if they all used the same programs. Naturally they standardized on those that had emerged as the new market leaders, and which worked so conveniently well with what had become the new standard in PC operating systems.

It was the most wonderful of double whammies for Microsoft – it now enjoyed powerful network effects in both operating systems and applications software. By 1994 it had 70 per cent of the markets for spreadsheets and word processing and its revenues from Office alone reached $4.5 billion. That was an exceptional year – companies soon realized that they did not need to upgrade these programs endlessly, but they had no inclination to move to other suppliers. They had by now made substantial investments in learning how to use this software, it already had more features than most people knew how to use, and they still needed to exchange documents with others. Regardless of whether they liked the programs or the supplier, they stuck with the devil they knew.

Like most cases where positive feedback kicks in, this was not entirely planned. Bill Gates acknowledged that he had been slow to spot the significance of computer networking in general and the Internet in particular, but he benefited enormously. This was one of those rare occasions when a discontinuity can produce a massive bonus rather than a disaster. Microsoft went on to pulverize Netscape and to extend its hegemony to the Internet browser market. It did most of the work itself, but feedback played a big part, once momentum built up behind Internet Explorer. When Netscape acquired the look of a loser, public expectations finished it off. It really did look for a while as if one winner was going to scoop up all the prizes.

The twenty-first century, however, has been less kind to Microsoft than the last two decades of the twentieth. Its victory over Netscape did not lead to anything like domination of the Web and triggered a long anti-trust suit. In 1999 a court ruled that Microsoft had used its 'prodigious market power' to 'harm any firm' that might threaten it. The revelations in court seriously tainted at Microsoft brand, and its share price has been sluggish ever since. Other browsers like Firefox and Safari have taken market share from Internet Explorer; Linux and the open source software movement have dented its hold on the server market; and Google has challenged its overall pre-eminence, and not just by making search its own. It has also put up credible alternatives to Microsofts's applications software products, offering 'free' word processing and other programs online that Microsoft still charges for. It is even threatening Windows obliquely by making the Web, rather than the desktop, the place to go both for information and for applications. Software as a service poses a deadly threat to Microsoft's business model.

Scalability

After Henry Ford blazed the trail, all manufacturers sought economies of scale through the largest possible scale of production.

New-economy entrepreneurs seek a different, more intangible quality, scalability, a concept borrowed from the world of software engineering. When applied to systems, it means that they are able to handle increasing amounts of work gracefully, that their performance continues to improve after more hardware is added, as happened spectacularly with Google. Its ability to grow exponentially with minimal human intervention was a major factor in its success. Yahoo's labour-intensive approach to classifying information did not scale.

During the Internet boom the term was frequently used to describe businesses and business models, and some now see it as a characteristic of all successful businesses in the networked economy. Amazon, Dell and Cisco were the most frequently cited examples of scalability in the 1990s, and their 'virtual' businesses did indeed grow astonishingly quickly, though not quite as easily as Google and eBay, who almost circumvented the physical world entirely.

Dell and Cisco avoided many of the constraints of supplying physical products to customers by relying on others to do most of the manufacturing. Conducting virtually all their communications with suppliers and customers over the Internet, they achieved enormous advances in productivity. Amazon presented itself as a virtual business but was not quite so scalable. Eventually it had to build warehouses to store books, and employ large numbers of people to despatch them to customers, even if it pretended that they were not really employees. Just as looking like the winner of a standards battle helps the company with a slight edge, looking scalable can be as important as the reality. Winning in tippy markets is partly a question of managing consumers' expectations.

One of the consequences of very scalable businesses is the 'long tail', Chris Andersen's term for the mass of small markets opened up on the Web. Book and record shops can only stock a tiny proportion of the titles available, but online retailers can offer infinite choice. Instead of concentrating on just the best-selling hits, as most retailers are forced to do, there is money to be made from selling the neglected titles in the long tail behind these. A small bookshop will stock a few thousand books, a giant store 100,000 or so, but over a quarter of Amazon's sales come from titles outside its top 100,000.

According to Andersen the long tail reduces the tyranny of the hit record and the best-seller in media markets, and means that many more niche markets can now be addressed economically He argues that it accounts for much of the success not just of Amazon but also of Google and eBay: Google gets most of its advertising revenues from large numbers of small advertisers, and eBay has got big by aggregating demand from tens of thousands of tiny markets. Apple's iTunes is another big beneficiary, but it is in the music business. Whether the long tail proves to be as significant as Andersen believes in markets beyond media remains to be seen.

Some of our market creators were more naturally scalable than others. A few benefited to the extent of enjoying 'positive returns' – instead of the diminishing returns that normally set in when the limits to economies of scale have been reached. However, bigger does not always mean better – some networks can get too big for comfort and reach 'complexity catastrophe'. AOL suffered from serious congestion on more than one occasion and permanently alienated millions of customers. In many large organizations, the hours people spend each day dealing with email has turned it into a bureaucratic quagmire.

Winner takes all?

Nassim Taleb laments the fact that a few lucky individuals and businesses in 'scalable' businesses and professions are now taking a grossly unfair proportion of the prizes. A tiny number of books are enormous best-sellers while 90 per cent are flops – similar ratios apply to movies and music. Taleb and Andersen, like many writers, take an intense interest in book sales. Sadly, books are not one of the biggest markets – and have always been skewed.

Several commentators have accused the new economy of increasing inequality – rewarding a few individuals and organizations out of all proportion to their apparent worth. Tiger Woods and David

Beckham, investment bankers and CEOs, make fortunes, the latter even when their companies hit big trouble, while average earnings in the US have stagnated. The share of national income of the richest 1 per cent of Americans doubled from 8 per cent in 1980 to 16 per cent by 2004, and that of the top 0.1 per cent tripled from 2 per cent to nearly 7 per cent.

Uneven distributions of all kinds of things are much more common than many people assume, and there is nothing new about them. Vilfredo Pareto discovered in the nineteenth century that incomes and wealth were not 'normally' distributed in a neat bell-shaped curve – he found that 20 per cent of the Italian population owned 80 per cent of the wealth and received 80 per cent of total income. The 80:20 'rule' seems to apply to many other groups.

Scientists in several fields have identified a mathematical structure to populations where power laws apply – where the relationship between variables is based on an exponent, or power. Metcalf's 'Law' says that the value of a network corresponds to the square of the number of members – it increases to the power of two. Power law populations are made up of a few giants, each one taller than the other, with lots of midgets at the bottom. Power laws apply particularly to social systems, where people are involved – to the popularity of websites and hit records, the frequency of certain words in languages, the prices of paintings and, most conspicuously, to incomes.

Enormous differences in incomes seem shocking because we became accustomed to the relative equality that was the norm between 1940 and 1980. But historically this was an exception – highly unequal shares were common for most of human history, long before the dawn of a new economy and even before capitalism raised its ugly head. Until about 1800 wealth mainly meant land, its ownership was concentrated in a few hands, and most people were subsistence farmers with no land, and no cash income at all.

Many reasons have been suggested for the recent increase in inequality: the deregulation of labour markets and fiscal policies; the elimination of many white-collar jobs by computerization; and global competition for all kinds of work. These, however, do not account for the extraordinary growth in incomes at the very top.

The most highly skewed rewards are found in occupations where differences in talent and popularity can have a massive effect on overall performance and success, where a few people become global superstars. Something like winner-takes-all applies to sport and entertainment, and to top executives, but not to teachers and accountants. Where a few people are perceived, rightly or wrongly, to be the very best, and capable of making a decisive difference to whether a film is a hit, a company's share price perks up, or a football team wins matches, businesses are willing to pay an enormous premium to obtain them. Demand for the A-listers exceeds supply, and they compete mostly only with other A-listers.

The only thing that the earnings of footballers and film stars have in common with network monopolies is that power laws and feedback loops apply in both cases. These occur much more frequently in the modern economy because of its complexity and interconnections – millions of virtual networks and feedback loops are interacting with each other, most of them invisibly. The reason why Microsoft and eBay have taken an enormous share of their core markets is because their strategic assets have given them monopolies, but they are unusual. In most markets competitive advantage is extremely difficult to maintain for long, as we shall see in the next two chapters. In fact, monopolies and oligopolies were more common before 1980, when markets were less volatile, market shares changed slowly and governments often protected businesses from foreign competition

Ultimately A-listers earn as much as they do because millions of investors and film and football fans are prepared to pay the price of obtaining their services. Something similar applies to market creators, who establish temporary monopolies because their capabilities are unique. Monopolies based on strategic assets mostly last longer, as competitors are locked out and customers locked in.

A-listers are less protected and have shorter shelf-lives. Footballers and film stars are lucky to play at the top for more than a few years, and the reigns of CEOs are getting briefer, which is why they demand contracts that reward them handsomely when they are terminated. Networks on the other hand tend to go on for many years, and when they die, do so slowly.

One of the industries where the leading incumbents have benefited from network effects for decades but now face decline is Yellow Pages. These directories were once immensely profitable, mainly because they delivered large audiences of consumers looking for suppliers, but also because many of the small businesses who advertised there had no obvious alternative. It was an affordable form of advertising for unsophisticated businesses and generated sales so long as consumers consulted the directory frequently to look for suppliers. It suited both parties to use just one directory, so this seemed like a benign monopoly.

Two things have changed: more and more consumers are conducting their searches for suppliers online, rather than using directories; and many small businesses are placing ads with Google and others. They find that this offers better value, as they only pay for sales leads that arrive at their website.

This does not mean that all Yellow Pages advertising will migrate to online methods, but a great deal of it will. This particular network is no longer as magnetic as it used to be, and positive feedback loops are accelerating the trend. The same decline hit AOL (and could one day hit eBay or Google). However, although AOL is a mere shadow of its former self, it still has millions of customers and remains a valued strategic partner for both Google and eBay. Even a gradually weakening magnetic force counts for a lot, so long as a rival does not have a devastatingly compelling new proposition.

Although they have become more common recently, because of the growth of various kinds of networking in the modern economy, really strong network effects are still rare. Scarcely any businesses benefit from them to anything like the extent that eBay and Microsoft currently do. They are generally discovered rather than planned. Virtually all of our market creators stumbled across these strategic assets, which quickly became their main defences against competitors.

The Maginot Line

There is a downside to relying too much on barriers to invaders – it can breed a dangerously defensive mindset. French military strategy in the 1930s relied on the heavily fortified Maginot Line running along its border with Germany and impossible to breach. However, when the Germans invaded in 1940, they avoided it by coming through weakly defended Belgium, just as they had in the First World War. Their 'blitzkrieg' technique of air-supported rapid tank movement was similar to a strategy that a certain Colonel De Gaulle had urged the French army to adopt before the war. Its defensive mentality and failure to develop strong mobile capabilities led to the most humiliating defeat in France's history.

Companies who rely on barriers to entry and strategic assets for competitive advantage, rather than developing new capabilities, may become similarly exposed when the rules of engagement change. Businesses like telephone companies and airlines that previously depended on national protection or monopoly have found competing in a deregulated market distinctly uncomfortable. Technological revolutions can have a similar effect on incumbents with previously enormous competitive advantage. Assets that they thought made them invulnerable can become irrelevant and even a handicap. Sales forces and distribution channels who fear being bypassed can cause fatal hesitation.

In the next chapter, we examine two venerable institutions who found that strategic assets they thought made them impregnable had become liabilities.

9

THE DISRUPTIVE PC

IBM and Encyclopædia Britannica, Inc. were two outstanding companies who came close to being destroyed by the most disruptive technology of the late twentieth century, the PC. IBM's was the most dramatic fall from grace in recent business history. It partly recovered, but the business model of printed encyclopedias like *Britannica* was blown away.

IBM

Ever Onward

IBM was the creation of two remarkable men, Thomas Watson and his son, 'Young Tom'. 'Old Tom' was a buccaneering salesman who rose to be general manager of NCR (National Cash Register), where his aggressive tactics earned him a jail sentence and the sack. In 1914 he became general manager of another business equipment company, CTR, and escaped serving his sentence. In 1924 he gave CTR a name to match his ambitions – International Business Machines. However, according to Young Tom, it was then 'still full of cigar-chomping guys selling coffee grinders and butcher's scales'. Its most successful products were machines for sorting and tabulating punched cards, crude precursors of the modern computer, but in their day the most sophisticated form of office machinery.

Watson created a paternalistic corporate culture, led by conservatively dressed, positive-thinking salesmen, with frequent singing of the company hymn, 'Ever Onward'. A spirit of boundless optimism and confidence prevailed. IBM offered generous benefits and a lifetime career to its employees, who reciprocated that loyalty. By 1932, IBM was the market leader in tabulators, and three years later it had 85 per cent of the market. By the outbreak of war, sales reached $40 million.

Young Tom joined the company as a salesman in 1937 and after war service rose rapidly. He realized that the days of mechanical tabulators were numbered and was the strongest advocate of investment in electronic computers. Old Tom is reputed to have declared that there was a 'world market for maybe five computers' and remained resistant to too radical a shift. On Young Tom's insistence IBM started building expertise in computing, and when he became president in 1952 hired large numbers of engineers. Other companies, however, were setting the pace in the marketplace and Young Tom quarrelled frequently with his father, who continued to rule the company with a rod of iron. He eventually backed his son's strategy of shaking things up and ceded control in 1956, at the age of eighty-two, when the company launched the IBM 704, the most powerful commercial computer then available.

Watson radically changed the company he inherited. 'The secret I learned early on from my father was to run scared and never think I had it made... I never felt I was completely adequate to the job.' He set up a new divisional structure, increased spending on R&D to 9 per cent of revenues, developed new capabilities in software development, and deployed IBM's well-honed marketing and distribution organization to sell the new products. He was conscious that he was building capabilities and know-how as much as sales. Early large orders for the US Air Force earned meagre margins but enabled IBM to build large automated factories and to train thousands of workers.

Watson decided to share authority with a Management Committee that would consider all important decisions and adjudicate disputes, instituting a 'contention system' to encourage executives to put the arguments for or against key initiatives so thoroughly that decisions were 'bound to be right'. IBM executives rose by giving winning presentations. To be 'good with foils' and exuding sunny optimism became core competences. Although Watson was as strong-willed and hot-tempered as his father, he liked occasionally to appoint 'harsh, scratchy people' and frequently listened to them. He welcomed the occasional 'wild duck' that did not always fly in strict company formation.

His real revolution though was effectively to bet the company on new technology, cannibalizing the old business and transforming the market for mainframe computers. (Mainframes took their names from the enormous metal boxes that housed them.) By 1960, Watson had shut down the punched-card division and made IBM not just the clear leader in the new industry but significantly bigger than its seven main rivals combined, dismissed as the Seven Dwarves. Nonetheless in the early 1960s he took a massive gamble on an entirely new range of computers, the 360 series, the first based entirely on integrated circuits rather than valves or transistors. IBM deployed enormous teams of people on the project who had to spend much of their time on coordination. The project took far longer than expected, ran way over budget, and took the company close to bankruptcy. It cost over $6 billion, then the biggest-ever privately financed commercial project, and more than three times IBM's annual revenues at the time. Announced in 1964, on the sales force's insistence, the full range was not available until 1967.

Nonetheless, the 360 series was a triumphant success – it was the first family of computers where each machine could use the same software, which meant that customers could easily trade up without incurring enormous costs. IBM's dominance was now complete: no other supplier could hope to supplant its leadership. For the next twenty years IBM was the most admired company in the world.

Watson retired early, having taken the company ever onward from sales of $1 billion in 1956 to well over $7 billion in 1971. His success owed much to his philosophy. 'If an organization is to meet the challenges of a changing world, it must be prepared to change everything about itself except its basic beliefs... The only sacred cow in an organization should be its basic philosophy of doing business.' Sadly his successors would lose sight of this truth.

Frank Cary took over as chief executive in 1972. Like every CEO until 1993, he had started his career as a salesman and worked his way up. But Cary was also a graduate of Stanford Business School and, unlike the Watsons, a cerebral, analytical manager. He had been prepared on several occasions to argue with Watson – and generally won his point. In his eight years at the helm, IBM reached revenues of $26 billion, and grew from 262,000 employees to 341,000. Cary reshaped the organization as the archetypal professionally managed American company, with extensive staff training programmes, the careful identification and nurturing of the 'hi-pos', heavy spending on R&D, and the intensive cultivation of customer relationships. The IBM salesman became a trusted adviser. The cliché that 'nobody was ever fired for buying IBM' became an enormous asset – IT managers had to be brave not to.

Cary surrounded himself with people who understood technology better than him and encouraged open debate. However, IBM was becoming increasingly bureaucratic and complacent and was taking longer and longer to get new initiatives moving. The contention system proved to be better at stifling ventures than allowing them to flourish. Although System 370 was a successful evolutionary extension of the 360, a task force to design the Future System was a major failure. Meanwhile, IBM's market share was gradually declining, as rivals opened up new markets like minicomputers and super-computers that IBM ignored at first. Others, notably Gene Amdahl, who had been the chief designer of the 360, set up his own company producing IBM 'plug compatibles', precursors of the clones that were to be the IBM PC's nemesis.

Cary worried about miniaturizing technologies which could eventually undermine mainframe markets, but much of his time and attention was absorbed by a twelve-year anti-trust suit, probing IBM's domination of the industry. Numerous analyses were made of the possible market for personal computers, but vast teams had failed to develop a marketable product. A frustrated Cary, as one of his last acts as CEO in 1980, ordered that a team of forty people be set up to develop a personal computer, isolated from the rest of the company, reporting to him personally, and led by someone outstanding. Cary remained as chairman until 1984 and on the board until 1991. His successor as CEO was John Opel, the man who had led the ill-fated Future System task force.

The accidental entrepreneur

The first leader of the PC team was Bill Lowe, whose initial strategy was bold and radical. He knew that IBM's elephantine teams could never produce anything quickly, and they needed to get a product launched within a year. He convinced Cary that IBM should set up a completely autonomous unit, not just to develop but to produce and market the new product, and recommended an open architecture that would allow customers to add other companies' peripherals. Even more revolutionary, the PC would not use proprietary IBM components, and if IBM units could not supply them immediately, the business would go elsewhere.

A small software company in Washington that knew something about PCs was keen to help, and in 1980 the fateful relationship with Microsoft began. At this point, Bill Gates's main goal was to build a close relationship with the biggest partner a business with only thirty-two employees could hope for. When IBM said it needed an operating system for the PC, he found one. The original core of what was to become the pivot of an entire new industry was known as QDOS, quick-and-dirty operating system. Microsoft bought the rights to this from Seattle Computer Products, hired the man who'd developed it, and set to work to make DOS, now called Disk Operating System, good enough to run the IBM PC. Accounts vary as to how much Microsoft paid for the original, but the highest estimate is $100,000, which made it quite a bargain.

Within a few months Lowe was promoted away from the PC group and replaced by Don Estridge. Estridge had not previously been considered a real high-flier or even a true IBMer. Although proud to work for what he considered the finest company in the world, he was frustrated by corporate in-fighting and procedures. He did not conform to the IBM stereotype, even to the extent of wearing cowboy boots. Unlike Lowe and nearly everyone else at IBM, he actually owned an Apple II and loved it.

He took full advantage of his direct line to Cary and refused to return the calls he got from other IBM executives or to attend the many meetings he was invited to. He concentrated not on being a good company man but solely on getting the PC project moving. Most of his team at the obscure office in Florida were equally happy to be free of the corporate straitjacket. Estridge proved an exceptionally effective and sympathetic leader of this team, happy to give others credit and encouragement.

One of his first tasks was to negotiate a proper contract with Microsoft. His main concern was whether it would deliver a good product on time – it never occurred to him or to anyone else at the time that control of the operating system would eventually enable Microsoft to wrest leadership of the new industry from IBM. Gates earned credit for not asking for a large payment for the work but suggested instead a small royalty on every sale. Even shrewder was his request to be able to sell the software to other manufacturers on similar terms. At the time this was a prudent insurance policy rather than a Machiavellian plot. Not even Bill Gates realized in 1980 how far the PC revolution would go.

Likewise Intel did not immediately appreciate the significance of IBM's order for its 8088 microprocessor and Estridge's stipulation that it should supply IBM's competitors too, to ensure a steady supply. Gordon Moore subsequently acknowledged that this decision 'changed the course of Intel's history'.

Lowe and Estridge knew that to create a serious market where normal customers could buy with confidence, as opposed to the near chaos that then prevailed in the mutually incompatible hobby computer world, standards were necessary. Using easily available, reliable components that other manufacturers could also obtain meant that the marketplace could develop on similar lines to hi-fi, where amplifiers, turntables and speakers from different manufacturers could work together. Naturally IBM would define that standard and dominate the market.

However, Estridge declined to place orders with other IBM units if they were not competitive with what he could obtain externally, which caused resentment. He also insisted, against considerable resistance, that PCs be sold through dealers as well as by the sales force.

Estridge knew that the PC would only be attractive to customers if there was good applications software available to run on it and he wanted a competitive supply. He encouraged the development of a software publishing industry for the PC, with IBM taking a share of the revenues. This was to play a significant part in stimulating demand, since without useful programs customers would have no need for the new machines.

Estridge's determination to behave like an autonomous entrepreneur offended the corporate planners. The biggest row was over manufacturing capacity. They heavily discounted his forecast of a million sales over the PC's first three years. Given that IBM only sold 2,500 mainframes in a year, a figure of 200,000 was surely much more realistic. Estridge's forecast in fact turned out to be accurate – which meant that he was short of manufacturing capacity for the first two years.

The hardware development was completed by March 1981 and the operating system by June. Against all expectations, in August 1981 the PC team had a product ready for launch, the IBM 5150. The Charlie Chaplin advertising campaign positioned it as quirky – and confirmed the prejudices of many IBMers who thought of it as a toy, but it was an instant hit, a computer with appeal beyond the ranks of hobbyists and technology enthusiasts. In 1982, IBM sold 240,000 PCs; in 1983, 800,000, which meant that it overtook Apple as the leading supplier. The basic version cost $1,565, but most customers were paying twice that. Estridge was suddenly the hero of the hour, bringing in billions of dollars in revenues nobody had dreamed of. Over the next five years, IBM sold 5 million PCs, many of them to middle managers not involved in IT, rather than the consumers originally envisaged as the target market.

Compared with present-day computers, the first IBM PC was a modest device, but by the standards of the early 1980s it was a significant breakthrough. It had a new Intel processor, allowed the use of floppy disks and had some graphics capabilities. Unlike any personal computer then available, it was a machine that businesses could feel comfortable buying, mainly owing to the halo effect from IBM's reputation. Most makers of home computers until now, even Apple, had looked distinctly flaky to business customers, but IBM conferred legitimacy on the PC. This product, unlike virtually all its predecessors, could cross the chasm from early adopters into the majority – and uncover a mass market.

Probably the key trigger for sales take-off was the appearance in January 1983 of Lotus 1-2-3. Like VisiCalc and the Apple II, this spreadsheet program quickly became the single most useful piece of software for the PC and the main reason for a business customer purchasing one. Mitch Kapor, who founded Lotus and developed 1-2-3, wanted to sell IBM exclusive rights to it but he could not get an appointment with Estridge. The executive he did get to see, who clearly had no idea of the potential of this program, discouraged Kapor from revealing too much, as IBM was worried about being sued for stealing other people's ideas.

Prometheus Bound

Estridge was finding it increasingly difficult to operate like a fast-moving entrepreneur. He soon found himself with thousands of staff he had not asked for – redeployed from other parts of the company. His own development people had to book appointments with him. Soon he was running, not a small team of mavericks, but a division with 10,000 people (named 'Entry Systems', to make clear its relationship with proper computers). He was now more senior and better paid, but he lost his direct line to the chairman that had ensured his independence. Many IBMers resented his sudden success and his reluctance to play the corporate game. Even those who wished him well thought he would benefit from the warm corporate embrace and could not see that he had succeeded precisely because he'd operated outside the system, running the PC group as if it were an independent business.

Estridge's priority, from which he was frequently distracted by twice-weekly visits to corporate headquarters, was building a range of new PC products. The first, the XT, was a big success. It was basically an enhanced version of the PC, but had a hard disk and much more memory. It came out in 1983 and accounted for a large proportion of sales that year and subsequently. The next was a disaster.

The PC Junior was conceived as low-cost machine for the home market. As originally planned it might have had a chance, but endless discussions in the Management Committee whittled it down, so that it would not compete with the PC product. The keyboard was ditched in favour of a calculator-like 'Chiclet' that pleased nobody. Peripheral attachments that might have allowed it to be enhanced into a PC equivalent were dropped. Distribution through K-Mart outlets was ruled out as undignified. And the price was set at $669, more than twice that of the Commodore and Atari models it was competing with. It was released too late for Christmas 1983 and, despite desperate efforts to correct things like the keyboard, sales never took off. In 1985 the company announced that the PC Junior 'would fulfil its manufacturing schedules', IBM-speak for death.

The important new product was the AT, based on a more powerful chip, the 286. This posed serious compatibility problems with the PC, which delayed the launch. Estridge's decision to buy up all available stocks of the chip proved not to be a smart move. Intel was able to churn out more chips, without the original flaws, for IBM's competitors, who benefited from the lower prices that came with higher volumes, but this error was masked by booming demand. Sales revenues for the PC division as a whole in 1984 hit $4 billion, making it the equivalent of a *Fortune* 500 company.

Estridge himself was courted by nearly everyone in the industry, but he turned them all down. Apple offered him $1 million a year to become their CEO, but he proudly declared, 'I work for IBM.' He had a radical plan for a new computer based on Intel's 386 chip. This would be so fast that it would marginalize every PC in the market, including IBM's own. His boss, Mike Armstrong, overruled him. What was the point of killing products like the PC and the AT that were earning good profits? The margins did not look attractive enough. What the PC business needed was more of 'the disciplines of the mainframe business'. Armstrong did not appear to have noticed that levels of innovation and competitive intensity in the new market were several orders of magnitude greater than in main-frames.

Armstrong and the Management Committee decided that the PC was now too important to be left in the care of a 'wild duck'. Early in 1985, Estridge was moved to a staff job at corporate headquarters, which he detested. Taking his first holiday in years, that August he and his wife were killed in a plane crash.

Estridge's replacement was Bill Lowe, a safe pair of IBM hands who had started the project. His view mirrored that of Armstrong: 'prevailing in the PC business is not that different from prevailing in the rest of the computer business'.

A new competitive landscape

What nobody at IBM, and very few people in the industry as a whole, had yet appreciated was that the competitive landscape was changing fundamentally, and that IBM's advantage in PCs was rapidly disap-pearing. The term 'IBM PC' was registered as a trademark, but PC was quickly adopted as a generic term for all computers made to IBM's specifications, regardless of who made them.

In 1983, Compaq had produced the first clone PC by copying IBM's design. Its first product was a portable, which only competed indirectly with IBM. It was an enormous success, selling 100,000 in six months. (Significantly, IBM did not bother to obtain one and took another three years to produce its own portable.) Compaq soon brought out other models and was followed by dozens of other suppliers, who also reverse-engineered IBM's designs to produce almost identical versions of IBM products which they sold at very much lower prices. The hardware of the PC was on its way to becoming a commodity. The critical success factors in the emerging marketplace would be relentless focus on cost, frantically fast product development and willingness to cannibalize existing products ruthlessly. IBM had none of these qualities and none of its key managers understood the new marketplace.

The industry's centre of gravity was shifting away from mainframes and towards chips and software. The significance of this took years to sink in at a company steeped in the traditions of Big Iron. More innovative companies carved out new markets that IBM spotted too late – DEC in minicomputers, Wang in word processing, Silicon Graphics and Apollo in workstations, Hewlett Packard in laser printers. In IBM's closed world, comforted by fat margins on Big Iron and continued overall growth, these seemed peripheral developments, as did the PC, despite its sensational success. Top management continued to regard Fujitsu, Hitachi and other Japanese mainframe manufacturers as their main rivals and failed to see that IBM's share of the overall IT market had been slowly but imperceptibly declining for years.

Ironically, IBM had pioneered many of the technologies that more entrepreneurial rivals turned into business success – Oracle with relational databases and Sun and Hewlett Packard with RISC architecture, and, of course, Microsoft with DOS. It was inconceivable to almost everyone at IBM that PCs could be anything other than an interesting complement to serious computing on mainframes, that IBM would not dominate this market in a similar way, and that a scruffy sub-contractor like Bill Gates might understand this business better than them.

By 1985, Gates was a 'partner' whose success and power IBM resented. Lowe was determined to cut him down to size and to put IBM's stamp on a future operating system for PCs. Since 1983 a growing number of IBM programmers had been working with a small Microsoft team on OS/2. Both IBM and Microsoft's growing band of worried rivals in the software industry hoped that OS/2 would marginalize the clunky but increasingly dominant DOS. Microsoft had been working in parallel on Windows, which would eventually give DOS the kind of graphical user interface that made the Apple Mac so attractive. To Gates's dismay, Lowe told him he had no plans to incorporate Windows features into OS/2. His plan was displace DOS or to make OS/2 the added-value layer on top of it, with Microsoft back where it belonged as a sub-contractor. This was the start of a long, complicated, immensely expensive and rancorous series of manoeuvres.

Gates was pursuing several strategic options and certainly wanted to maintain a relationship with IBM. At several points he tried to persuade Lowe to make Windows a joint project. In 1986, when Microsoft went public, he proposed that IBM should buy 10 per cent of Microsoft's stock, in order to bind the two companies more closely together. There was a good precedent – IBM had recently acquired 20 per cent of Intel and the cost of 10 per cent of Microsoft would have been a very affordable $100 million (which by early 2000 would have been worth $29 billion). Lowe, however, declined, saying that IBM did not want to be seen to be dominating the PC market.

In fact, by 1986, IBM was barely holding its own against Compaq and the pack of clone makers, who were not behaving remotely like dwarves. Its share of the business market for PCs had plummeted to 40 per cent and Compaq seemed on the verge of overtaking it. As well as the portable, Compaq produced the first AT clone, the first PC with the 386 chip in it, and later the first laptop and notebook computers. In 1986, only four years after its foundation, it entered the *Fortune* 500.

Lowe and Armstrong, though, still looked on the PC marketplace as a minor adjunct to the real computer world, where everyone knew their place. The place of the PC, in the kind of large company they felt comfortable selling to, would be as part of an integrated network, the hub of which would of course be a mainframe. Their underlying assumption was that they could bind their PC customers to them as closely as they had their mainframe ones and insulate themselves from impertinent competition. In fact, the opposite was about to happen – the PC vandals were soon to disrupt their cosy, mainframe monopoly.

Most of Lowe's PC initiatives either got bogged down in endless IBM decision-making or arrived too late and underpowered. The gestation of the Convertible laptop was so elephantine that, when it arrived in April 1986, it was pathetically inferior to others already available – too heavy, underpowered and without a modem. It quickly became a standing joke.

Lowe's big idea for the next range of PCs, the PS/2, was to try to make them proprietary, to deviate significantly from the open-architecture approach he had pioneered in 1980. He decided to make a big play of a minor technical feature, a fast 'bus', which he called the Micro Channel Architecture. All this did was eliminate some trivial radio interference problems, but it would be difficult for competitors to copy: to do so they would have to obtain a licence from IBM. Changing the plumbing frequently, to keep competitors behind, had been an IBM wheeze in the mainframe world. It now promised to inconvenience some customers, whose previous equipment would not be compatible with this, in order to provide rather spurious differentiation.

The PS/2 was launched with a big fanfare in 1987. On the chairman's insistence, these were called Personal Systems, to make them sound more like serious computers. They had some mildly interesting features (but not a new operating system – OS/2 was still in development), and although sales were respectable they were hardly sensational. And no other manufacturer bothered to ask for a licence to replicate the Micro Channel. Compaq and several others publicly argued that it was an irrelevance that offered no benefits to customers. The *Wall Street Journal* compared it to the 'mystery ingredient' in toothpaste ads. IBM was in danger of appearing ridiculous. It was becoming clear that it no longer controlled the PC standard.

Trouble at the top

IBM was led in the 1980s by two men who failed to realize that the computer industry was going through a revolution. They assumed that the mainframe would always be king and that the company's continued success was preordained. The organization as a whole was suffering from a serious case of hubris, compounded by stifling bureaucracy and blindness to new realities.

John Opel, who succeeded Cary as CEO in 1980, presided over IBM's period of greatest profitability. This was partly due to his decision to encourage customers to purchase rather than lease their computers. New accounting policies deferred expenses on big software projects but treated new leases as sales. These measures and the early success of the PC gave an enormous boost to profits in the short term. They doubled from $3.3 billion in 1981 to $6.6 billion in 1985, which was more than any company had ever earned before. Opel retired in that year and handed over to John Akers, feeling that he'd done a great job.

In fact, he'd stored up problems for the future that IBM could ill afford in a turbulent new environment. IBM now had a staggering 405,000 employees. It had lost its comforting annuity income from leasing. Its salesmen, instead of concentrating on their long-established role of advising customers and keeping them happy, were now targeted on new sales, often regardless of need, but had little incentive to evangelize on behalf of new and unfamiliar products. Like most people in the company they had got used to having an easy life.

Life was about to get very difficult indeed. The growth of personal computing and of minicomputers was starting to have an effect on mainframe sales. Customers who had previously been happy to take the advice of IBM salesmen on trust were deciding for themselves that they did not need to build their PC networks as if they were dumb terminals governed by mainframe hubs. Already, high-powered workstations could handle applications like payroll as effectively as many mainframes, and at a fraction of the cost. Typically, IBM was slow to produce a competitive workstation of its own.

The new CEO, John Akers, was frequently described as the perfect IBMer, 'the CEO from central casting'. A great salesman, a former fighter pilot, handsome, apparently decisive, he had been groomed for the role for years. On his appointment, Akers declared, 'IBM's prospects have never been brighter than they are today.' Until 1989, he insisted that mainframe sales revenues could continue to grow at 15 per cent a year – stagnation in 1987–8 had just been an aberration. In fact, Amdahl and Hitachi reduced IBM's mainframe market share from 90 to 80 per cent in the 1980s. A detailed study by a high-powered internal task force concluded that the days of effortless sales growth were over – 5 per cent was the best that could be expected. In fact this turned out to be far too sanguine an estimate – mainframe sales would soon start to decline sharply. To have anticipated this would have been to think the unthinkable.

In this situation, sunny optimism, which all good IBMers liked to exude, was a serious handicap. Akers had never had to deal with challenges like this before and took what appeared to be decisive steps. He declared that IBM would shift away from hardware and should aim to earn half of its revenues from software and services, without specifying how. A large proportion of IBM's software sales were to captive customers, developing customized programs to run on their mainframes. It did not have to compete too hard for this business and could get away with being slow and inefficient, but in fast-changing areas like PC software its competitive weaknesses were much more apparent. Its enormous teams of people were much less efficient than Microsoft's, yet on joint projects IBM executives would frequently complain that they were producing more lines of code than their now detested partner, as if that were a good measure of productivity.

Akers tried various reorganizations and took the first timid steps to reduce IBM's bloated workforce, announcing a voluntary severance package. Unfortunately, of the first 10,000 who volunteered, 8,000 came from that 10 per cent of the workforce marked 'One' on their annual reports – the company lost precisely the people it most needed to hold on to. IBM's commendable concern for the welfare of its employees now became a serious handicap – staff released from manufacturing and programming jobs were assigned to the sales force, which was already probably too large and certainly did not need unqualified people. Many business units, notably PCs and minicomputers, objected to contributing to a large overhead that did not have the expertise or the enthusiasm to sell their products effectively. Akers, however, could not bring himself to decentralize the sales force – every IBMer knew in his bones that it was the company's greatest asset.

OS/2 v Windows

At the end of 1988, Jim Cannavino took over from Lowe as head of the mess that was now the PC business. Cannavino was a blunt, aggressive trouble-shooter who had started at IBM as a technician and through enormous determination and application had risen to be head of the mainframe division. He quickly decided he did not trust Bill Gates. He resented the fact that Microsoft was making millions from DOS licences on the back of IBM and suspected him of being half-hearted in his support of OS/2. He told Akers that Gates did not feel like a partner to him. Given that, it might have been better to make a clean break, but he found that Microsoft people were working on key parts of OS/2 development and divorce now would mean losing yet more time on a project which was turning into a nightmare. Since 1987, IBM had had more than 1,000 people working on it, and was spending close to $125 million a year. To kill the project would have meant a big write-off at a difficult time financially.

And so the 'partnership', which was coming to resemble an acrimonious marriage in its final stages, staggered on for another two years. Although it is difficult not to sympathize with Cannavino's predicament, there is no doubt that Gates was considerably more astute. At the Comdex show in Las Vegas in November 1989, they negotiated a deal whereby each would 'endorse' each other's pet operating system – Windows was to be for the low end of the market and OS/2 for more powerful machines. Both of them were publicly vague and evasive under questioning, but the software industry, most of whom were hoping to see Microsoft checked, concluded that Cannavino had acknowledged that Windows had a future in the mass market, without getting much out of Gates. They decided that if the issue was which one would prevail, it was going to be Windows.

And they were right. When Windows 3.0 appeared in 1990, despite still being markedly inferior to the Macintosh operating system, it sold 10 million copies in two years – OS/2 struggled to reach 300,000 in three. Some customers buying machines with OS/2 already installed actually paid their dealers to switch them to Windows. OS/2 did eventually become quite a respected system for powerful servers, but Windows was good enough and affordable enough for the mass market. It was also benefiting from powerful network effects. Cannavino, with his mainframe mindset, had not really believed that a flimsy product, as Windows certainly was in 1990, could actually be a stunning commercial success. By the end of 1991, IBM had spent $2.5 billion on OS/2 and earned revenues of only $200 million. Microsoft's revenues more than doubled between 1989 and 1991, to $1.84 billion – and were to reach $6 billion by 1995.

Cannavino was scarcely more successful with his other ventures. The PS/1, a home computer launched in 1990, repeated virtually all the mistakes made on the PC Junior. Because of concerns that it might compete with other IBM products, it used the now six-year-old AT processor and was markedly less powerful than other models already on the market and was of course priced above them too. All hopes were pinned on the brand. Within a year PS/1 was being sold at a 50 per cent discount.

In 1991 a new laptop appeared. Cannavino wanted to price it at $5,000, the then going rate, but Marketing insisted an IBM model would sell at $6,000. The Management Committee agreed – IBM did not engage in aggressive pricing. Rivals immediately dropped their prices well below $5,000. IBM was forced to respond, but it was now positioned as a $6,000 machine. Within months, Compaq brought out a model which greatly surpassed it, and IBM's price dropped to $3,500, but this was widely viewed as a sign of desperation.

Endgame

After four years of stagnant or declining earnings, Akers won a reprieve in 1990 with the success of the AS/400. This was a minicomputer, developed like Estridge's first PC by a maverick team, kept apart from the main business. After years of coming up with products that could not match Digital's, the AS/400 won a knockout victory. It went on to become a $14 billion business – considerably bigger than the PC had been. The results in 1990 also looked better because some losses had been anticipated in 1989. Akers told the board that IBM's troubles were now over and secured for himself and his senior team raises of 35 per cent plus stock bonuses.

Analysts were predicting record profits of $7 billion for 1991, but, in March, Akers had to announce that the year as a whole would show a loss. Cary, who having reached the age of seventy was obliged to resign from the board that month, was livid and wanted to fire Akers. The most senior board member now was Opel and he decided not to rock the boat. Over the rest of the year, Akers managed to convince the board that the problems were primarily due to the state of the world economy and the Gulf War.

Akers called a meeting of his top twenty managers and roundly criticized each of them in turn. 'Stem the tide!' he shouted. 'I used to think my job as a sales rep was at risk if I lost a sale. Tell salesmen theirs is at risk if they lose one.' His remarks were widely reported and had a devastating effect on morale and respect for the CEO.

As 1991 got worse, Akers decided that underperforming people had to go and instituted a new, much harsher rating system. People in the bottom category were told they were on their way out. This deeply shocked most IBMers, long accustomed to the comfort of lifetime employment. In November he announced another, this time truly radical, reorganization. This dispensed with tens of thousands of people and broke the company up into thirteen independent business units. He was prepared to sell off some of these, but could not bring himself to abandon the Management Committee or give the business units their own sales forces. 'We think 1991 was an aberration,' he declared; in 1992, IBM would return to 'normal growth'.

At first it appeared as if this might be the case, as a new range of mainframes was launched. Demand stayed buoyant during the first half of the year but quickly petered out. This had never happened so quickly with a new range and soon a price war broke out which badly hit all manufacturers, from Amdahl to Fujitsu. Akers had been telling analysts that gross margins would stabilize somewhat lower than the 60–65 per cent of the good old days, but still at a comfortable 52 per cent. In fact they were now closer to 40 per cent, and falling. The lay-offs rose to 40,000, but still the overheads were too high for such drastically lower margins. For the first time in living memory, IBM cut expenditure on R&D and reduced the dividend to shareholders.

In January 1993, Akers finally 'resigned'. Two months later, Lou Gerstner, then CEO of RJR Nabisco and formerly with American Express and McKinsey, was appointed to replace him. A new era had begun.

Postscript

The story of how Gerstner rescued IBM from disaster is a remarkable one. Briefly, he rejected the option of breaking up the company, cut costs drastically and dismantled some of IBM's bureaucratic culture. He proceeded to reinvent the company, building in particular on the growing success of the division that provided computing services to businesses. Gerstner realized that what most business customers needed, as they concentrated on their core businesses and capabilities, was for technology providers to provide them not with complex products and challenges but simple solutions.

With the unparalleled depth of its research, its range of technical knowledge and capabilities, and its still strong sales culture, IBM was better placed than anyone to cater for a wide range of clients. Its service revenues rose from $200 million in 1984, when it was only the tenth-largest supplier, to $16.6 billion in 1994, when the Global Services division overtook EDS as the leading provider of outsourced computer services. By 1997 a third of IBM's revenues came from services: overall revenues reached $78.5 billion and earnings $6 billion. By 2004, Global Services, which now included the largest global IT consulting operation, accounted for nearly half of IBM's

$96 billion revenues.

Gerstner's mantra was to put the customer first, IBM second and the business unit third. If third party products were better for the customer, IBM would offer them rather than its own. It turned its back on the old reliance on proprietary systems and the obsession with hardware and monopoly. It embraced wholeheartedly the open standards it had inadvertently done so much to bring about. In the second half of the 1990s IBM benefited enormously from the explosive growth of networked computing and the Internet, positioned itself as the leading provider of e-business solutions and hosting, and Internet-enabled thousands of companies.

The main question that concerns us is how IBM could have fallen so badly, so quickly. Why, after such previous success, was it suddenly on the verge of extinction? Part of the answer lies in the consequences of previous success.

It was firstly the victim of an extraordinary set of discontinuities that came close to destroying one computer industry and replaced it with a completely different, brutally competitive and volatile one. IBM was merely the most illustrious casualty of this revolution that dislodged the leaders of most sectors of the industry, from Digital to Wang, Apple to Bull, Apollo to Silicon Graphics. The PC was one of the most disruptive technologies since the printing press.

IBM was suffering in the 1980s from an advanced state of hubris, complacency and organizational sclerosis. Until the crisis had almost destroyed it, it was incapable of recognizing the nature and scale of the challenges and of adjusting to a radically new environment. Its past success and present eminence played a big part in this paralysis. It had had more competitive advantage than was healthy for any firm.

In the mainframe world this owed much more than it realized to strategic assets like customer switching costs and fear of the unfamiliar, rather than distinctive capabilities. Once the enormous price/performance advantages of PCs were clearly established, and it was obvious that Compaq and Dell made machines at least as reliable as IBM, frequently better and always cheaper, its old world was bound to collapse.

IBM simply did not have the capabilities to be an effective or efficient equipment supplier in the new mass markets that emerged for PCs. Its initial success would almost certainly not have happened without the accidental entrepreneur, Don Estridge. IBM enjoyed the success it did in the early eighties mainly because of the efforts of his team and because its enormous reputation made the PC seem respectable and safe for millions of customers. But in the long run, however heroically it might try, IBM was never going to be able to compete with companies like Dell on speed of movement, customer responsiveness, rapid product development and ruthless cannibalization of old products.

Many have blamed IBM's failure on management stupidity, but this is too glib. They certainly made plenty of mistakes, and Opel and Akers were particularly inadequate leaders, but given IBM's history it must be doubtful whether any management could have coped with a catastrophe of these dimensions very much better. Cary, or a good outsider, would have made a difference; Akers should have taken drastic action much earlier than he did; and the board should have fired him much sooner, but from the inside of a cocooned, mainframe-centric, inward-looking world, it was next to impossible to see just how much and how fast the world was changing. It was not that easy from the outside either: hardly anyone even in 1990 was predicting the near collapse of the mainframe market or the extent of the onward march of the ever-mightier PC.

The remarkable thing about IBM is that it was not destroyed by these events but had the capacity to reinvent itself. The redoubtable Mr Gerstner cannot claim all the credit. He was only able to do what he did because the company had deep reservoirs of talent, knowledge and capabilities that it was able to extend and redeploy. Though we have not considered it in detail, IBM's redefinition of the role of computer services, in particular the provision of e-business capabilities to its many clients, was an achievement every bit as impressive as those of the other market creators described here.

Encyclopædia Britannica

The Legacy

Encyclopædia Britannica was a product of the eighteenth-century Enlightenment, a noble attempt to encapsulate the sum of human knowledge. It was founded in 1768 by Colin Macfarquhar and Andrew Bell in Edinburgh, home to Adam Smith and David Hume, partly as a response to Diderot's radical French *Encyclopédie*. *Britannica's* patriotic first edition was dedicated to King George III, had three volumes and was a huge success. The third edition, in 1801, had twenty volumes. Contributors over the years included some of the most distinguished authorities on their subjects – Sir Walter Scott, David Ricardo, Sigmund Freud, Albert Einstein, Henry Ford and Milton Friedman. The quality and comprehensiveness of its content and its unparalleled reputation set it clearly apart from all other general works of reference.

The business belonged for most of the nineteenth century to the Scottish publishers A & C Black, who sold it to Americans in 1901. All its American owners invested in new editions and strengthened the brand, making the articles shorter and more accessible to the non-academic reader, and introducing more aggressive sales and marketing methods. The eleventh edition in 1911, produced in conjunction with Cambridge University, was universally praised as definitive. In the 1930s a permanent editorial department was established for continuous revision of content – the first encyclopedia to do this.

In 1941, William Benton, founder of the Benton and Bowles advertising agency, acquired *Britannica* from Sears Roebuck and presided over it until his death in 1973. Benton dedicated himself to the University of Chicago, to politics and to the organization and communication of knowledge. He wanted the encyclopedia to be a kind of open university, a place where the brains of the world would meet to synthesize their knowledge and make it available to everyone. He committed $32 million, an enormous sum then, to the development of the fifteenth edition of *Britannica*, which appeared in 1974.

This edition, which took a radical approach to the systemization of knowledge, has been updated but never replaced.

Benton's other legacy was a marketing operation, with a budget larger than the editorial department's. Its core was a large, carefully trained direct sales force, highly skilled at persuading parents that buying a full set of encyclopedias would be an investment in their children's education. Unlike most door-to-door salesmen, *Britannica's* had an aura of professionalism and authority. National advertising generated a steady flow of sales leads and a high proportion of them were converted into instalment plan contracts. The salesman collected a commission on each sale, as did his district manager, the regional manager and the sales vice-president. By 1990 there were 2,300 people in the sales organization, and top management regarded it as an asset as valuable as the product itself.

There was frequently tension with the high-minded editorial department. The guardian of its independence was the philosopher Mortimer Adler, the part-time chairman of *Britannica's* editorial board. It was an article of faith across the company that it was the quality, breadth and depth of the encyclopedia that made it such a success and justified a selling price of $1,500. *Britannica* had not just the best product, the best brand and the best sales force, it was the only encyclopedia whose margins allowed it to afford such a business model. Competitors' prices were much lower and had to rely on more conventional sales and distribution methods.

However, a large proportion of sales were made to families on modest incomes who possessed scarcely any other books. For the most part, they made little use of the encyclopedias after purchase: *Britannica's* market research showed that scarcely any customers or their children opened the books a year after purchase. They mostly bought them because of a vague feeling of anxiety about their children's education and in the hope that such a magnificent set of volumes would give them an edge. Spending $1,500 reassured them that this was a serious investment.

Britannica's sales reached a peak of 400,000 copies in 1990, with revenues of $650 million. They then started to decline, as the market became saturated and many alternative reference books and part-works were being published. Much worse was to come.

New media

During the 1980s a number of technology companies, including Microsoft, approached Encyclopædia Britannica, Inc. with the idea of putting the encyclopedia or parts of it on to CD-ROM. In 1987 it started a serious programme of research and development in new media. It was not, however, prepared to make anything like the whole text of the encyclopedia available to personal computer users. One reason was that this would have been technically difficult, other than for the most powerful workstations, given the capacity of PCs at the time. A more fundamental objection was to the principle of selling the content separately from the books – the sales force was viscerally opposed.

In 1961, Britannica had acquired *Compton's*, an illustrated encyclopedia, aimed at children, which it sometimes bundled free with purchases of the flagship product. In 1987, work started on an electronic version of this. Britannica also acquired a sophisticated text search tool and built a strong team of electronic publishers and technologists. In 1989 it demonstrated *Compton's Multimedia Encyclopaedia* on a CD-ROM. Encouraged by its initial reception, it started work on developing a range of CD-ROM products.

In 1992 the first electronic product based on *Britannica* appeared. Aimed at publishers and professional users, this was sold as a 'fact checker' but, priced at $10,000, only sold a hundred copies. When the price dropped to $2,500, a few thousand were sold. In 1993 a networked version was developed and, in September 1994, Britannica Online was launched for universities, as a gateway to knowledge on the Web. Considerable work was also done on new search techniques.

But in the main marketplace sales revenues were slipping badly even before disruptive competition came along. In 1993, Britannica felt obliged to sell the Compton business for $53 million, shortly after Microsoft entered the fray. Having been rebuffed by Britannica, it decided to take another tack. What Microsoft wanted was a product that would make PCs more attractive to ordinary consumers, and it decided to obtain its own one. The Funk & Wagnalls encyclopedia was at the opposite end of the quality spectrum from *Britannica* and was sold cheaply, mainly in supermarkets. For Microsoft, however, it was the 'good enough' equivalent of a quick-and-dirty operating system, something that could be turned into an electronic product quickly. It acquired the electronic rights and in March 1993 launched the multimedia version of this on CD. *Encarta* was an instant success: hundreds of thousands of copies were sold at $50 a time and almost as many were given away with PC purchases. The market was suddenly flooded with cheap electronic encyclopedias.

This had a devastating effect on *Britannica's* sagging sales. In 1994 revenues were down to half the levels of four years before, at only $325 million. Britannica decided it must bring out its own CD version, but the sales force protested vigorously: this would destroy their livelihood. So when the CDs arrived, they were given free to customers who bought the full set of encyclopedias but priced at a forbidding $1,200 for everyone else. Even when the price was reduced to $895 a few months later, there were scarcely any takers. Meanwhile *Encarta* continued to sell like hot cakes and *Britannica* to flounder.

In desperation, the Benton Foundation put the business up for sale, but nobody was interested in buying an obviously dying enterprise. Eventually, in 1996, a Swiss financier, Jacob Safra, with no previous publishing experience, bought the business for $135 million, considerably less than its book value. Shortly afterwards he sacked what was left of the sales force and tried to turn it into a mainly electronic publishing business.

The business model has changed several times under the new owner. For a brief period an online version was available free, supported only by advertising. Now the full encyclopedia is sold at *Encarta*-like prices on CD and DVD, and is available online as a subscription service for $69.95 a year. It is now a very much smaller business, and faces even more devastating competition from

Wikipedia. The entire encyclopedia industry has shrunk massively in value, and it seems unlikely that anyone will ever again finance the kind of investment Britannica once made in new editions.

Diagnosis

Like IBM, Britannica was hit by an enormous discontinuity that changed everything. Once a 'good enough' alternative was available at a massively lower price, its value proposition and business model simply collapsed. That proposition was based to a considerable extent on the ignorance of many customers, who were buying the equivalent of a Rolls-Royce when all they needed was an old Ford. Most of those who might have made use of a 32-volume set, other than libraries, would never pay $1,500, since they could obtain much of the content or its equivalent from other sources. What was remarkable was the enormous number of copies that had been sold to people who had no real use for them.

The real competitor to *Britannica* was not so much the CD-ROM or *Encarta* as the PC itself. When consumers in the 1990s were looking for a learning aid for their children, a PC, rather than an encyclopedia, seemed like the ideal investment. They were probably deluding themselves almost as much, as solitary learning is mostly only done by highly motivated, well-disciplined adults.

Microsoft's business model bore no relation to that of encyclopedia publishers: its goal was to make the PC attractive to mass markets, rather than to make money directly from *Encarta*, but its pricing strategy commoditized the whole industry and ended the days of expensive, multi-volume encyclopedias. Like mainframe computers, they suddenly looked to most people like very uneconomic ways of obtaining and processing information. The CD-ROM version might not have had the breadth or the depth, but it was adequate for the needs of most people most of the time.

Since then, Microsoft has invested in improving the quality of *Encarta* considerably. More significantly, Wikipedia has produced an online collaborative encyclopedia, now one of the most popular sites on the Web, and much bigger than *Britannica*, if less authoritative (see Chapter 12). More fundamentally, Google has revolutionized the way in which we obtain information.

With the benefit of hindsight, the only viable strategy for any encyclopedia publisher was to learn how to use electronic media as quickly as possible, and to develop a radically different business model. In Britannica's case drastic shrinkage was unavoidable. Premium pricing on the Britannica scale would not be possible and a large direct sales force unaffordable. Leaving the initiative to others was fatal, as was the refusal to 'cannibalize' Britannica's own product. The revolt of the sales force was a classic case of channel conflict at a time of dramatic change. It would have taken a very brave and far-seeing management to recognize that what was once a prime source of competitive advantage was turning into a serious liability. Ditching sales methods that had served Britannica so well in the past seemed as unthinkable as the notion that a work as serious, as scholarly and as esteemed as the *Encyclopædia Britannica* could be wiped out by something as plainly inferior as *Encarta*.

Not only did Britannica's management fail to recognize how profoundly their competitive environment was changing, they over-estimated the potency of their value proposition. Their more cynical salesmen probably had a more accurate notion. It was perhaps too galling to acknowledge that the pride of the Scottish Enlightenment was being sold to a mass market by means of a confidence trick.

The fundamental problem was not the CD-ROM as a technology: far from being the future of publishing, it was largely bypassed by the Internet by the end of the decade. What this sad story shows is that the finest products with the proudest pedigrees can easily fail if they are no longer providing real value to their customers.

10

CREATIVE DESTRUCTION

*Competition from the new commodity, the new technology,
the new source of supply, the new type of organization...
competition which commands a decisive cost or quality
advantage and which strikes not at the margins of the profits
and the outputs of the existing firms but at their foundations.*
Joseph Schumpeter, *Capitalism, Socialism and Democracy*

*As many more individuals of each species are born than can
possibly survive; and as, consequently, there is a frequently
recurring struggle for existence...*
Charles Darwin, *On the Origin of Species*

*Competition, competition – new inventions, new inventions
– alteration, alteration – the world's gone past me. I hardly
know where I am myself; much less where my customers are.*
Charles Dickens, *Dombey and Son*

What blew both IBM and Encyclopædia Britannica off
course was an entirely unfamiliar external challenge to which they
had no effective response. Something similar happened to Kodak

when it found the market for photographic film evaporating, to traditional airlines when no-frills airlines established themselves, to telephone companies when first discount operators slashed the price of international calls and then Skype made them free.

These are all examples of disruptive competition, the dark side of creative destruction for incumbent businesses. Almost every innovation by one business presents a competitive challenge for another. In extreme cases it can totally undermine the economics of incumbents and effectively wipe them out. The most dramatic instances are where geography or regulation previously protected them from serious competition. In many small towns, food, clothing and hardware shops enjoyed a modest but sheltered existence that lasted for decades, but was suddenly shattered when giant supermarkets appeared with a vastly greater range of merchandise and much lower prices. Manufacturers have similar experiences when governments lift tariffs on imports, but often the process is gradual and indirect – in the period after the Second World War, the availability of cheap, ready-made clothing put dressmakers out of business, record players made dance bands obsolete, and mass ownership of cars drew shoppers away from high streets and city centres.

The man who identified the phenomenon of creative destruction and realized that it was what generated economic progress was one of the most remarkable thinkers of the twentieth century, Joseph Schumpeter.

The prophet

Schumpeter was a man of many parts. His declared goals in life were to be the greatest lover in Vienna, the greatest horseman in Europe and the greatest economist in the world. Modesty compelled him to acknowledge that he had not succeeded in the second of these, but he occasionally attended faculty meetings clad in jodhpurs and wielding a riding crop. He also dabbled in politics and finance, which almost ruined him. In his brief term as Austria's finance minister after the First World War, he was unable to halt hyperinflation, and the bank he ran in the 1920s folded because of bad debts. These dramatic experiences taught him two things: to see economics from the point of view of the businessman, and just how precarious and unpredictable life, particularly business life, could be.

He returned to the more congenial world of academic economics and in 1932 went to Harvard, where he became the most celebrated and flamboyant member of the faculty and remained until his death in 1950. He had developed the core of his most famous idea much earlier: in 1911, at the age of twenty-eight, he had published his *Theory of Economic Development*, which attributed economic progress primarily to the entrepreneur, 'the pivot on which everything turns'. This idea he subsequently developed into a theory of capitalism as a continuous evolutionary process, with creative destruction as its driving force.

Prior to Schumpeter, economists had tended to think of entrepreneurs primarily as inventors. He showed that they innovate just as significantly by introducing new means of production, new products and new forms of organization, and by making old methods obsolete. They are society's most useful revolutionaries, because their innovations are quickly imitated and widely diffused. Economists were also inclined to ignore the role of individuals in economic development and to base their models on assumed states of equilibrium. Indeed, the concept of 'perfect competition' is one where all suppliers in an industry produce the same good, sell it for the same price, and use the same technology.

Schumpeter thought this was nonsense – equilibrium was only ever a brief lull in the normal state of affairs, which was constant change. Progress depended on innovation, disequilibrium and temporary monopolies: 'Without innovations, no entrepreneurs; without entrepreneurial achievement, no capitalist returns and no capitalist propulsion. The atmosphere of industrial revolutions – of "progress" – is the only one in which capitalism can survive.'

He was concerned about its survival in 1942 when he wrote his most famous work, *Capitalism, Socialism and Democracy*. Two world wars and the Great Depression had destroyed public confidence in capitalism, but he argued that, contrary to Marx's predictions, and despite temporary dislocations, it had steadily increased the standard of living of the masses. 'The capitalist achievement does not typically consist in providing more silk stockings for queens, but in bringing them within the reach of factory girls in return for steadily decreasing amounts of effort.'

It was in *Capitalism, Socialism and Democracy* that he elaborated the idea of 'creative destruction':

> But in capitalist reality as distinguished from its textbook picture, it is not that kind of [price] competition which counts but the competition from the new commodity, the new technology, the new source of supply, the new type of organization... competition which commands a decisive cost or quality advantage and which strikes not at the margins of the profits and the outputs of the existing firms but at their foundations.
>
> The opening up of new markets... illustrates the same process of industrial mutation – that incessantly revolutionizes the economic structure from within, incessantly destroying the old one, incessantly creating a new one. This process of Creative Destruction is the essential fact about capitalism.

When Schumpeter wrote this, most people feared that economic conditions after the war would be much like those before it. After a decade of mass unemployment, his ideas were less appealing than those of his great rival, John Maynard Keynes, who offered a way to break the vicious circle of low consumer confidence, lower spending, lower production and, above all, higher unemployment. Avoiding unemployment was the main concern of policymakers in the post-war era and Keynesian ideas dominated economic thinking. While respecting Keynes's achievement, Schumpeter felt that his philosophy was essentially 'stagnationist': it sought to restore the economy to a state of equilibrium and took no account of the restless, dynamic nature of capitalism. Schumpeter only really came into vogue in the 1980s, thirty years after his death, in a United States newly taken with the virtues of entrepreneurialism and innovation.

Schumpeter's immediate influence was less on the economics profession than on sociology, history and the new academic field of business management. It was Schumpeter who coined the phrase 'business strategy' and argued that it needed to be one of the primary concerns of managers. The notion of management's role being to keep the ship on an even keel was hopelessly inadequate. 'Mere husbandry of already existing resources, no matter how painstaking, is always characteristic of a declining position.' The reality – not widely understood even today – was that almost all businesses are eventually overtaken by competitors of one kind or another. Only through innovation and entrepreneurship could any business other than a monopoly survive in the long term, but this did not make for the easy, comfortable life successful businessmen aspired to. 'The introduction of new production methods, the opening up of new markets – indeed the successful carrying forward of business combinations in general – all these imply risk, trial and error, the overcoming of resistance, factors lacking in routine.'

Schumpeter did not claim to be a prophet – for him capitalism was a continuous evolutionary process without any end point – but he described remarkably accurately many facets of today's economy. We do not need to accept all of his ideas, or those of his most enthusiastic evangelists, to see that they provide profound insights into how the world of business actually works. As he himself put it, 'Every piece of business strategy... must be seen in its role in the perennial gale of creative destruction.'

Dickens had an inkling of this when he described the arrival of the first passenger train in the 1830s. 'Burrowing among the dwellings of men, flashing out into the meadows with a shriek and a roar!' – the railway was the most disruptive invasion of England since the Norman Conquest. It transformed the landscape and society, put stagecoaches and turnpikes out of business, and made many of the recently constructed canals obsolete. Yet by the middle of the twentieth century, railways all around the world faced fierce competition themselves from the motor car, the bus and latterly the aeroplane. In Europe, they mainly survive thanks to subsidies from the state – in large parts of the US, they have disappeared.

The same remorseless pattern of new businesses and markets undermining old ones has wiped out entire industries. Labour-intensive coal mining, which employed tens of millions for much of the twentieth century, contracted massively in Western countries when faced with the more sophisticated technology and lower costs of the oil and natural gas industries. The theatre has been in a long decline since the arrival of the talking pictures in 1929. Music halls and vaudeville theatres disappeared by the middle of the twentieth century. Movies and cinemas themselves faced an enormous challenge from television from the 1950s onwards, as did radio. In the 1970s cable television channels started to compete seriously with broadcast television, and satellite and the Internet have cut further into its share of the audience.

Most industry leaders enjoy only a brief tenure at the top – of the biggest 100 American companies in 1917, sixty-one had ceased to exist by 1987 and only eighteen of the remainder were still in the top 100. A third of the firms in the *Fortune* 500 in 1970 had disappeared entirely by 1983. Of the top 100 British companies in the FTSE index in 1984, seventy-seven had dropped out twenty years later. Few large organizations last as long as forty years. According to some calculations the average lifespan of a business is seven.

Disruptive technology

Richard Foster and Clayton Christensen have illuminated one of the commonest ways in which creative destruction undermines incumbent businesses. It was Foster who first described the disruptive technology, though Christensen later coined the phrase. In *Innovation, the Attacker's Advantage*, Foster argued that when the underlying technology in an industry changes, new entrants have the advantage over incumbents, and market leaders are displaced. It was outsiders in the electronics industry who developed the transistors that replaced valves, and a generation later it was a new set of outsiders who developed the integrated circuit.

In 1997, Christensen produced a more elaborate version of this thesis in *The Innovator's Dilemma*. This was based largely on his research on the disk drive industry, where he noted the frequent repetition of certain patterns: the leading firms invested heavily in new technology, but not in radically different ones. They failed to take them seriously because initially they did not perform as well as the old one and their customers had little interest in them. It was new firms, with no commitment to the old technology, who developed the new one, typically addressing a different group of customers. Gradually performance improved and costs fell, and eventually it became attractive to mainstream customers – it had migrated upmarket. It was only then that incumbent businesses started to take it seriously, but it was too late to develop the necessary capabilities as the challengers had vastly more expertise.

This pattern was true of the early history of the personal computer, which at first seemed irrelevant to the needs of businesses and to the firms supplying them with powerful mainframes and minicomputers. Most of the innovation came from outsiders like Apple and Microsoft. What was unusual about IBM was that it, or parts of it, had a good stab at being entrepreneurial – for a while. But it could not cope with the pace of innovation that led to the PC eventually overtaking the mainframe.

Established businesses are organized for what Christensen calls sustaining innovation, which improves the performance of existing products and sometimes over-complicates them, not the kind of disruptive innovation that creates a new market. That requires much more than technical skills and invention.

The disruptive technology is a very important idea, especially in industries like IT. However, Christensen went on to apply the term more broadly to disruptive phenomena that most people would not think of as technology at all. Four main qualifications need to be made to his thesis:

1. It is a common pattern, not a universal rule. Ironically, IBM in the 1960s was an incumbent that successfully bet the company on a radically new technology; Intel in the

1970s abandoned memory chips in favour of integrated circuits. And now that 'the dilemma' is well known in Silicon Valley, companies are constantly debating how best to cannibalize themselves.

2. Not all new technologies are initially underperforming. Some, like fibre optics, DVD and HDTV, clearly performed better than the alternatives from birth. There may be other barriers to their adoption, but not performance.

3. This pattern of oblique substitution is not the only, or necessarily the most important, way in which technologies disrupt existing markets and lay the foundations for new ones.

4. Technology, however broadly defined, is not the only source of disruption.

New technology shakes things up in several ways. The one that eventually supplants another is common in industries where the main driver of change is technological advance, but more often a new technology marginalizes, rather than replaces, an old one. The PC toppled mainframe computers from their primacy in IT but did not eliminate them entirely. Likewise radio survived, in attenuated form, the onslaught of TV. The rapid adoption of digital photography made film redundant for most consumers, but not all.

Some technologies, like the microprocessor, have transformational consequences. They do not just spawn entirely new industries – in that case several – they change society and generate waves of disruption that go on for generations. The greatest of them all, Gutenberg's invention of moveable type printing in 1450, led to the mass-produced book and played a crucial part in the flourishing of the Renaissance and the Reformation. The disruption was not simply of the religious institutions who had previously enjoyed a monopoly in the making of books. Protestantism and the questioning of authority spread when people were able to read the Bible and think for themselves, rather than hear the word of God only through the mediation of priests. Without the printed book, the scientific and industrial revolutions, and most subsequent political ones, could scarcely have happened. The book, more than any other single invention, made the modern age.

Another transformational technology was Marconi's 'wireless telegraph', which first sent signals through the ether in 1895. This quickly displaced one of the marvels of the second half of the nineteenth century, fixed-line telegraphy, 'the world's system of electrical nerves', and went on to provide one of the foundations not just of radio and television broadcasting and radar but, a century later, of mobile telephony and a whole new generation of wireless technologies and products, like wi-fi and WiMAX.

Technological innovation affects most firms when a new infrastructure enables new ways of doing business and new social patterns. Steamships and railways in the nineteenth century, transporting goods faster and more cheaply, vastly expanded trade, both national and international. Motor cars and buses made ostlers, grooms and farriers redundant, but entirely new fields of economic activity were opened up. The modern corporation and virtually all the industries created in the twentieth century would not have been possible without the invention of electricity and telephony. Lamplighters and messengers were two immediate casualties, but the number of white-collar and technical jobs mushroomed.

The technology that has transformed the infrastructure of our globalized economy is networked computing, and in particular the Internet, transporting not people and goods but information, instantly, and at zero marginal cost, disrupting telephone companies, travel agents, bookshops and video stores. The single most popular Internet application, email, has greatly accelerated the decline of private letters. The number of emails sent each day in 2003, 31 billion, equalled the total number of letters posted in the year. A rise in parcel deliveries, thanks to all the goods ordered online, has partly compensated the postal services, though more technically sophisticated private carriers like FedEx have captured much of this new traffic.

Sometimes technology can be disruptive without there being any obvious winners.

Teenage pirates

The music industry for a long time viewed technological change as a menace to be resisted at all costs, suing even the makers of equipment that could be used for 'illegal' copying. This was not a winnable war and the strategy verged at times on the hysterical, but the fears were not entirely misplaced. Digitization fundamentally undermined what had been a lucrative business model built around hit records with negligle marginal costs. Ironically the industry benefited in the 1980s from customers replacing their LP collections with CDs, but that was a one-off event. The CD, as some Cassandras had foreseen, put the equivalent of a master disc into the hands of every customer. It was now easy for anyone to produce perfect copies of records. A little over a decade later, the Internet and broadband access made it equally easy to distribute them.

In 1999, Napster, Gnutella and a host of other file-sharing networks sprang up. Members, mostly teenagers, could find almost any song they wanted on the hard disc of someone else's computer and take a free copy. None of them felt they were hurting anybody. The record companies, however, called them pirates and pursued them remorselessly through the courts. Napster was unfortunate in becoming the best-known, attracting 26 million members and ferocious retaliation. The record companies eventually managed to close it down in 2002, but this was a many-headed monster. Dozens of file-sharing programs are still freely available, and thousands of networks, all carefully keeping a lower profile than Napster.

Between 2001 and 2005 the value of record sales fell by a quarter, a combination of lower volumes and desperate discounting. File sharing may not be the only cause of this but it is almost certainly the most important. Those aged under twenty-five, previously the most important demographic segment, have largely stopped buying records. A related discontinuity, accentuated by online music stores like Apple's iTunes that sell songs individually, is the unravelling of the logic of the album that bundles together a dozen songs of variable quality. Another is that the young now mostly listen to their personalized selection of music on iPods and mobile phones.

Technology was the source of most of this disruption but other forces helped to displace the hit record from its central role in the musical firmament. From the 1950s to the 1970s there had been something like a homogeneous mass market for popular music. Most people heard much the same songs on the same radio stations, bought a fairly narrow range of hits on LPs, and listened to them, generally with others, on hi-fi equipment in their homes. That mass market has fragmented into hundreds of niches and genres, from acid house and hip-hop to roots reggae and zouklove.

Another technological revolution lowered the barriers to entry for small record companies. The arrival of very much cheaper recording and mixing equipment, and sophisticated software that can run on ordinary PCs, makes tiny studios economically feasible. Many musicians have become their own producers, and are much less dependent than previously on record companies.

Technology has commoditized the markets for recorded music and hi-fi equipment. According to David Bowie, music will soon be free 'like running water and electricity'. However, fans will pay a lot to see and hear artists like him – more money is now spent on live performances than on CDs. Where once concerts were used to promote records, it is now the other way round. Some bands would rather sell merchandising like T-shirts at concerts than records, from which they make less money. The disc has been dislodged from its central position, and it is now mainly older adults who buy them.

All this has left the major record companies reeling and facing the prospect of further painful contraction. Some may continue to blame teenage pirates for their fate, but they are the victims of a set of discontinuities that was bound to destroy their old profitable, rent-seeking business. They would certainly have done better to come to terms with change earlier, but there was no way of avoiding the fact that the days of plenty were over.

Unknown unknowns

The fundamental challenge that faces every business, from the mighty corporation to the corner shop, is how to make sense of and adapt to apparently sudden changes in its environment. In the long run discontinuities of one kind or another are inevitable. Technological innovation and creative destruction may be the most important sources, but they mostly arise from the unpredictable, uncontrollable nature of life itself.

Discontinuities are changes in the environment so radical that things can never be the same again, but whose implications we only dimly understand. The new situation works by different rules, demands new capabilities and frequently new people. Andy Grove of Intel called them 'strategic inflection points', but that implies we know what to do about them – which fortunately he did. Whether a change represents a discontinuity for an individual firm, person or place depends on their particular circumstances, but it can only be understood in a broader context.

There is no systematic way of spotting them – the point about discontinuities is that they come mainly as surprises. Some trends seem clear and amenable to tracking, but they rarely continue in straight lines and their consequences are not always obvious. Nobody really knows, for example, what will be the long-term effects of global warming or the unprecedented fall in birth rates in Western countries. Yet these are 'known unknowns'. The really challenging discontinuities, though, are the 'unknown unknowns' that take us completely by surprise. The change may appear to be a sudden event, but more typically it is the culmination of a gradual process which we do not notice until it has acquired irreversible momentum.

The banking crisis of 2008 came as a shock to almost everybody, but its origins were long-standing: unprecedented increases in personal and corporate debt; an unsustainable property boom; frantic selling of mortgages and derivative products so complex that nobody understood them, and the growth of an enormous shadow banking system; bonus schemes that encouraged penalty-free risk-taking; and unwillingness on the part of governments to regulate an industry that appeared to be generating so much wealth. Pundits like Gillian Tett were warning from 2006 that disaster was looming, but not many people were listening.

We will consider in Chapter 12 the difficulties, both cognitive and psychological, of making sense of change and facing up to its implications. What all businesses have to come to terms with is the fact that their environment is constantly changing, sometimes slowly and imperceptibly, sometimes dramatically quickly. Yet most organizations and people cling to the hope that most things can remain the same and work on the implicit assumption of continuity.

Even the word discontinuity implies that continuity is the norm but in the long run it is continuity that is the exception. What appears to be permanent is mostly merely long-lasting. Everything dies eventually – and is succeeded by new forms of life and new ideas. Those who last longest, like all successful species, are those that adapt best to a changing environment.

Some changes can be anticipated and planned for, but the most important often cannot. They pop up in the most surprising ways.

Craigslist

Craig Newmark is an unlikely subverter of the newspaper industry. A self-confessed nerd, he has difficulty dealing with people, and prefers email to any other form of communication. He is wary of corporations and has little interest in making money. Yet his creation, Craigslist, now run by only twenty-four people from a house in Cole Street, San Francisco, has become the largest medium for classified advertising in the world and is credited by some with destroying the economics of local newspapers.

Craig came to San Francisco in 1993, after seventeen not very happy years as a software engineer at IBM. He became a devotee of The Well, the most idealistic and counter-cultural of online communities, and enthusiastically embraced leftish libertarian values – open source software, complete freedom of speech and members of the community helping each other out. In 1995, he started sending out a regular email to friends, listing interesting local events. Some of them started sending him other items to include, and the number of recipients quickly grew to 240, the maximum number of emails that can be sent using cc. So it became a listserv, a popular way of disseminating information in the days before the Web.

When word got around among businesses that this was a good way of finding technical people, jobs postings started flooding in. A separate list was created, along with several others. Many San Franciscans found Craigslist the answer to some of their most vital needs – apartments, lonely hearts and, sometimes most poignantly, lost and found. One lady who found her beloved dog after days of impassioned postings described Craigslist as the place 'where everyone goes when they're looking for something'. A young woman who thought she'd never meet a nice vegetarian man found the love of her life after addressing a posting 'to the cute blond guy on the L train'.

Craig's idea was not to build a business but to 'try to give people a break' and to support a community. The Internet has a purpose, he believes, 'to connect people to make our lives better'. When the community asked for a Web interface, he built it with the help of unpaid volunteers, using, naturally, Linux and other free software. He wanted to call it SF Events, but his friends persuaded him to stick with the name they had all got attached to. The site kept the same frugal look of the email lists, much cherished by the community. The production process remains largely automated from entries posted by email. Editorial intervention is mainly limited to cutting out the more outrageous (and hilarious) entries under Casual Encounters, which caters for those seeking strings-free sex. Traffic grew constantly, fuelled by powerful feedback loops and network effects.

It was not until 1999 that Craig finally decided to give up his day job and make this a business. Well, sort of. He only charged for job ads – everything else was free, but in 1999, the height of the dot.com boom, there were many more techie jobs than people in the Bay Area, and the money rolled in. It slowed down the following year with the slump in the Net economy, but the shake-out also brought

Jim Buckmaster to Cole Street. Jim was a bright programmer with the social and organizational skills that Craig lacked, so he made Jim the CEO. Jim quickly worked out ways to extend the site beyond San Francisco – within a few months it was in New York, Boston, Seattle, Washington and Chicago. Charging brokers $10 for every apartment listed in New York brought in revenues that compensated for the dip in job ads.

These revenues and the $25–75 Craigslist charges for job ads made the business comfortably profitable. In 2007 it was the ninth most popular site in the US, serving over 7 billion pages a month to 25 million visitors, and covering 450 cities around the world. It was receiving 10 million new ads each month, 500,000 of them for jobs, and its (undisclosed) revenues were in the tens of millions.

Jim makes it plain that maximizing revenues and profits is not on the agenda. They're doing this primarily to help people find jobs, flats, dates, whatever they're looking for. 'We've had the luxury of doing well and being able to follow a moral compass and not have much conflict between the two.' He and Craig have refused all attempts to carry conventional advertising or to float the company on the stock market. That would completely change the nature of Craigslist. It works, Craig believes, because 'it gives people a voice, a sense of community, trust and even intimacy'.

But these hippies have dealt a deadly blow to local newspapers, particularly the *San Francisco Chronicle*. It believes it has lost tens of millions of dollars in revenues from classified ads, the mainstream of its business. Al Saracevic, its business columnist, is indignant.

> You shouldn't take the money and run... You need to give some-
> thing back to society other than cheap apartment ads and funny,
> dirty personals... There has been a social contract for hundreds of
> years – news-gathering organizations derive revenue from
> community advertising. Well, Craigslist is changing that equation.

We can take the rhetoric about a social contract with a pinch of salt. The *Chronicle* was never exactly essential reading – like most local papers it had become a complacent monopoly – but Saracevic was right that the business model of newspapers is unravelling. Readers have never had to pay much for expensive news content. It has always been subsidized by advertising, which accounts for the vast majority of local paper revenues – and a good chunk of those of national titles – but very few people now rely on papers as their main source of news. Advertisers are starting to move online and readers are doing so even faster, accelerating a long decline in newspaper circulations. In 1970 one daily paper was sold for every American household; by 2000 that number had halved. In 2007 alone sales fell by 3.5 per cent in the US, in Britain by 2.7 per cent and by similar amounts in all Western countries. Newspaper advertising revenues are falling even faster.

This is becoming a classic vicious circle, the obverse of Craigslist's virtuous one.

Age of discontinuity

Why do discontinuities like the slump in classified advertising and record sales seem so much more prevalent recently? Why is there so much creative destruction now? How has our world become so volatile?

This is partly a question of perception. To Europeans like Schumpeter living through the first half of the twentieth century, two world wars and the Great Depression, these would seem strange questions. Our lives in the West are infinitely more stable, secure and prosperous than they ever were for our great-grandparents. What is undeniably true, however, is that the economic order has become more volatile in the thirty years since 1979 than in the immediate post-war decades.

The term discontinuity was popularized for business readers by Peter Drucker, while the post-war boom was still going strong. In 1968 he pointed out that most of the then leading industries – steel, motors, tyres, chemicals, oil – were based on technical innovations made in the fifty years before 1914. Drucker was struck by how comparatively little the structure of industry had changed since then.

With remarkable prescience he predicted that technologies then being developed would lead to a new age of turbulence. Several of the leading firms in 1968, like Firestone and International Harvester, have indeed since disappeared entirely, most have declined in importance, while major companies like Microsoft, Intel, Cisco and Dell did not even exist at the time he was writing.

Discontinuities entered mainstream thinking in the 1970s with two major shocks: the end of fixed exchange rates and a massive hike in the price of oil. Economic growth spluttered, and both inflation and unemployment soared. The Keynsian consensus crumbled and a flood of best-selling books forecast yet more turbulence.

What few people realized was that the post-war era in the West had been unlike any other period of history, in combining record rates of economic growth with low levels of industrial volatility and virtually zero unemployment. Creative destruction seemed to have been banished. For those whose ideal was a lifetime career at the same organization, *Les Trente Glorieuses* (1945–75), as the French fondly recall them, seemed like a golden age of prosperity and security from which we have sadly strayed.

In fact this era saw big, but comparatively painless structural changes in most Western economies. In Western Europe there was a massive shift of manpower away from farming and into manufacturing, entirely new mass markets for cars, refrigerators and televisions mushroomed without any obvious collateral damage, and international trade grew at an unprecedented rate. Old industries like textiles and shipbuilding faced fierce foreign competition, but a booming economy soaked up the displaced manpower.

Within most European countries competition was muted and often tightly controlled by government regulation. Mergers and acquisitions, particularly hostile ones, were frowned on as likely to mean job losses. Even in the US, oligopolies were the norm in mature markets. European governments fondly believed that their 'management' of the economy and prudent planning had played the key part in the steady rise in post-war prosperity. Strategic planning was also taken seriously by large businesses. Most executives, politicians and bureaucrats believed that the future could be planned and controlled.

That confidence was shattered by the shocks of the seventies and the even greater volatility, both economic and political, that followed. The 1980s saw a rash of hostile takeovers, the collapse of smokestack industries, and booming stock markets, but this was a revolutionary decade in a more profound sense. Two revolutions occurred in the 1980s, one technological, the other political-economic. Arguably there was a third in terms of attitudes towards free markets. In combination they led directly to our present globally integrated, electronically connected, much more competitive economy.

The long decade between 1979 and 1991 saw enormous leaps forward in the application of information and communications technology. In 1980 there were 720,00 personal computers in the world; by 1990 there were 50 million, and their average power had increased a hundredfold. Those in use within businesses were connected over local area networks, and those in universities over the Arpanet. After the Arpanet had adopted a communications standard, TCP/IP, that enabled any computer to connect to it, it metamorphosed into the Internet. By 1990 the number of host computers on this network of networks had grown 800-fold, from 231 to 180,000. The Internet was soon to be transformed even more dramatically by Tim Berners-Lee's invention of the World Wide Web. Not many people outside academia noticed until 1994, when Netscape brought out its browser.

Three other giant digital strides were taken in the 1980s: Sony launched the CD, and the media industry reluctantly started to come to terms with digital formats; a group of technologists known as Moving Pictures Expert Group 3 (MPEG3) agreed a standard for compressing music and video into digital files which later became known as MP3; and a digital standard for mobile telephone networks, GSM, was agreed that was subsequently adopted across Europe and Asia.

The world also became significantly smaller: the first fibre optic cable was laid underneath the Atlantic, at a stroke massively increasing international telecommunications capacity and heralding a sharp drop in the cost of phone calls; and communications satellites started broadcasting multi-channel television across national boundaries.

The technological foundations were thus laid for the take-off of mass markets in personal computing, the Internet, mobile telephony and pay television, the automation of millions of jobs, and hundreds of thousands of businesses operating on a global scale. Politics, however, played as big a part as technology in making the world economy more open, more connected and very much more competitive. And the key steps were totally unexpected and mostly taken independently.

The first was the Chinese government's decision to introduce free markets to its previously closed, heavily regulated economy, and subsequently to open it up to trade with the rest of the world. Its example was soon followed by India. The second step was the dawn of Thatcherism and Reaganism. Starting with the abolition of exchange controls in Britain, both governments pursued parallel programmes of market liberalization, tax cutting and reducing the role of the state in business. Even the governments of the European Union decided to create a single market across all their countries by 1992.

The most dramatic event was the collapse of the Soviet Union and its empire in 1989, largely due to economic failure. With it ended both the Cold War and any lingering faith in planned economies, and Russia and the countries of Eastern Europe joined the global economy. Finally, after decades of suspicion of economic liberalism, intellectual and governing elites, in almost all countries and across the political spectrum, came to accept the arguments for free markets open to global competition.

The most important consequence of this transformation was the establishment of a globally integrated economy. Between 1982 and 2001, the effective global workforce increased fivefold, levels of foreign direct investment grew almost tenfold, and world GDP tripled. The networking of the world made it smaller, more connected and more complex, and feedback loops fuelled the growth of thousands of virtual networks. Information, once scarce and costly, became instantly available, and mostly free.

Capital also became very much more plentiful. Between 1996 and 2005 new companies coming to the world's stock markets raised $1.2 trillion in initial private offerings, more than $600 billion poured into venture capital, and funds in poor countries offering micro-credit to farmers and small businesses reached $7.5 billion. Capital markets had become the most global of all. Largely free from govern-ment control, funds moved rapidly between countries, rewarding companies that performed well, and ruthlessly punishing those that did not. The value of mergers and acquisitions reached a record $850 billion in 1995 and soared to $3.8 trillion in 2006.

It had never been easier to start a business, or to expand or acquire an existing one, but the new environment has not been entirely business-friendly. Competition has never been so intense, nor competitive advantage so elusive and fleeting. Many more industries are now global, and the barriers to entry that arose from distance and government regulation have crumbled. Monopolies in long-protected industries like airlines, banking and telecommunications have been ended. Differences in wage costs have shifted vast amounts of manu-facturing production, software development and call-centre work from rich countries to China, India and Eastern Europe.

Customers, both businesses and consumers, have very much more information, choice and power, and can switch suppliers easily. The profits of non-financial companies in the US, which were 18 per cent of GDP in 1950, had fallen by 2003 to 6 per cent.

Most publicly quoted companies in the West re-engineered their businesses to concentrate on their core capabilities, stripping out layers of management, and streamlining their supply chains. Many of the new businesses challenged, and sometimes displaced, incum-bents. More new technologies and business models have been developed and more widely adopted, more new markets created and industries disrupted than ever before.

This was a world in which creative destruction worked on a scale and at a pace that would have amazed even Schumpeter.

Disruptive competition

For incumbent firms, creative destruction often hits the fan in the form of a new kind of competitor who invades their markets obliquely. In many cases this is harmless – fast-food joints do not take business from good restaurants, and distance-learning courses do not threaten top-flight universities. In two of the new markets considered here, the disruptive competition was downright destructive. The ever growing power and performance, and ever lower prices, of the personal computer effectively wiped out the markets for typewriters, computer terminals, word processors and workstations. Low-cost airlines like Southwest and Ryanair have destroyed the business models of traditional airlines on short-haul routes, and precipitated the bankruptcy of many.

On a smaller scale, eBay has had a devastating effect on some segments of the antiques business and other second-hand markets, and Starbucks has made life uncomfortable for independent cafés. New forms of advertising, notably those of Google and Craigslist, are doing serious damage to newspapers and directory publishers.

What makes competition disruptive is not that it comes from an unfamiliar source, but that it offers something the incumbent cannot: it changes the basis on which incumbent businesses compete, and undermines their competitive advantage. Challengers come up with new propositions that incumbents cannot match, generally because of new capabilities. IBM did not have the capabilities to be a nimble, low-cost PC maker; Encyclopædia Britannica could not come remotely close to the price of *Encarta*.

Disruptive competition typically arrives in one of three forms – a radically different business model, encroachment from another industry or market, or an entirely new product or service concept. Sometimes, and most devastatingly, it has elements of all three.

The most important immediate cause of disruptive competition in the Age of the Internet is an original business model. The Net has been their most fertile breeding ground, but there have been many others. Dell's rise to leadership of the PC industry was almost entirely due to a brilliantly streamlined model that others could not easily imitate. It was not the CD-ROM that destroyed Encyclopædia Britannica, but the fact that Microsoft had a completely different model and objectives. Its content production costs were negligible, and it was not trying to make money directly from *Encarta*, so it could afford to give it away. Britannica's business model, with its enormous sales force and margins, was not sustainable in the long term.

What most recent business models offer customers is a combination of low prices, greater choice and more convenience. Typically they simplify the sales and marketing process, cutting out expensive sales people and intermediaries, and ask customers to do more of the work – in return they get something more tailored to their needs. This can be devastating for companies organized differently, particularly those with large sales forces or long-established marketing channels. The new model often exposes how little value these were adding.

A variation on the new business model is invading or encroaching on an adjacent market. The trespasser is not an unconventional start-up, but an established company offering a wider range of products to its existing customers. The marginal cost and risk are low, it does not have to develop major new capabilities, and benefits from economies of scale and scope. Supermarkets are the most conspicuous example, endlessly extending their range into areas like books, music and clothes. Their buying power means that they can offer exceptionally low prices on selected items, and customers benefit from the convenience of one-stop shopping, leading to the disappearance of specialist shops in many areas. Likewise Microsoft was able to expand from its base in operating systems to invade one area of desktop software after another and is now getting a taste of its own medicine from Google.

The disruptive consequences of new product and service concepts ought to be obvious – an iPod is a clear alternative to a Walkman, Naxos to a full-priced CD, a DVD to a videocassette. Typically though the disruptive effect is initially either under- or overestimated.

Nobody at first thought that the PC posed a threat to the typewriter industry, though the movie industry in the 1970s regarded the video-cassette as a deadly danger. In fact it has proved to be its salvation –

videocassettes and their successors, DVDs, have greatly increased total revenues from films and reduced dependence on receipts from declining ticket sales. However, they started to change the way people used their televisions and weakened the hold of broadcast networks.

DVDs have been less disruptive, because of the way that the consumer electronics industry collectively introduced them, knowing that they would quickly replace the VCR. Partly for this reason, but mainly because of better sound and picture quality, and more sophisticated pricing, the DVD found a mass market faster than any other consumer electronics device has done. For some consumers they have become the main way of watching TV series, further weakening the hold of the broadcasters.

The physical characteristics of DVDs provided the platform for a new business model that proved highly disruptive to video rental stores. In 1997 Reed Hastings realized that DVDs were small, light and robust enough to send through the post, and that the Internet made online ordering easy. He set up Netflix, the first company to allow customers to order their movies online, without the bother of going out to stores. Netflix simply posts the DVD to the customer and as soon as it comes back, sends out another one from her list. For people with limited time, this proved a more convenient way of renting movies, but Netflix could also offer vastly greater choice than was available in most local stores. The model was quickly imitated by hordes of other businesses across the world, and video rental shops are fast disappearing.

In some cases it takes years to see how a new product will compete with existing ones, and with which. When Sony launched the original Walkman, nobody, least of all Sony, saw it as a threat to the recorded music industry, let alone to conventional hi-fi. Its only obvious competitor was the portable cassette player. In the long run, however, the new market for personalized, portable music weakened demand for all kinds of audio equipment and paved the way for the MP3 player. In conjunction with file-sharing on the Web, this has had a devastating effect on sales of CD players, not to mention CDs.

Quite often competition is only mildly disruptive. PCs have not destroyed mainframe computers, though they have marginalized them; eBay has not eliminated antiques dealers; nor Amazon, bookshops. Television and the wealth of modern home entertainment systems have not destroyed cinemas. Some of the incumbents in these industries had capabilities and assets that challengers did not, and their propositions were still relevant. They have been able to emphasize the things that they do better than the challenger.

At good antiques shops customers can see and touch the objects, there is ambience, service, personal relationships – particularly important at upmarket shops and auctioneers like Sotheby's and Christie's. People who like books like to see and touch them, to leaf through, to browse, which they can't do on Amazon. Cinemas offer a more intense, exciting, communal experience than home entertainment.

None of these industries is about to disappear entirely. They may have lost market share but they are a long way from dying.

The critical questions for an incumbent are: What does the challenger have that we do not? How strong is their proposition? What capabilities and assets do they have? How strong is our proposition, and for which customers? How strong are our relationships with our customers? What else do we have that they do not?

If an incumbent business has a strong proposition of its own, and good relationships with its customers, it takes an exceptionally strong proposition to prise them away. Proprietary games consoles have resisted the seemingly irresistible march of the PC, thanks to customer switching costs and frantic levels of competitive innovation by manufacturers and their networks of developers. Webvan was a business whose proposition was nothing like strong enough to tempt customers away from supermarkets. All it really had to offer was convenience – to penetrate a mass market quickly, it would have needed much lower prices.

Embracing creative destruction

For those on the receiving end of disruptive competition, the experience can be painful and sometimes catastrophic. The trauma associated with business failure and the devastation caused to local communities when a major employer disappears explain why most people dislike creative destruction intensely. They may welcome the creation, but rarely the destruction: Chinese factories, Indian call centres, Polish plumbers, not to mention Microsoft, Wal-Mart and Tesco, have been demonized as cruel predators destroying incumbent businesses and jobs. There is no denying that competition can be a brutal business. There are always losers as well as winners, but the winners' gains are widely dispersed, largely among consumers, and less pronounced than the losers' pain.

In the long run the process is irresistible in open societies. Only in the poorest countries do blacksmiths, millers and cobblers still ply their trade. Societies that place more emphasis on guaranteeing employment than on encouraging new enterprise risk economic stagnation. Feudalism and the medieval guild system produced centuries of stability but virtually no growth. The centrally planned socialist economies eventually seized up.

So far as businesses are concerned, creative destruction is an unavoidable fact of life. Richard Foster and Sarah Kaplan have argued that, now that discontinuities are the norm, companies should adopt creative destruction as a business philosophy and apply it to themselves. Most businesses, they say, are still organized on the false assumption of continuity and are unable to change and create value at the pace and scale of markets. They argue that sustainable competitive advantage is an unattainable ideal, predicting that no more than a third of today's major corporations will survive over the next twenty-five years: the inability to adapt will account for the slow demise of the rest. The main reason they say that corporations find it difficult to respond to the messages of the marketplace is 'cultural lock-in' – the inability to change corporate culture in the face of clear market threats.

The fundamental problem is the conflict between the need for corporations to manage their present operations and the conditions that permit new ideas to flourish and old ones to die. Companies are generally not good at the divergent thinking required for innovation, and can only match or outperform the market by abandoning the assumption of continuity. Foster and Kaplan contrast this with capital markets, which do not show the sentimentality and attachment to the past of mature corporations. Private equity firms can change at the pace and scale of the market, engaging continuously in creative destruction, thinking of businesses as a revolving portfolio.

The diagnosis is irrefutable, but their arguments need some qualification. Discontinuities may hit every business sooner or later, but some firms manage to surmount modest ones. Competitive advantage may always be eroded eventually, but like human life it is precious while it lasts. A few companies are able to enjoy decades of good returns from competitive advantages based on distinctive capabilities, strong brands and the ability to maintain modest levels of innovation. These companies may not offer the shareholder returns of rapidly growing new companies, but stability and predictability are valued by some investors as well as by managers, not to mention employees and customers. They also have marked advantages over new entrants whose strength lies in a single innovation. Innovations can be quickly imitated or surpassed and it is frequently mature companies that do this.

The comparatively small number of companies who have been able to innovate repeatedly over long periods, such as HP, 3M, Glaxo and GE (not to mention Apple, Sony and some parts of IBM), tend to have had strong relationships with their employees, customers and suppliers. They are, among other things, strong social networks of people who have learned how to collaborate effectively, and to do so repeatedly – not easily achievable in an environment where failure is instantly punished by the death of the business. The most startling transformational innovations nearly always come from outsiders, but it generally takes large companies, as Apple and Sony quickly became, to build on these and produce the waves of innovations that earn years of profitable revenues.

The near-collapse of the global financial system in late 2008 shows how badly capital markets can get things wrong. Quite apart from periodic boom and bust crises and grossly underestimating risk on this occasion, it has been clear since the mid-1990s that stock markets had been assigning unrealistically high valuations to technology stocks and reinforcing an unhealthy obsession with short-term financial performance. The revolving portfolio approach to business does not encourage managers to think about the long term or to develop new capabilities and opportunities. It contributes to the failure of many firms to take note of subtle changes in their environment and the challenges they represent, and produces creative destroyers like Enron, singled out for praise by both Foster and Kaplan and Gary Hamel.

At the time of writing some commentators were predicting the end of the era of unbridled capitalism. Almost certainly commercial banking will be more tightly controlled, not least because governments now own so much of it. But banking has always been a special case, since the failure of just one bank can precipitate the collapse of the entire financial system. However, even if more regulation spreads to other markets and the tide of globalisation ebbs, creative destruction will always be with us.

As we shall see in the next chapter, although Nokia successfully applied a form of creative destruction to itself, it is very much the exception, and the circumstances were extreme. It is questionable whether more than a tiny handful of super-companies could really embrace it as a guiding principle, and whether it would be desirable for most to try. Companies cannot transform themselves quickly like Superman in the phone booth: they have to start with the capabilities they have. As Foster and Kaplan establish conclusively, it is exceptionally difficult for an organization that is optimized for operational efficiency to create new markets. To combine both with a ruthless willingness to cull one's young is asking a lot.

Acceptance

We do not have to rejoice in creative destruction, but we cannot ignore or deny it. What all businesses must come to terms with is the fact that their competitive environment is constantly changing, sometimes slowly and imperceptibly, sometimes dramatically quickly.

The most positive response is to embrace change and become an active agent in shaping the future. Only a few immensely talented and fortunate organizations can create entirely new markets. However, many more learn how to adapt to change, cultivate new capabilities, build new businesses and products, and counter disruptive competition.

What is not an option is hoping that the status quo can be the way forward or denying that change is happening. That is the road to ruin. Sooner or later, a discontinuity will hit every business, very likely from a new kind of competitor. To have any chance of anticipating it and developing an adequate response, it must seek to understand its business environment and how it could change – from a broader perspective than that of the firm and its present customers and competitors. That means constantly scanning the horizon, preparing itself for a range of possible futures, good and bad, and thinking of continuity as a medium-term exception, not the rule. And most importantly, to keep the organization's capabilities distinctive, it must seek to extend them and to learn new ones.

Most businesses fail to do these things, which largely accounts for their high mortality rate. Schumpeter would say that business failure is healthy as it makes way for thrusting newcomers. Most businesses, however, would like to survive and prosper for as long as they can. The only ones who have any hope of doing so are those who learn how to adapt to disruptive change.

11

WIRELESS WINNERS

In the early 1990s BSkyB and Nokia were on the brink of bankruptcy yet both went on to create enormous and immensely profitable new markets in pay television and mobile phones. Sky smashed the cosy world of British broadcasting and made Premier League football a global industry. Nokia applied creative destruction to itself and to Ericsson and Motorola, who had opened up the markets for mobile telephones.

BSkyB

The best television in the world

Until Sky shook it up in the 1990s, broadcasting in Britain was not so much a competitive industry as a highly regulated public service. There were no paying customers, but all homes with a television set were obliged to pay a licence fee to the government, the proceeds of which financed the BBC. Although there had been a commercial alternative to the BBC's monopoly since 1955, all ITV franchisees were themselves monopoly suppliers of television advertising in their regions. As one of them carelessly remarked, it was a licence to print money.

All concerned agreed that television was too important to be left to unbridled market forces. ITV franchises were awarded, and occasionally not renewed, on the basis of the quality of programming, as determined by the Independent Broadcasting Authority (IBA). Simply giving viewers what they wanted, everyone believed, would lead to the 'crass commercialism' of American television. Although ITV programming was more populist than the BBC's, the incumbent companies absorbed many of the values of public sector broadcasting defined by 'Auntie', and the overall tone of British television was thoroughly respectable.

The BBC's mission since the days of John Reith, its high-minded founder, had been 'to inform, educate and entertain'. It was not alone in believing that the unique British hybrid of public and private had produced 'the best television in the world'. This much-repeated claim was not as self-satisfied as it would sound today: the BBC had repeatedly produced superb drama and documentaries, and entertainment programmes that won enormous audiences; the output of ITV companies like Granada was comparable. When Channel 4 was introduced in 1982, it too was given a public service remit.

Most people working in British television thought there was no need for more television channels – viewers were perfectly happy with the ones they had. Nobody had seriously considered the possibility of very different programming, aimed, for example, at the millions of people who enjoyed unashamedly vulgar newspapers like the *Sun*.

One person who had was the *Sun*'s proprietor, Rupert Murdoch, familiar with a more raucous style of broadcasting in Australia. He was convinced that there was untapped demand for more varieties of programming, but knew that what he saw as the stuck-up British Establishment would never admit a rude Aussie to its ranks. In 1969, the same year his company, News International, acquired the *Sun*, he had also bought a stake in London Weekend Television, then in desperate trouble. His money and hands-on involvement probably saved the company from bankruptcy, but the television world was appalled at his taking LWT downmarket. The IBA barred him from playing any executive role and tightened its rules to ensure that anyone with extensive press interests could not hold an ITV franchise. So Murdoch sold his stake in LWT and looked elsewhere.

The *Sun* had built its circulation with topless models on page three and headlines like 'Freddie Starr Ate My Hamster'. News International's acquisition in 1981 of the most prestigious titles in British journalism, *The Times* and *Sunday Times*, prompted howls of indignation in Parliament and other papers. The *Sun* covered the Falklands War with gusto, and its front page of 'Gotcha', over a picture of the sinking Argentinian ship the *Belgrano*, caused outrage. Murdoch was the most hated figure in the British media, but he admired Margaret Thatcher and she him. They both loathed the BBC, his papers loudly supported her government and she helped him at critical junctures.

Murdoch pursued his television ambitions in the US, where he bought the 20th Century Fox film studio in 1985 and used this to create a fourth American television network, Fox Broadcasting. Fox's combination of innovative comedy like *The Simpsons* with tabloid-style journalism defied the expectations of most commentators and eventually became a serious competitor to the ABC, CBC and CBS oligopoly.

Spectrum scarcity

The fundamental reason for restrictions on the number of television channels was 'spectrum scarcity'. At a time when all broadcasting was done in analogue rather than digital mode, each channel required enormous amounts of bandwidth. The allocation of this finite resource was controlled directly by governments in most European countries.

Until about 1990 the only way to offer more channels was cable television, a huge success in the US, where more than 60 per cent of homes subscribed. The cable industry, mostly American-owned, did not expect to reach similar levels in Britain but hoped to do better than the 5 per cent of British homes passed by its cables that it was achieving. Cable penetration in the US was high partly because of the poor reception of terrestrially broadcast channels. British viewers could get good pictures over the air and did not find the additional channels on offer particularly enticing. The cable industry, however, continued to concentrate on the logistics of laying cable and gave little thought to finding more compelling programming.

Another way around spectrum scarcity was satellite broadcasting, but in the 1980s satellites were not powerful enough to broadcast directly to homes – only dishes several feet wide could pick up signals. Satellites were mostly used to distribute television programming to cable operators, but a new generation of higher-powered satellites was coming along, capable of direct-to-home (DTH) broadcasting that could be received on small dishes. The economics of DTH promised to be more attractive than cable, which required the enormous up-front cost of digging up streets and laying the cables. Signals from a satellite could in principle reach an infinite number of homes within its geographical footprint. The challenge, as with cable, was to find attractive programming.

Murdoch had long suspected that satellite would be the way for him to break into British television. In 1983 he came across an obscure, loss-making business, SATV, whose footprint included the UK but whose only source of income was supplying cheap American programming to cable TV operators in Holland and Germany. News International acquired control of SATV in 1983 and renamed it Sky. It continued to make losses for the rest of the decade, as Murdoch had only the haziest idea of how to make a successful business of it.

Business considerations did not greatly trouble the technologists and regulators who were blithely planning the future of satellite broadcasting across Europe. Curiously, their pan-European committees took little account of satellite's ability to ignore national boundaries, assuming that governments would continue to control all broadcasting in their own territories. The Eurocrats were more interested in the technology, where they hoped to define a new world standard. At the urging of the IBA, all satellite broadcasters would be obliged to adopt a sophisticated, expensive new system, D-MAC.

The government decided that two of the five satellite channels allocated to the UK should go to the BBC and the other three to a single winner of the kind of beauty contest the IBA organized for the award of ITV franchises. The BBC prudently decided that it did not have the financial resources for a speculative venture of this kind. Several more intrepid organizations, including News International, applied for this new monopoly. Predictably, and with the benefit of hindsight fortunately, its application was rejected in favour of that of British Satellite Broadcasting (BSB), a consortium of respectable British media companies like Pearson, Granada and Reed, plus the French group Chargeurs.

Murdoch, however, spotted another way in. A Luxembourg company, Société Européenne des Satellites (SES), was planning the launch of a 'medium-powered' satellite whose technology would be less sophisticated than D-MAC, but well tried and adequate for DTH broadcasting to the UK. If Murdoch went with SES, and the signals were uplinked to the satellite from Luxembourg it would be a 'non-domestic service' and escape British regulation. Simply leasing capacity on SES's satellite, Astra, also avoided the expense and hassle of launching and operating Sky's own one. In these pre-digital days, a whole transponder was needed for each channel, and Murdoch ordered four of Astra's sixteen for ten years and told the Sky team to start developing three new channels.

SES launched Astra in December 1988 and Sky unveiled its new channels two months later. It hoped to reach an audience of 2.5 million in its first year, but Sky's team had little television experience and limited resources and the quality of its programming was mediocre. The non-Murdoch British media crowed with delight at pathetic 'tabloid television', clearly doomed to failure.

Just before the launch, Murdoch had come close to obtaining some uniquely compelling programming, offering the Football League £47 million for the rights to live coverage of matches. This was three times as much as the League was then receiving from the BBC and ITV, but it was reluctant to commit itself to an unproven medium and a flaky business. Instead it used Sky's bid to squeeze more money out of ITV, where Greg Dyke was determined to 'strangle satellite at birth'.

BSB was also inclined not to take Sky seriously, and assumed that it would be the only player in satellite broadcasting. This was to prove a big mistake, the first of many.

Other people's money

BSB's management suffered from three delusions – that its success was assured when it won the franchise, that it need not worry about competition, and that its shareholders had unlimited funds.

Its managing director, Anthony Simmons-Gooding, like most of his senior colleagues, had no experience of television, competitive or otherwise. His background was in marketing and advertising, and he had spent the last three years trying to make Saatchi and Saatchi a global marketing conglomerate. He had a relaxed, well-fed air, and unwisely told an investment banker that he was having fun 'spending other people's money'. All the work on preparing the business plan was done by consultants, who like everyone else at BSB travelled first class everywhere. When BSB's Marco Polo satellite was eventually launched, a year late in April 1990, a team of executives flew out to Florida to witness it, while their colleagues moved into a swanky new building in central London, Marco Polo House.

According to Michael Grade, then chief executive of Channel 4, 'If you went to the BSB car park, everyone had a BMW and a chauffeur. They were all there on a gravy train, and they weren't focused at all.' The contrast with Sky was stark. 'Murdoch's people were on a mission, a mission to destroy the BBC, a mission to destroy ITV. A mission to destroy the old order. They were zealots.'

When it started to look as though Sky might pose a real problem, Simmons-Gooding complained to the government about 'unfair' competition:

> The government insisted on BSB marketing a pioneering
> technology, complex, high-risk, but if successful, of great long-
> term benefit, not only to the consumer but also to the European
> manufacturing and retail community as a whole. The

government then allowed BSB to be bypassed by a powerful competitor unregulated, with no technology demands, and promoted by the most powerful media group in the UK.

Extensive lobbying of Parliament came close to subjecting 'non-domestic' satellite services to restrictions on cross-media ownership. But the government was now in the process of changing the way that television franchises were awarded, not least because Mrs Thatcher had decided that the top ITV executives were cosseted fat cats and the unions 'one of the last bastions of restrictive practices'. She was determined to subject them to commercial pressures, and would not countenance hobbling her most loyal supporter in the media, particularly not to prop up yet another fat cat.

BSB decided that the key to beating Sky was signing up the rights to as many recent movies as possible: content would be king, and movies the killer application. (They did indeed turn out to be almost fatal.) Executives were despatched to Hollywood to sign up as many studios as possible and to outbid Sky, now pursuing an only slightly less reckless strategy. Like Sony's acquisition of Columbia this was one of those opportunities that Hollywood relished – taking money from star-struck suckers. It convinced BSB that to win the rights it would have to guarantee payment, even if the movies were never shown. There was not enough cash in the coffers to meet these demands, so Simmons-Gooding persuaded Pearson, Granada and Chargeurs to guarantee minimum payments to the studios of £700 million over the first five years.

BSB was already spending £170 million on building and launching two satellites – one just for back-up – and had to shell out another £400 million on encryption technology and 'squarials'. These strange dishes promised to be not just smaller and more unobtrusive, but to perform better than the conventional round ones. However, none had yet been manufactured, and scarcely any were available even by 1990. The scarcity of squarials was one of the reasons why BSB signed up many fewer subscribers than Sky – by early 1990 it was losing £7 million a week.

Struggling at Sky

Life was not much easier down at Isleworth, Sky's defiantly unglamorous headquarters next to a razor-blade factory. There were not enough dishes available in shops, and many of the consumers who were vaguely interested were inclined to wait and see whether BSB might be better. Memories of the VHS–Betamax war made customers wary of committing to what might become a losing system.

By mid-1989 only 10,000 dishes had been sold, against a revised forecast of a million for the year. Sky was losing £2 million a week and sucking money out of the rest of News International. At just this juncture, Disney withdrew from a movie deal and disaster loomed.

Murdoch made frequent changes to the management team. None of the chief executives had any television experience, though he did bring in a sprinkling of Australian broadcasters to support them. For a few months Andrew Neil, the editor of the *Sunday Times*, was part-time chairman, but Murdoch interfered frequently himself. Neil found much of his time taken up with the operational problems of getting studios and broadcasting facilities working in what was effectively a building site.

In March 1990, Murdoch took personal command, but the strain was beginning to show – he frequently reminded everyone that it was his money they were spending. Sky's losses were part of a much bigger cash flow problem in the Murdoch empire.

The most conspicuous sign of failure was the slow sales of satellite dishes – all of Sky's efforts had gone into getting schedules together and retailers were as hesitant as consumers about making a commitment to either of the rivals. It seemed to Murdoch that the business's biggest problem was marketing and distribution. Until subscriptions were sold more aggressively, they would never have enough viewers. In desperation he decided to get into the business of selling dishes directly to consumers. This meant competing with retailers, as BSB was quick to point out, and also begged the question of whether Sky had the capabilities to run an operation like this.

It certainly had a talent for improvising, but the sales operation was amateurish and clumsy. By offering customers the chance to rent a dish for £4.99 a week, rather than splash out £300 on buying one, it was only winning them temporarily. If BSB were to come along with a more attractive proposition, those customers could leave in a week, as their switching costs were tiny. Most prospective customers were still holding back.

Recruiting thousands of salespeople in a few weeks meant that their overall quality was dire – many of them came from the unscrupulous end of the double-glazing industry and several of them simply fiddled the numbers to earn commission. However, this initiative did generate 50,000 customers a month, most of whom remained. As dish supply and installation could not keep up with this surge, Sky went into the installation business too – the overall exercise cost £140 million.

Sky did at least have the priceless promotional asset of its newspapers. They helped it win many more subscribers than BSB, but nothing like enough to take it to break-even. The cash Sky was burning contributed to a massive debt crisis at News Corp, Murdoch's international holding company, which owed $8.7 billion. Sky's losses took the British businesses into the red and News Corp's survival was in the balance. Things got so bad that it failed to make a payment on some bank loans in 1990. Murdoch managed to squeeze more money out of his bankers for the core newspaper businesses, but they insisted that none of it be diverted to Sky. He realized he had to find someone special to turn the company around, someone who really knew about television and beating competitors.

The tough guy

Sky's saviour was a short, pugnacious New Zealander, with a striking resemblance to James Cagney, in both physical appearance and tough-guy persona. In Sam Chisholm's case, no acting was required – he was genuinely ferocious and such was his reputation that his mere presence at a meeting could swing negotiations. Unlike everybody else in British television, Chisholm had experience of fierce competition, thrived on it and despised most of the whinging poms he came across. He had spent most of his working life in Australian television, where he had made Kerry Packer's Channel Nine number one. He had a sharp eye for promising programming, and his policy was that if he couldn't buy it or grow it, he would kill it, to make sure that rivals didn't get their hands on it. Like his new boss, Chisholm was ruthless and domineering but could also turn on the charm.

Chisholm arrived at Isleworth in September 1990, never having worked in Britain before, and found an almost completely inexperienced team. He was impressed with what they'd achieved, but felt that most of the energy had been misdirected. 'These people were a brilliant start-up team, but they needed someone to carry the ball the next 100 yards, and to define and direct the energy of the company in a few more sensible areas.' He did not hesitate to give them that direction, and to bully and humiliate anyone who fell short of his expectations. Chisholm's brutal management style almost lost him the best executive he inherited, David Chance, who later became his valued right-hand man and deputy. In this highly effective relationship, Chance was the strategic thinker and master of detail, while Chisholm conveyed a sense of urgency, embattlement and supreme self-confidence.

Chisholm's approach to management was summed up in some much-repeated aphorisms: 'To err is human; to forgive is not my policy. Delegate, then interfere.' It did not suit everybody, and he was quick to punish poor performers, but he certainly motivated the rest and rewarded those who served him well. Unlike British television executives, who sometimes boasted about how little TV they watched themselves, Chisholm devoured it, and no detail of Sky's programming and presentation escaped his attention. A major priority in the early days was cutting costs. He insisted on signing all cheques for more than £10,000 himself, and mostly only did so when a writ was attached to the demand.

By the summer of 1991 disaster was looming for both Sky and BSB. Sky had a mere 750,000 subscribers, but BSB a pathetic 110,000. Both were bleeding money and were in danger of ruining their owners. It was clear to both sets of shareholders that a merger was unavoidable. The terms of the deal that were thrashed out by

September gave Murdoch 50 per cent and management control. Chisholm would remain in command of the merged company, to be called British Sky Broadcasting (BSkyB). It would broadcast solely from Astra; Marco Polo would remain in now pointless geostationary orbit and eventually be sold off, along with the eponymous building. There were howls of protests in Parliament and the non-Murdoch press, but the government declined to intervene. Its relationship with Murdoch was irrelevant – if it had held up the merger both companies would have gone bust.

Removing the uncertainty over which satellite service would prevail solved only one of the problems. The combined company still lacked a customer proposition compelling enough to attract millions of subscribers. It was also so strapped for cash that it was verging on insolvency – no bank would advance any more loans, and some directors were concerned that they could face criminal prosecution for trading with insufficient funds.

Chisholm's immediate priority had to be saving the business from ruin. He got Arthur Andersen to conduct an audit (payment deferred), and they advised him that the combined business was losing £14 million a week. He immediately swung into even more ruthless cost-cutting mode, firing all 580 former BSB staff in the first two days and sixty from the old Sky operation.

BSkyB's biggest burden was the enormous commitments the two rivals had made to the Hollywood studios, amounting to $1.2 billion over five years. Chisholm spent eighty-five days in California trying to renegotiate these deals, while running the company in the time left over. His philosophy on negotiating was simple: 'You think about your own position, that's what you consider all or most of the time. You don't waste a lot of time thinking about the other guy's position. You concentrate on your own business.'

This was the studios' philosophy too. As they had capitalized the minimum commitments as precious assets in their balance sheets, they were not inclined to make any concessions, especially as Murdoch's Fox was a direct competitor. But Chisholm proved even more determined, doggedly sticking to his line that they were in danger of losing everything. 'Look, if we go under the next guy who comes along is going to offer you five per cent of what we're offering. This is a high-risk business, so you'd better go along with us and keep the business alive.' It was a long, hard struggle but eventually Sony's Columbia crumbled, and that broke the resistance of the others. The minimum commitments were cut by 30 per cent, more money than anyone had ever won back from Hollywood.

By the end of 1991, there was light at the end of the tunnel. Sky now had 1.8 million subscribers and losses were down to £1.5 million a week. By April 1992, the business was breaking even operationally, though still making big losses because of interest payments on its £2 billion of debts. What it needed now was an equally miraculous improvement in its appeal to customers.

Premier League

By a happy coincidence, it was at this point that the television rights for English football came into play. The top twenty-two clubs were breaking away from the moribund Football League to form a new Premier League and were determined to get the best possible deal from television rights. The old Football League had been a curious association of 92 rival clubs united mainly by resistance to change, suspicion of outsiders, particularly foreigners, and fear of television. They saw it less as a potential source of new revenue than a threat to attendances at grounds, which had been declining since the 1960s. Most clubs were poverty-stricken and none were run as real businesses, though the bigger ones like Liverpool and Manchester United were aware that their brands could be making them serious money.

David Dein of Arsenal had been the prime mover behind the deal with ITV in 1988. Prior to that the BBC and ITV had operated a cartel for dealing with the Football League, to keep the cost of the rights down. In 1985 they had jointly agreed to pay £16 million over four years and shared the matches between them. When Sky had made its unsuccessful bid in 1988, ITV had broken ranks and won an exclusive four-year contract worth £44 million for an unprecedented eighteen live matches per season, though none of them on sacred Saturday afternoons. This had tripled the Football League's TV revenues, and given the top clubs even more. As the BBC could never afford that kind of money, ITV looked like holding on indefinitely, despite its unexciting coverage. The main point of the top clubs forming the Premier League in 1992 was to win very much more money from television.

The arrival of BSkyB as a credible alternative raised their expectations, but ITV remained in pole position: it was the incumbent, had close relationships with top clubs like Arsenal and Manchester United, and was now willing to go a great deal higher than before.

Dein, and most of the top clubs, were reluctant to put all their eggs into Sky's basket. He suggested either splitting terrestrial and satellite rights and selling them separately, or the Premier League starting its own channel with a transponder on Astra. Failing that, Arsenal, Liverpool and Manchester United favoured sticking with ITV: the audiences on Sky would be much smaller, which would slash the substantial revenues they gained from ads on billboards seen on TV screens. Most clubs, however, were wary of Dein's schemes and resentful of the greater share the big clubs had won for themselves in 1988. They were determined that the crucial decision should be taken by majority voting where the big clubs would always be in a minority.

Sky targeted the other clubs and the chief executive of the Premier League, Rick Parry, whom Murdoch himself joined Chisholm in courting. By this stage they were convinced that securing these rights would be the making of Sky and were willing to bet the business to get them. Chisholm spent hours every day monitoring ITV's discussions with the clubs and planning his next move. Drawing on his experience of televised sport in Australia, Sky thought hard about how football could be presented and promoted more excitingly.

Aware that many club chairmen felt more comfortable with the familiar faces at the BBC and ITV and looked askance at aggressive antipodeans, Murdoch and Chisholm devised a brilliant ploy, a joint bid with the venerable BBC. To the amazement of its top management they proposed the unlikeliest partnership in media history, and an offer too tempting for Auntie to refuse. The BBC had given up all serious hope of bringing back its immensely popular Saturday evening *Match of the Day* programme of recorded highlights, but now this could be done at modest cost. Sky would put up the serious money for live coverage, but the alliance with the BBC would give the bid respectability and allay the misgivings of the clubs.

The key consideration was, of course, money. In April, Dyke agreed to top Sky's offer of £30 million a season and eventually put £235 million over five years on the table. When he realized that Chisholm's talk of pay-per-view was swaying Parry, he decided at the last moment that ITV must raise this even further and bypass Parry. As each club chairman walked into the meeting that would make the crucial decision, he would be handed an envelope containing ITV's final bid of £262 million, dangerously close to Sky's last offer. Given ITV's relationship with the clubs, that ought to have clinched it for them, but Sky's courting of Parry now paid off. He tipped off Chisholm, who convinced Murdoch that they simply had to raise their own bid substantially to blow ITV out. Arthur Andersen and several Sky executives felt that, given their tight finances, the amounts could not be justified, but Murdoch knew that this was a bet he had to make. On the day of the meeting, Chisholm faxed the Premier League his final offer of £302 million. Although most of the top clubs still favoured ITV, a narrow majority voted for Sky. Labour MPs and the non-Murdoch press were indignant at this 'theft' from the public of access to a national treasure.

This was the crucial turning point that secured Sky's future. In the next four months it signed up a million subscribers for its Sports Channel at £5.99 a month, and put itself on course for full profitability the following year. That was only the beginning – Sky would win many more subscribers and get them to pay much more. The new programming, with more lively presentation and more use of different camera angles and slow motion, was an immediate hit.

The deal also transformed the economics of football in Britain. The infusion of unprecedented amounts of money, subsequently to become much greater, helped to make the Premier League the richest in the world, and attract the very best players. The top clubs were on their way to becoming multibillion-pound businesses, and English football became a globally attractive product, with large television audiences in countries from Australia to China. Contrary to the fears of club directors, attendances at games actually rose in the 1990s, thanks to much greater exposure.

New business model

Sky's coup represented a fundamental shift in the balance of power in British broadcasting. It demonstrated that in a bidding war like this, multi-channel pay-TV had an advantage over conventional broadcasters, whose business models relied either on advertising or the licence fee. Sky could provide many more hours of coverage than ITV, which had to cater for other viewers, notably women, on its single channel. Every penny Sky spent on rights could be recouped from viewers willing eventually to pay considerably more than £5.99 a month. Subscriptions were becoming the core of Sky's business model, which was proving be a much more profitable one than most people had ever imagined.

John Birt, the BBC's Director General, was quick to recognize the new reality. 'All of a sudden, something very attractive indeed was created on subscription services, which materially altered the balance between ourselves and the satellite broadcasters.' Sir George Russell, the chairman of the ITC, the successor organization to the IBA, also saw that things had changed fundamentally.

> Once you lose your major sports, and have to go to Parliament to beg for the little bit left, you can see already the long trail of those who argue, 'Why are we paying the licence fee when we don't get the things we want to see?' Everything stems from the moment when the BBC did the Premier League deal with BSkyB.

In the wake of this revolution a distinguished mainstream broadcaster, David Elstein, decided to move to Sky. Elstein had been offered the job of programme director at ITV Network Centre, and his friends were amazed that he would go somewhere as peripheral to British broadcasting:

> But that was just not understanding what Sky was about and not beginning to get a sense of how far Sky had moved between 1990 and 1992.

He felt more and more uncomfortable with the idea that 'terrestrial TV was so good that nothing else could possibly attract the consumer'. Elstein took charge of acquisitions and scheduling, and of reorganizing Sky 1, the general entertainment channel. Although programme making was not yet a strong capability, Sky was now able to attract other talented broadcasters.

For most of Sky's life, it had been assumed that advertising would be its main source of revenue. The only successful pay television channels so far, HBO in the US and Canal Plus in France, had the advantage of competing with mediocre programming on terrestrial channels. Murdoch and Chisholm had come to realize, well before the Premier League deal, that when they had attractive programming customers could not obtain elsewhere, subscription would become by far the more important source of revenue. Capabilities that Sky had started to develop back in the darkest days enabled it to turn subscription management into a powerful strategic asset.

In 1989, Sky had established the skeleton of what became its subscriber management system in Livingston in Scotland. Initially it consisted of twelve people with a few telephone lines, but its manager, David Wheeler, and some innovative technology turned Livingston into Sky's main channel of communication with its customers, and its most powerful marketing arm. Telesales staff at Livingston became adept at persuading customers to upgrade to premium channels.

In 1990, News International came across Adi Shamir, a brilliant Israeli who had invented a sophisticated algorithm for encrypting streams of data. It quickly decided to obtain an exclusive licence to this technology and set up a new subsidiary, News Datacom (NDC), to exploit it. NDC had only one customer, Sky, and provided it with a conditional access system for premium programming. Starting in 1991, the Movie Channel was transmitted in encrypted form – the pictures could only be viewed by subscribers whose set-top boxes contained the right smart cards. Livingston was able to reprogramme smart cards remotely and instantaneously, giving subscribers access to the channels they had paid for, and sell them upgrades to more. By August 1992 two movie channels and sport were encrypted, but that was just the start of something much bigger.

The next stage was to get customers to pay for non-premium programming too, not just Sky's channels but others like MTV, CNN and Nickleodeon that were currently available free on Astra. According to Chance, the architect of the new strategy, the initial impetus for developing a multi-channel package was anxiety that someone else might do it first. TCI and Viacom had tied up multi-channel television in the American cable industry, and Chance feared that they could do something similar in the UK, with their own subscriber management systems and set-top boxes.

If Sky moved quickly, it could not only make its package the standard and head off potential rivals but would make it easier to sell subscriptions to its premium channels while also generating revenues from the basic channels. It then found it could put up prices steadily by introducing new channels which very few people watched, while convincing premium customers that they were getting the basic package free. It launched the multi-channel package in August 1993 at the 'introductory' price of £6.99 a month, but only £2.99 for the first six months. Customers swallowed that without complaint, and in 1994 the price for the basic service jumped to £9.99–£14.99 with one premium service, £19.99 for two and £21.99 for all three. Sports fans who had only become subscribers in 1992 for the football were now paying £9 more than the £5.99 they had signed up for. This transformed Sky's finances. It was finally making profits – £93 million, or 17 per cent of its gross revenues of £550 million – and these showed every sign of going higher still.

With Sky now the standard, it became a wholesaler as well as a broadcaster – both content owners and cable operators were obliged to deal with it. Sky's package became the dominant one in cable as well as satellite and nobody could compete with it without something equally compelling in movies and sport, but Sky had snapped up the most attractive premium content available. The cable companies were paying the penalty for their neglect of programming and marketing. Now that Sky had become the gatekeeper in pay television, it could determine the terms on which it would do business with others. Cable needed the sport and movies that only Sky had, and other channels needed to swim in its slipstream, so all of them had to accept Sky's prices.

Sir George Russell did not see this as a problem, as Sky 'only had 3 per cent of the total television market'. The IBA's (now ITC's) concept of competition had always been simplistic. At least the Office of Fair Trading (OFT), which investigated the pay television market in 1996, could see that it was an entirely new one where Sky had established big barriers to competitive entry. It imposed some mild restrictions on Sky's terms of business with cable companies but did not suggest any capping of the prices it charged its subscribers.

Sky presented its stupendous growth in revenues and profits as due to a similar rise in subscriber numbers, but this was misleading. The number of viewers did double between 1994 and 1999, but most of these new 'subscribers' were actually the customers of cable companies. By 1999 they accounted for slightly more than half, 4 million out of a total of 7.4 million viewers. In the same period, total revenues more than doubled, from £550 million to £1.27 billion, making Sky the most profitable television company in Britain.

It was getting most of its revenues not from the new customers but from the 3.5 million DTH subscribers, whose numbers increased very little during this period. They were the hard core of sports and movies fans with whom Sky had a direct relationship and it extracted nearly £1 billion from them in 1999, an average of £23.50 a month, compared to £13.50 in 1994, and £5.25 each from cable customers.

The skills of the 3,000 people who now worked at Livingston had much to do with this, but the main explanation was that these customers were hopelessly addicted.

The over-mighty subject

Sky's long-suffering owners decided to go for a stock market flotation in 1994. This sold 20 per cent of the shares and raised £867 million, valuing the company at over £4 billion. Murdoch reduced News Corp's stake to 40 per cent, without weakening his control, as the other shareholders remained happy to defer to him. The only shareholder that decided not to cash some chips was Chargeurs. Chisholm greatly appreciated the steadfast support of his French investors over the years and took his team to Paris to thank them publicly. He also wanted the Sky management team to be rewarded with a bonus scheme worth £24 million. The nominal chairman, Frank Barlow of Pearson, strongly opposed this and Murdoch used the bad blood that the row generated to persuade the other shareholders that it was time for Barlow to go. He was replaced by Gerry Robinson of Granada, but Murdoch remained the real power. In 1995, Pearson sold the rest of its stake, which meant that BSkyB now had enough tradable shares to qualify for membership of the FTSE 100. Chargeurs' judgement proved to be better than Pearson's – a year later BSkyB was valued at £12 billion.

As the chief executive of a major public company, Chisholm now had an independent power base and was starting to behave like the boss. Murdoch was merely the largest of his shareholders, and BSkyB's market capitalization actually exceeded that of News Corp. According to some accounts, Murdoch was jealous of his over-mighty subject and the acclaim Chisholm was getting. Murdoch had a history of quarrelling with executives who threatened to eclipse the monarch's majesty – Barry Diller at Fox and Andrew Neil at the *Sunday Times* had also fallen from grace after notable success.

Murdoch himself was endlessly juggling his many business interests, of which Sky was just one. Just as Google in its early days always needed more computers than it could easily lay its hands on, the monarch always needed more money for his ever-expanding empire. In October 1996, News Corp raised another billion dollars and used some of its stake in BSkyB as collateral. This alarmed the stock market and led to a fall of £1.4 billion in the company's value, to the annoyance of his fellow shareholders. Murdoch shrugged this off and subsequently told the *Sunday Times* that, at forty times earnings, he felt the shares were overvalued.

We can only imagine what Chisholm thought, but there is little doubt that he was irked by Murdoch continuing to treat Sky as if he owned it. In 1996, Murdoch tried to tempt David Chance away to run his satellite venture in the US, ASkyB, and parachuted in his ambitious, 27-year-old daughter, Elizabeth, as General Manager of BSkyB. Many commentators and some insiders concluded that she was being groomed to succeed Chisholm. Chance decided for himself, correctly, that ASkyB would not fly, but Elizabeth's arrival upset Chisholm. He told journalists that Chance was his natural successor, and that Elizabeth was not yet ready for a big job. Like Murdoch, Chisholm also sometimes forgot that he did not own BSkyB, but when he spoke of it as 'the success story of the decade', the implication was that its success was largely attributable to him. The fact that Murdoch had not managed to replicate it in the many other parts of the world where he tried may have rankled.

Murdoch knew that Chisholm would be needed to win one last battle, holding on to the company's most valuable asset, the one supplier over whom Sky did not have a stranglehold. Sky's contract with the Premier League would expire in 1997, and the clubs knew that Sky simply had to win a renewal of the lease. They were now very interested in getting a direct share of pay-per-view revenues, which they thought could eventually be worth as much as £2 billion. In the long run they might start their own television channels, but in the meantime they were hoping to double their money.

The only serious challengers to Sky in the auction that took place early in 1996 were United News and Media and Carlton Communications. United offered the clubs a deal worth £1 billion over ten years, but Rick Parry was determined that the League should not to be locked into a long-term contract. Carlton was Sky's closest rival, offering £650 million over five years. However, it did not have transponders on Astra, and redirecting all existing dishes to another satellite was obviously unthinkable. Sky's hold on Astra was an even bigger strategic asset than it then realized. But the rival bids had served Parry's purpose: to make certain of holding on, Sky put £670 million on the table for the next five years. That and Chisholm's personal pitch clinched it.

In June 1997, Chisholm announced that he would bow out at the end of the year, and Chance decided to leave at the same time. Both men had health problems and were looking for a quieter life. Their departure hit the share price, but it soon became apparent that they had laid solid foundations for their successors. Sky now had capabilities in depth, both men stayed on the board to give advice, and Chisholm's successor, Mark Booth, had long experience of pay television in Australia and Japan. In 1999, Murdoch formally became chairman of the board and subsequently made his son, James, chief executive.

Sam Chisholm's main legacy was to have turned Sky from aggressive, amateurish outsider to the well-managed and confident incumbent that has dominated pay television in Britain ever since. It was able to see off subsequent challenges, mainly as a result of capabilities built during the Chisholm–Chance era and of Sky continuing to think like an outsider, never taking its pre-eminence for granted.

Digital dilemmas

The most conspicuous threat Sky faced in 1997 was the 'digital revolution', though quite what this meant was not clear. It had been understood for some time that the answer to spectrum scarcity would be to transmit programmes, both terrestial and satellite, digitally. Sophisticated compression techniques meant that several channels could now be squeezed on to one transponder and that one satellite could now carry hundreds of channels. Saving money on communications was not the issue – digital broadcasting could make it easier for new players to get into pay television. Up to now, Sky's hold on

Astra and its capabilities in subscriber management had been big barriers to competitors.

The great unknown was the impact of digital transmission on terrestrial television. Dozens of new channels might soon be broadcast to millions of homes. As they could be encrypted, it looked as though the barriers to entry in pay television were about to fall. How serious a threat this might represent depended on how quickly consumers acquired the set-top boxes needed to receive digital signals and on whether new players would come up with fresh ideas.

In order to remove the bottleneck on channel capacity and make the allocation of spectrum less contentious, the government had decided that digital conversion was a national priority and warned that analogue transmission would cease by 2012. However, it was not clear who would pay for the set-top boxes, since many consumers showed no interest in having more channels but would be aggrieved if terrestrial channels disappeared, particularly if they still had to pay a licence fee.

Sky started making contingency plans in 1996 and two years later, excluded from terrestrial digital, decided on a pre-emptive strategy. It would launch 200 free channels (mostly variations on existing ones, with different starting times for movies) and entice customers to make the switch to digital by offering them free set-top boxes. In 1998–9 it spent a whopping £450 million on the transition, which briefly took it into the red again. However, this killed three birds with one stone: it ensured that all its subscribers switched over quickly, made Sky the undisputed champion of digital broadcasting, and completely outflanked its first serious challenger. This turned out to be overkill, as the rival was yet another headless chicken.

The old ITV companies were desperate. They had been the biggest losers from Sky's transformation of the television landscape. They had lost football, seemingly for ever, and the BBC and Channel 4, both now with more populist programming, were winning larger shares of the main terrestrial television audience. A wave of mergers had consolidated the industry but had not led to any fresh thinking. The last two ITV companies left standing, Carlton and Granada, formed yet another cartel to attack the digital opportunity. Their strategy was based more on aspiration than real capabilities: not only would they launch a pay television channel, ONdigital, but also an internet television service, ONnetwork, and a pay-per-view platform, ONrequest. Ominously, the new venture took over the space in Marco Polo House vacated by BSB, in whose footsteps they were now faithfully following.

They were offering what they called 'manageable choice' – only 'sad people living in lofts wanted 200 channels'. Their biggest miscalculation, though, was to imagine that they could carve out a new market made up of the fans of the second-tier clubs left in the old Football League. There may have been an audience for this, but it was not a large one nor prepared to pay much. Sky and the BBC were giving the League £16 million a year for some limited coverage, but knew it was not worth much more. Feeling lucky, in 2000, ONdigital bid £315 million over five years for the rights to eighty-eight Football League matches. They were careful to insert in their offer weasel words to the effect that some of this money was conditional on them winning sufficient subscribers. As their proposition to potential subscribers was equally complicated, the take-up was predictably poor. By 2001 what was now called ITV Digital was unable to make even its minimum payments to the Football League. Before the end of the year, it went bust and an undignified court case ensued. Carlton and Granada managed to extricate themselves from further payments, but their reputation in football was mud.

Undisputed champion

No serious competitive threats appeared in the next few years. Weakened cable TV companies like NTL became takeover targets and ITV's share of the television audience plummeted. From 42 per cent in 1991, it dropped to a mere 20 per cent by 2006, when its revenues were half of Sky's.

The other beneficiary of the ITV Digital debacle was the BBC, Sky's ally in the first Premier League battle. It acquired the unused transmission capacity and became the cheerleader for the adoption of

Freeview – a growing band of channels like its own BBC3 and BBC 4 that had previously languished on Sky's package. Cheap set-top boxes attracted more viewers than anyone had predicted, and were in a quarter of British homes by 2005. This persuaded Channel 4 and ITV that they should throw in their lot with Auntie and concentrate on advertising rather than subscription revenues. By the end of 2007 more than 8 million homes were using Freeview.

This was only slightly fewer than Sky's 8.8 million satellite subscribers, who were paying an average of £33 per month for the privilege. Sky had managed to gain 5 million new satellite subscribers since 1999 and to persuade virtually all of them to take a premium package. Total revenues in 2007 were £4.5 billion, on which it made an operating profit of £815 million. It was also earning revenues from its new broadband and mobile telephone services, and from Sky Plus, its version of TiVo, that enables easy recording of programmes.

The main cloud on the horizon was the possible loss of its grip on Premier League rights. The European Commission had forced the Premier League to sell the rights to at least some games to other broadcasters, and in 2006 both its monopoly and Sky's were breached, when the Irish satellite broadcaster Setanta won the rights to cover forty-six matches in 2007.

Sky remains the dominant player in pay television in Britain. Strategic assets like sports rights and Sky's position on Astra have played a major part in its continued success, but so have its capabilities in marketing, presentation, subscriber management and scheduling. It has also been the shrewdest strategist in the industry and arguably the best-managed television business in the world.

Nokia

From lumber to mobiles

In 1865, Fredrik Idestam, a Finnish mining engineer, built a lumber mill on the banks of the Nokiavirta river and called his business after it. When Nokia was acquired by Finnish Rubber Works in 1918, the name stuck and the company diversified into cables and more than a hundred other fields. Until the 1990s, Nokia was best known in the Nordic countries for its toilet paper and Wellington boots and its biggest market was the Soviet Union, with which Finland had long historical ties. Although it had been quoted on the Finnish Stock Exchange since 1915, most of the shares stayed in a few hands and effective control was exercised by two rival banks, to the considerable frustration of senior managers.

In the 1960s the company branched into electronics, telecommunications and radiotelephony, but twenty years later more than 80 per cent of its revenues still came from rubber, paper and cables. Its expansive CEO, Kari Kakkonen, knew that these offered little prospect of growth, and that it was in electronics and telecommunications that the future of the group must lie. He invested heavily in research and development, made many acquisitions, and gave Nokia a more international outlook, but he did not go so far as to get rid of any of the old businesses. With 187 of them, the unwieldy group made Sony look like a paragon of focus.

The small radiotelephony business had by the 1980s moved beyond walkie-talkies for fire brigades and police forces and was supplying Nordic Mobile Telecom (NMT), the first cellular network in Europe, with mobile handsets. These were clunky, utilitarian tools, weighing 10 kg, not fashion items: nobody saw this as the forerunner of an enormous consumer market. Despite steadily growing sales, the business lost money in 1987, along with many others in the group. Kakkonen's biggest and most unfortunate investments had been in TV set production, where Nokia became the third-largest supplier in Europe, but the losses were catastrophic. The bankers wanted him to sell off some of the newer businesses, and in December 1987, deeply depressed and seeing no way through the morass, Kakkonen killed himself.

Kakkonen may have been over-ambitious, but he had been far-sighted: he knew that if Nokia was to conquer new markets, it would have to develop distinctive capabilities in digital technologies. As early as 1979 he had started a research centre which focused on them and on new telephone networks.

David v Goliath

The Goliath of telecoms, not just in the Nordic countries but across Europe, was the Swedish company Ericsson. With its stellar reputation for engineering excellence, it sold networking equipment to nearly all the state-owned telephone companies. In 1956 it had conducted the first experiment with mobile telephony, when the handsets weighed a crippling 40 kg. It also became the main technology supplier to NMT, which was launched in Norway, Sweden and Finland in 1981.

Although not a huge commercial success, NMT broke new ground: because the network was made up of many radio cells, between which callers could roam freely, it allowed much more extensive geographical coverage than any previous network. There were soon many cellular networks in Europe and in the US. However, they all transmitted analogue signals, which required considerable bandwidth, and handsets so bulky and expensive that they were only usable in cars and only affordable by businessmen.

The Nokia mobile buffs saw more clearly than most that digital technology could transform these networks, and hoped that the expertise they had been building would enable them to break into network provision. In 1986 they got their first order for an (analogue) NMT exchange, which broke Ericsson's monopoly, won Nokia considerable prestige, and helped it to sell for handsets in countries like Thailand that had adopted the NMT system. Sales of handsets grew at an annual rate of between 30 and 50 per cent through the 1980s, and Nokia's global market share got close to that of Ericsson and Motorola. Its priority, though, was preparing for the shift to digital.

What exactly a digital network should look like was a matter of heated debate. Technical experts had been arguing about a pan-European 'Groupe Système Mobile' (GSM) in European forums since 1982, and its complexity and opacity earned it the label 'Great Software Muddle'. It eventually became known as the Global System for Mobile, which made it more inviting to Asian countries, and GSM subsequently became the standard in Europe and Asia. Nokia's years of research and development paid off when it played the leading role in defining the new standard. The Finnish government was the first to deregulate telecommunications in 1987, a year later a new operator, Radiolinja Oy, was formed, and in 1989 Nokia was well placed to win an order for the first GSM network.

However, Nokia's entrepreneurial, engineering-led, but slightly chaotic mobile division had enormous difficulty delivering the first network and meeting rapidly rising demand for phones. Nokia's board seriously considered selling the business, before deciding to see whether a new manager could sort it out.

The saviour

Jorma Ollila was not a technologist but a highly analytical finance specialist, who had been Nokia's CFO since 1986. 'What I was told by my superiors was, "Look, you've got six months to make a proposal on whether we sell it or what we do with this business." After four months I said, "No, we're not going to sell this one."' But he had his work cut out organizing production and supply chains and later acknowledged that the 'GSM project was in disarray. There was a lot of disillusionment with the spec and the difficulty of the technology.'

In 1991, however, Radiolinja Oy's network was ready to go live when the Finnish Prime Minister made the first GSM call, using a Nokia phone. By then Nokia had also won orders from Cellnet, Vodafone and Orange in the UK, Sonafon in Denmark and Europolitan in Sweden. It had emerged as the undisputed leader in the new market, having stolen a march on Ericsson and Motorola. That year it sold 800,000 phones.

Although networks were more prestigious, and Nokia's technological edge strategically crucial, Nokia was to make its fortune selling phones in numbers nobody had ever dreamed of. The shift to digital led directly to the development of a massive consumer market: the new technology made possible much smaller, lighter, more portable handsets and lower prices, which made them immediately attractive and affordable for hundreds of thousands of individuals spending their own money.

Nokia's board of directors was delighted, but had other things on its mind. Since Kakkonen's tragic end, the infighting among share-holders and directors had intensified, with the chairman, Mika Tivola, manoeuvring to make his son-in-law the CEO. In 1989 the Soviet Union – Finland's and Nokia's biggest trading partner – had collapsed and two years later its economy imploded. This turned Nokia's crisis into a catastrophe. The Cables and Telecoms businesses were particularly hard hit by the abrupt loss of these markets, but large inventories of toilet paper also had to be written off and, to cap it all, Europe was now in recession, with Finland hit particularly hard. In 1991 every Nokia business other than mobile phones reported sales lower than budget, and even Telecoms with its GSM orders had been badly hit by the loss of Russian markets. The group as a whole made a loss for the first time, and banks would not extend credit. One of them wanted to sell the company to Ericsson.

It was at this low ebb in the company's fortunes that the board decided that the miracle worker at Mobile Phones might be the man to save Nokia. In January 1992, at the age of forty-one, Ollila, virtu-ally unknown outside the company, became group CEO. The next day, the share price fell by 10 per cent. Analysts were expecting him to take the company back to basics and to rein in grandiose expan-sion plans.

The immediate challenge was the state of the company's finances and its credibility with investors, but Ollila never wavered in his conviction that Telecoms and Mobile Phones were growth businesses with outstanding prospects. They would need substantial working capital, which would be difficult to raise externally: the funding would have to come from cash flow. Cost savings could only make a modest contribution – the unavoidable conclusion was that some businesses would have to go. Consumer Electronics, now the largest business in the group, but with losses of FIM303 million in the first half of the year, was the main candidate. If the worst came to the worst, even a core business like Telecoms might have to be sacrificed if it could not become more profitable – it would certainly fetch a good price. However, when Siemens, the German engineering giant, expressed an interest in buying it, Ollila refused without hesitation.

The financial results in the first quarter of 1992 were grim, with an overall loss of FIM 178 million, but fortunately both Telecoms and Mobile Phones comfortably exceeded their sales targets. Two immediate tasks were to establish a better relationship with shareholders and to resolve the interminable boardroom squabbles. As part of the shake-up that brought Ollila in as CEO, several board members, including the controversial chairman, Tivola, left. His replacement, Casimir Ehrnrooth, was no longer a direct representative of his bank, and he and Ollila established an effective working relationship.

Ollila also sought to further internationalize the company's shareholder base. Rather than seeing Nokia as a Finnish conglomerate with some technology add-ons, he wanted US investors to think: 'Focus, global, telecom-oriented, value-added.' In 1993, Nokia raised FIM 1 billion in international capital markets. The following year it was listed on the New York Stock Exchange and raised a further FIM 2.5 billion (€400 million). In the bullish American stock market, it was not difficult to position Nokia as equivalent to a start-up in an exploding new market. These investors were not alarmed by the short-term negative cash flow, particularly as operating profits at Telecoms and Mobile Phones were so good.

With the crisis over and a healthier balance sheet, the company could now concentrate on the long term. At a board meeting held in Hong Kong in May 1994, Ollila presented his radical strategy: Nokia was far too widely diversified, several businesses were unhealthy, and mobile communications required strict focus if Nokia was to realize its potential. It should concentrate almost exclusively on the Mobile Phones and Telecoms divisions and dispose of most of the rest. Ollila expressed it with characteristic understatement: 'It is inappropriate to devote the company's financial and intellectual resources to the development of all current businesses.' He predicted that a restructured company could increase sales twelvefold over the next six years, to FIM 65 billion (€11.9 billion), and profits more than sevenfold to FIM 5 billion.

The board could not quite believe that this was achievable or that Mobile Phones could generate enough cash to finance this expansion, and engaged McKinsey, the consulting firm, to evaluate it. McKinsey concluded that the plan was achievable, and it was largely implemented over the next three years.

What Ollila and his colleagues had realized was that deregulation and new technology were changing the telecommunications industry fundamentally and that Nokia was in an excellent position to benefit from this. Because GSM was much better than analogue systems at accommodating rapid growth, it could soon overtake the old networks, and Nokia had more expertise in digital technology than any other supplier.

Fixed telephony was a mature, slow-moving industry, accustomed to monopoly and cosy relationships between operators and national suppliers. Deregulation gave birth to a new breed of operators like Radiolinja Oy who needed to generate cash flow quickly to pay for their massive investments. It was new telecoms operators, and not just in mobile communications, who would become Nokia's most important customers. They looked to their suppliers for speed of installation and technical advice and Nokia stood out by the innovativeness of its technology and its agility.

The most extraordinary growth, however, took place in the market for mobile phones rather than networks. In 1992 Nokia doubled its sales to 1.6 million units, and they came close to doubling every year throughout the decade. However, even in 1994, most of these customers were early adopters: only in three countries, Finland, Norway and Sweden, was market penetration greater than 10 per cent. By 1998 it had reached 50 per cent in the Nordic countries, 22 per cent in Britain, 26 per cent in the US and 36 per cent in Italy. Three years later, it was more than 70 per cent in all of these countries, apart from the US – in Italy and the UK it was more than 80 per cent. These explosive rates of take-up were unprecedented for any new communications device and in Europe greatly exceeded those for PCs and the Internet. The company that benefited more than any other from this was Nokia. By 1995 its mobile phone production reached 5 million, and 128 million by 2000, when the last of its old businesses was divested.

Meeting that runaway demand was immensely difficult, one of several ways in which Nokia distinguished itself from its competitors, though not without periodic crises. Throughout the 1990s one of the key success factors was, in the understated words of Ala-Pietila, the head of the Mobile Phones division, 'adapting production to unpredictable and rapidly changing demand'.

Nokia had to overhaul its approach to logistics and the supply chain radically on several occasions after Ollila's first shake-up in 1991. When sales really took off in the mid-1990s, there was a major crisis in logistics and delivery. In 1995 sales growth for once failed to meet the forecast, and the company had to issue a profits warning. Pertti Korhonen was named 'The Man Who Saved Nokia' after he introduced Nokia's Integrated Supply Chain. This transformed its ability to deal with peaks and troughs in volume, drastically reduced its overall inventory levels and halved the costs of holding finished products. Although it sub-contracted some of its production, most of it was still done by its own increasingly international workforce.

New industry, new company

Nokia's other most distinctive capabilities were its deep knowledge of the underlying technology and in design and marketing. Its main rivals, Motorola and Ericsson, were focused, like most incumbents, on the technologies, capabilities and processes that had served them well in the past and were less ready for the changes in the 1990s. They were particularly unprepared for the new consumer markets.

In the old world of telecommunications, dominated by male engineers, customers had little or no choice of handset. Like the venture capitalists who told Howard Schultz, when he was seeking funding for Starbucks, that a cup of coffee was simply a commodity, most people in the industry thought that a phone was just a phone. A mobile telephone, however, turned out to be a very different beast, something that was more useful, more versatile and more intimate than the old device in the kitchen or the hall, shared between the whole family. The generic value proposition of mobile telephony was radically different and more compelling. It was meeting a market need that fixed-line telephony could not, and one that turned out to be enormous.

The market structure was also different. It was mainly network operators who sold phones and service as a bundle, but customers increasingly made their purchases in retail outlets, and demanded the phone of their choice. Although the contractual relationship was with the operator, increasingly the emotional attachment was to the maker of the handset. Nokia's innovative products and advertising made it the coolest brand and one to which customers developed real loyalty, few of them knowing that it was Finnish.

Ollila and his colleagues soon realized that this new market was made up of several segments with different tastes and priorities. For many customers the mobile phone became one of their most prized personal possessions, something they carried with them everywhere and that reflected their personality and lifestyle. They did not want just any phone, but the phone that was right for them.

Nokia's approach to design was integrally related to its thinking about customers and their needs. Different products were specifically designed for particular kinds of customer. The segment Nokia calls 'experiencers' are interested in fashion, so the key attributes of a phone for them are appearance and individuality. 'Controllers' on the other hand are most interested in efficiency and personal productivity. (Nokia's other four main segments were impressers, maintainers, balancers and sharers.)

Producing phones whose size and appearance were fine-tuned to different customer segments was basically applying well-established marketing principles to a new industry. What was truly radical about Nokia's approach was extending this thinking to the user interface, the way in which the user communicates with the device. This turned out to be the most important way of making different phones fit different needs and perform different functions. To create an appealing 'expression' phone for balancer customers who also value simplicity, they needed to limit the number of control keys. Technically demanding customers, like impressers, are prepared to work harder to use more sophisticated features.

One of the major constraints in design was the small size of these phones. That limited both the screen size, how information and choices could be presented and how users could understand and interact with the device. These constraints became greater as the devices got smaller and smaller, and more and more functions and features were introduced. The big challenge was how to offer these options without making them impossibly difficult to use.

Nokia's approach was to present functions in a menu that was so well organized that even first-time users could understand it. They did not need any advance knowledge – they merely had to learn how to browse the menu and to select items from it. Since 1997 users have chosen functions using what Nokia calls soft keys, 'dynamic function' keys that perform different functions depending on what the user is doing. In phones with two soft keys, the left one is used basically for saying 'yes', for going forward or deeper into the menu hierarchy, and the right one for saying 'no', going back and cancelling. Other phones use a single 'Navi-key' in conjunction with a Clear key.

This simple, intuitive approach has been stunningly successful. Millions of Nokia phones are now everyday objects, used regularly by unsophisticated customers who cannot operate their VCRs and would find a PC intimidating. Nokia has been careful to 'evolve' these user interface styles very gradually. If customers felt that changes were challenging, they would have failed. The aim, largely achieved, was for customers to see changes as improvements.

Many people contributed to these achievements. Good design had been important in the company for some time, and has deep roots in Nordic culture. Nokia, however, was not hesitant about bringing in other talent and set up its main design centre in Los Angeles. Frank Nuovo, an American, whose previous experience was in car interiors and medical instruments, became an adviser in the early nineties and in 1995 joined as chief designer.

Nokia's design philosophy is effectively part of its approach to marketing – seeing it not just as a way of promoting or selling goods to customers but as a holistic process by which the firm understands the needs of customers and organizes itself to satisfy them. Under Ollila it took branding very seriously. The old firm had sold television sets, electronics equipment and phones under a variety of names but used the Nokia brand for its least glamorous products. Even in 1990 at the Mobile Phones division, Ollila set the goal of making Nokia's brand awareness and image in Europe and Asia stronger than Motorola's. He was not immediately successful, as brands take time to build, but by the middle of the decade Nokia had clearly achieved this goal in Europe.

The brand concept was first expressed as 'Hi-tech with a human touch', later changed to the more succinct, 'Connecting People', which came to be used in its English-language version in most countries. By 1999, Nokia was the best-known mobile brand in Europe, but still behind Motorola and Ericsson in Asia-Pacific and America. However, Nokia's brand was about much more than name recognition. It represented a much closer relationship with its customers, stood for cutting-edge technology, modern styling, ease of use, freedom, personalization and quality. At the end of the century, Interbrand made it the only non-American brand in its list of the ten most valuable in the world.

Global leader

By 1998, Nokia was the world leader in mobile phones with sales of €8.4 billion and a market share of 24 per cent. In 2000, Nokia achieved another cherished goal – it became one and a half times bigger than its nearest rival and reached a market share of 32 per cent. The bold financial targets Ollila had set in 1994 proved to be conservative: group sales in 2001 were almost three times higher than forecast at €31.1 billion and profits six times, at €3.4 billion.

Acts of creation like this are normally only achieved by outsiders, but Nokia was a very mature company. Like Sky, the new business was effectively a start-up. Ollila disposed of its baggage with a lack of sentimentality that Schumpeter would have applauded.

According to Ollila, the years 1992–6 were its 'entrepreneurial growth period', calling for 'value-based management', tolerant of a certain amount of chaos. By the end of the decade, this had given way to what he called 'fact-based management', with a much stronger emphasis on efficiency and professionalism. Economies of scale were now more important than the agility of earlier years. It sought continuous, incremental, measurable improvement in the key foundations of its competitive advantage: products, branding and logistics.

Ollila introduced a new metric during this phase: the ratio of added value to total employee costs. All external purchases are deducted from sales and divided by the total costs of staff – a good way of measuring overall productivity in a networked economy, where so many elements of production are outsourced. Nokia's ratio was around 1.5 in the early 1990s – at the end of 1994 it surged to 2 and climbed steadily to 3.5 by 2000.

In 1992, Ollila described the company's culture in terms that are still largely applicable:

> The Nokia way of operating is based on a flat, decentralized organization, with emphasis on efficient teamwork and entrepreneurial spirit. Flexibility, being able to innovate, react speedily and make fast and effective decisions are the central features of Nokia's way of operating... continuous learning, improving skills and quality, ambitious goals and respect for the individual

– a pretty good definition of disciplined entrepreneurialism.

Seven years later he summed up the reasons for Nokia's success in typically low-key, rational terms:

> If one can manage growth of this magnitude, and the organization operates sharply, that alone facilitates the increase in the margins. It appears that we have succeeded in our operating philosophy in volume production and in organizing logistics. This paves the way for improved productivity. In addition, when things go smoothly employees are happier in their jobs. All that

shows in the operating profit. And when you add successful marketing and strong products, that's it.

That is a fair if rather flat summary, but not a complete explanation. Other factors are also evident: the deep roots of Nokia's capabilities in digital technology and design; the originality and boldness of Ollila's strategic vision; minute concern for the needs and tastes of individual customers; ruthless focus on mobile communications markets and abandonment of its old core businesses; crystal-clear, consistent leadership; highly differentiated strategy and brand; and acute awareness of its broad competitive environment and ability to adapt to change.

Disruptive competition

At the start of Nokia's rise to superstardom it had been a classic disruptive competitor, first to Ericsson, and then even more so to Motorola, the global leader in mobile communications in the 1980s. Motorola's main customers, American network operators, had no great interest in digital technology, so it remained a low priority. Nokia, in an obscure, faraway country, with its bizarre, troubled history, and led by someone with no background in technology, did not look like a serious threat until well into the 1990s. By 1998 it was the clear global leader, and both Motorola and Ericsson were obliged to issue profits warnings and lay off thousands of staff.

Now it was Nokia's turn to be the blindsided incumbent. In that triumphant year, a former sub-contractor in an even more faraway country, Korea, made 4.7 million phones of its own: suddenly Samsung had 2.7 per cent of the global market, but as nearly all of them were sold in Korea not many people noticed. Then Samsung spread its wings and had the bright idea of adding photographic capabilities to phones: total sales rocketed to 21 million in 2001. Several other Asian manufacturers entered the fray, and not just the obvious ones: Japanese consumer electronics giants like Sony, now in alliance with Ericsson, were bringing different capabilities to the party.

It was in 2001 that Ollila declared, with just a hint of smugness, that the greatest danger Nokia faced was complacency. Nokia did not have the kind of binding relationships with its customers, based on technology standards, that industry leaders like Cisco, Intel and Microsoft enjoyed. Network operators became distinctly wary of becoming too reliant on Nokia. The barriers to entry in mobile phone production were much lower now that components and software could be bought off the shelf. Mastery of the network technology was no longer essential to being a credible phone maker, the market was now much more open, and Nokia was no longer the leading innovator.

In the early 2000s, Nokia was slow to recognize consumer demand for new kinds of phones, particularly those with clamshell designs, a trend which Motorola and Sony-Ericsson were quicker to spot. For a while Nokia started to look old-fashioned in a very fashion-conscious marketplace, where young consumers changed their phones as often as their outfits. It found that it did not own its customers – the only way to keep them was to out-innovate its competitors.

Nokia bounced back with a new generation of sophisticated models. Since 2002, it has produced forty-eight models of camera phones, and in 2004 became the biggest manufacturer of digital cameras globally, selling 12 million of them. It moved aggressively into music, making its phones the cheap alternative to the iPod. It also made itself the largest supplier by volume of smart phones, the multi-purpose communications devices that now threaten the PC's primacy. However, Research In Motion's Blackberry still dominated the business market, and Apple's iPhone made Web access much easier for consumers than Nokia's cumbersome Symbian software. In 2008, even Google announced an operating system for phones. Disruption in this market is still going strong.

Nokia clearly cannot take its continued leadership for granted, but has held onto it rather well to date. In 2007, its sales reached €51 million, up 25 per cent on 2006 and 64 per cent on 2001.

Nokia's momentary blindness to changes in its environment in 2001–3 was nothing compared with that of Motorola in the 1990s, but its wobbles are a reminder that superstars can find it as hard as the rest of us to see the big picture. Previous success can actually make it more difficult.

12

MISSING THE BIG PICTURE

We are all capable of believing things we know not to be true.
George Orwell, *In Front of Your Nose*

How to understand it all! How to understand the deceptions she had been thus practising on herself, and living under! The blunders, the blindness of her own head and heart!
Jane Austen, *Emma*

The real voyage of discovery consists not in seeking new landscapes but in having new eyes.
Marcel Proust, *The Captive*

Clear-sightedness is easier for challengers like Sky and Nokia focused on the new market in their mind's eye. Incumbents have to cover an infinitely broader landscape: new opportunities are hazy at best and threats could be lurking anywhere. It is not just new technologies, new competitors and business models that could change everything, but so many things that don't seem to make sense.

This chapter describes some initially bewildering features of an ever more complex business landscape and some attempts to make sense of them. It also considers the greatest obstacle to seeing the big picture – human frailty. We are nothing like as rational as we would like to believe, crave simple explanations for complex phenomena, and hate admitting we were wrong.

First impressions

Our first impressions of a new technology are almost invariably inaccurate. We try to get our minds around it by reference to something familiar. The first motor cars were called horseless carriages. Alexander Bell named his most famous invention the harmonious telegraph – telephones were originally sold in pairs, so that people would have one in the home and another in the office, and use them like walkie-talkies. The idea of networks and exchanges only came later when there were many more phone owners to connect. Scarcely anybody anticipated until well into the twentieth century how ubiquitous the telephone would become as a medium for personal communications. Marcel Proust used to lie in bed with his phone clasped to his ear, listening to live relays of performances from the Paris Opéra, but it never occurred to Thomas Edison that his phonograph invention might play recorded music and be the springboard for a massive new industry – he envisaged that its main function would be for dictating business memos. Fifty years later Harry Warner thought silent movies were as good as cinema could get. 'Who the hell wants to hear actors talk?'

Giuseppe Marconi and other early developers of radio technology never dreamed of the possibilities of broadcasting, let alone mobile phones and wi-fi, but were rightly confident that it would transform telegraphy. The board of the Anglo-American Cable Company knew better. 'The possibilities of wireless telegraphy commercially,' it declared in 1901, 'are so remote that we can look forward to the future without alarm.'

One of the reasons that the computer industry took so long to understand the importance of the PC was that it compared it to the mighty mainframe and the minicomputer – and found it rather puny. Even Gordon Moore of the famous law whose company, Intel, had made the first microprocessor, was initially unimpressed. 'There is no reason,' declared Ken Olsen of DEC, 'for any individual to have a computer in their house.' In fact, very few PCs, other than those that are networked together, are used for the kind of large-scale data processing or number crunching that mainframes performed – they have evolved into something completely different: tools for personal productivity and communications. However, it was years before even word processing emerged as a standard application on PCs, let alone email or web surfing.

When the Open University was launched in 1971 it was almost invariably referred to as the University of the Air. Because television broadcasts were its most publicly visible feature, most people tended to think of it as an adjunct to television, while conventional academics dismissed it as a gimmick. In fact, even in the 1970s, television lectures made up only a tiny proportion of the OU's teaching material – print was a far more important medium and the Internet is now the main one for many subjects. The OU's most important pedagogical innovations – course development by multidisciplinary teams and student support through a network of part-time lecturers – are still not widely understood. This has to some extent worked to the OU's competitive advantage, as other universities offering online courses have tended to overemphasize technology and content rather than learning.

Solutions looking for problems

It is rare that a new technology or medium immediately finds a market – Cisco's Internet router was one of the few to be an instant hit. More often a new technology or the prototype for a new product is a 'solution looking for problems' – its practical applications (and disruptive implications) are not yet clear, and nobody really knows whether it has commercial potential. The phrase is often used dismissively, but that can be foolish. It was first applied to the laser when it appeared in 1960, and then to the two most successful new products of the last quarter-century – the personal computer and the mobile phone.

Before the arrival of VisiCalc, the first spreadsheet program, there was scarcely any generally useful application for the personal computer, other than for technical enthusiasts who were basically buying Apple IIs as toys. Hundreds of thousands of computers, bought by consumers in the early 1980s in a flush of enthusiasm, ended up in the backs of cupboards, never having proved of any real use. Yet the PC went on to become one of the most versatile, ubiquitous and disruptive of all technologies, as it became ever more powerful and affordable, as more and more software applications were developed for it, and as more and more people learned to use them.

The mobile phone evolved equally dramatically, from the expensive, barely portable, brick-like objects that business executives lugged around in the 1980s, to the tiny, stylish emblems of personal identity virtually all of us carry today, and that the young in particular use to take photos, tell the time, play music, pay for small purchases and act as all-round remote control devices for their lives.

The Internet followed a similar path, mutating from its academic, techie, anti-commercial, low-bandwidth origins to something that is now used by so many people and organizations for so many purposes as to be almost impossible to categorize as a medium. It is gradually melting into the background infrastructure that we simply take for granted, just as electricity and telephone networks did in the twentieth century. Yet well into the boom that started in the mid-nineties, scarcely anyone realized what a powerful platform it was becoming for shopping and doing business, let alone advertising and exchanging videos. And business models were mostly mysteries until well into this century.

The solution looking for problems that is best avoided is the elaborate kind designed by strategic planners who think they know what the future holds and what customers will want, or rather what they think customers need. Webvan is a classic example, as were the Prodigy service developed by IBM and Sears Roebuck and Time Warner's Full Service Network. The D-MAC technology that bureaucrats, boffins and politicians saddled British Satellite Broadcasting with ensured its ignominious failure and absorption by Sky.

In 1998 a consortium led by Motorola launched the world's 'first global mobile telephone service', Iridium. Over the previous few years it had spent $5 billion building and launching a constellation of sixty-six low-orbiting satellites which would give telephone coverage to customers just about wherever they went in the world. There was no shortage of investors and the share price soared, as banks like Merrill Lynch and Kleinwort Benson joined Iridium in predicting that global services should win 2.5 per cent of the total mobile market. Iridium's management, which had conducted extensive market research, forecast between 32 million and 42 million customers by 2007.

The obstacles were formidable. The telephone handsets were large and heavy, cost $3,000 each, and unfortunately did not work inside buildings or in many parts of urban areas, as satellites require a direct line of sight to receivers. Calls were astronomically expensive – as much as $7 a minute. Meanwhile, the GSM standard was making roaming with ordinary mobile phones easy, and relatively affordable, in large parts of Asia and Europe. An advertising campaign costing $180 million was expected to bring in 500,000 customers by March 1998 but only managed 10,000. A crisis of confidence among investors and customers quickly spiralled and in July, Iridium filed for bankruptcy.

Compared to some of the other investments being made in 1998, Iridium did not seem particularly foolish. The warning sign, however, was in that '2.5 per cent of the total market'. The line of reasoning that all we have to do is win a tiny share of an enormous market is a close relative to that of the Greater Fool, and equally specious. Iridium's proposition was feeble, its prices ludicrous, and even the technology, for all its innovative brilliance, not fit for purpose. This was a wild gamble that made enormous, unknowable assumptions about demand and the costs of alternative networks. Yet hundreds of experienced investors and businessmen were convinced that this was going to be a glorious winner.

The only way to find out if a radically new product or service solves any serious problems or meets a real market need is by a process of discovery and trial and error, modifying initial hypotheses in the light of fresh evidence. Eric Beinhocker has described it in *The Origin of Wealth*, his masterly account of how our economy has evolved, as 'deductive tinkering'. People play around with new technologies serendipitously, but apply deductive logic to shape them into useful products.

Most new technologies fail to find a substantial market – or indeed any market at all. Even those that justify interest and investment mostly fall by the wayside, because the need they might address is not sufficiently great for a compelling customer proposition to emerge. They are nice, rather than need-to-have.

The CD-i (Compact Disc Interactive) player was launched by Philips in 1991 as the first multimedia device for a mass market. It was immensely versatile – according to Philips visionaries, it could be used for playing games and karaoke, for watching short movies, but also for training and sales presentations. If anything it was too versatile – most consumers could not see what it would do for them that would justify spending $700. It also found itself competing with the ever-evolving PC. Even though the PC in the early 1990s had far fewer multimedia capabilities, it could do more immediately useful things, like word processing and spreadsheets. The problem was that the CD-i was a smorgasbord of impressive technical features rather than a product that meant much to ordinary people. Sales were modest and soon tailed off. Most of its features were eventually replicated by the CD-ROM, the DVD and the Net. This failure cost Philips over a billion dollars, and so crippled it financially that it was no longer able to compete effectively with the Japanese majors in consumer electronics.

Another technology that never realized its digital destiny was videoconferencing, despite being loudly hyped in the 1990s by specialists like Picturetel, telecom operators eager to find bandwidth-hungry applications, and companies like Intel looking for reasons for users to upgrade to more powerful PCs. The trouble with video-conferencing and video-telephony was that they were only valuable in certain very limited circumstances and not really adequate substitutes for most face-to-face meetings, with their much richer exchanges of subtle signals and emotions, and often added little value to a simple telephone conversation or to collaborative software tools: the proposition was simply not strong enough to convert most customers. Even now that webcams and Skype make video-telephony virtually free, few people use it regularly.

E-learning was another new industry that tried to reduce complex human interactions to a semi-automated process. Learning is about much more than presenting information to passive recipients, as some technology suppliers seemed to think. Most people only learn successfully when they want to do so and when they are actively involved themselves. They generally do it much more effectively when they learn with others, which is why classrooms still have their uses. Some e-learning projects, which took account of these principles, notably at Cisco and IBM, have been outstandingly successful, but they are the exceptions. Most businesses who bought 'e-learning solutions' did so to save money, but most employees assigned to take e-learning courses were so bored they never completed them.

Even after this was well established, the consensus in Silicon Valley and much of Wall Street was still that e-learning would soon replace conventional training. Few people thought that John Chambers was exaggerating when he said, 'The volume of Internet traffic generated by e-learning will make email look like a rounding error.' And Cisco was not even a (direct) supplier.

In all these cases, the customer proposition was almost entirely concerned with efficiency, of a rather reductive kind, with little attention to the needs of real users. This highlights two of the most challenging questions for market pioneers: who exactly is the customer, and who pays? We have come a long way from the innocent days when a supplier sold a product to a customer who paid money for it. Frequently the most important 'customers' do not pay anything at all, which can make profitability problematical.

Something for nothing

Most Internet users pay nothing for the services they use, other than a flat fee for broadband access and anything they purchase online. The most popular application from the earliest days has been email, and that has almost always been free or bundled in with the price of Internet access, Blackberry being a notable exception – its customers, or their employers, pay handsomely to feed their addiction. Even low-speed dial-up access has been widely offered free since 2000, in return for expensive customer service. Search, thanks to buoyant revenues from unobtrusive advertising, enables users to delve into the enormous treasure trove of the Web to their hearts' content, without the comfort of expensive walled gardens.

And almost all of that content is free too – from newspapers and magazines, which put much of their print editions online, to archived radio programmes, music videos and thousands of Internet TV channels. Some believe that this is because 'information wants to be free' – but that is to conflate different meanings of the word.

The reasons why so many information products are now free are fundamentally technological and economic: most forms of digital information are now easy to copy, the marginal cost of transmitting a web page (or an email) is zero, competition between suppliers is intense, and barriers to entry minimal for new suppliers like bloggers. Information providers who try to charge quickly find their audiences evaporating. Only those with uniquely valuable content and exceptionally strong brands like the *Wall Street Journal* and the *Financial Times* succeed.

'Free' business models are not confined to the Net. In other industries where marginal costs are low and competition intense, we can see similar patterns. No-frills airlines sell flights for pennies; many newspapers are now distributed free or only charge a nominal cover price; telephone calls are bundled with rental; much software is now funded by advertising.

This is very different from the economy of thirty years ago, when most economic activity consisted of manufacturing and labour-intensive services, and variable costs, mainly manpower and materials, accounted for the lion's share of business expenditure. What determined profitability was the margin between the cost of making each good and the price at which it could be sold. In large parts of the new economy, however, the important costs are fixed or sunk, and variable costs negligible.

If the market is highly competitive, the prevailing price is likely to be close to the marginal cost, which is quite often zero. This makes profitable business models elusive: eBay has pulled it off through overwhelming network effects and virtually no marginal costs; Ryanair through maximizing seat occupancy, having the lowest operating costs and high margins on extras like food; Google through selling advertising to a previously under-served market at a negligible cost of sales. They are among the happy few – Skype is more typical in not yet having succeeded.

Very few content owners have satisfactory business models for the Web – only rarely do they enjoy network effects. They must either find ways of 'versioning' their content, so that those for whom it has high value pay a premium, normally by subscription, or they survive on what is often only a trickle of advertising. The free version is often good enough for most users. Publishers and broadcasters have to rely on the Web acting as a shop window for the 'real' product, but margins are meagre. Entertainment content like BSkyB's, however, can be sold at a substantial premium, even when consumers were previously accustomed to free television. It all depends on the power of the proposition.

Free extracts from books and records generally stimulate purchases of the physical product, but liberating entire works can have the opposite effect. For some libertarians, it is an article of faith that information should always be free and that musicians, writers and software developers should earn their livings indirectly through the boost to their brands. That may profit a fortunate few, but fame is no guarantee of fortune, as Netscape learned.

The value of information and what its recipients will pay for it depends mainly on what they will do with it. If it is merely interesting, like most news, they will pay very little, since they can get it free elsewhere. If on the other hand information is critical to business decisions, some customers will pay a great deal for it. Consequently the most valuable markets for information are narrow ones for specialized financial services like Reuters and Bloomberg. Most content owners on the Net find it difficult to make money from information as such, unless it is bundled with something else, whereas new services like MySpace, YouTube and Wikipedia have flourished through 'user-generated content'.

Web 2.0?

Some have hailed these as part of a new paradigm, Web 2.0, an elastic abstraction that encompasses many voluntary 'collaborative' activities. However, several quite distinct phenomena are being bundled together here, and all go back to before even Web 1.0.

One is the explosion of blogs, which encourage open dialogue with their readers and fellow bloggers, and are in a direct line of descent from Usenet groups and chat rooms, that predate the Internet. Blogs represent part of the Web's threat to newspapers and may carry some advertising, but typically they are written for self-expression, not profit.

Another is a new set of businesses enabling social networking, mainly by young people. MySpace and Facebook are following in the footsteps of AOL and Friends Reunited, seeking to build businesses out of connecting communities and selling advertising to them. MySpace was sold to News Corp in 2005 for $585 million, Google paid $1.65 billion for YouTube in 2006, and Microsoft a desperate $240 million a year later to acquire 1.6 per cent of Facebook. These currently immensely popular sites are intriguing social epidemics but as yet unproven business models and are less about collaboration than self-promotion. Older networking sites have lost many of their adherents, and the same could happen to these.

All these are quite distinct from serious collaborative work conducted by 'self-organized', often altruistic groups, defying conventional thinking about the need for financial incentive, hierarchical organization and the efficient division of labour.

Open source software was born in Bell Labs in the 1960s along with Unix, 'an operating system around which a fellowship could develop'. A culture of close collaboration and sharing of software enhancements emerged which went way beyond the fixing of bugs. Fellows were motivated more by the desire to do interesting work and to impress each other than any notion of business efficiency. There was no concept of intellectual property – software became a common good, like water or air, and its producers craftsmen, constructing the virtual cathedrals of the modern age.

This philosophy was anathema to the man who was to become the most powerful figure in the software world. In 1976, when Microsoft was a tiny company, Bill Gates wrote an open letter to the Homebrew Club where Wozniak and Jobs used to hang out: 'As the majority of hobbyists must be aware, most of you steal your software.' Microsoft decided to withhold its source code, distributing its software strictly as a product, for use by single users, who were prohibited from copying it, outraging the 'open' camp, alarmed also by the PC's eclipsing of Unix.

Unix, however, was chosen as the platform for TCP/IP, the new communications protocol for the Arpanet. TCP/IP made it possible for any computer to connect to the network and helped Arpanet to morph into the Internet. As Unix became the core operating system of most host computers, it seemed as though its hour had finally come.

But there were three clouds on the horizon: there was no easily affordable way for freelance craftsmen to use Unix – the mini-computers and workstations that it ran on were still expensive; the adoption of Unix by computer manufacturers was leading to fragmentation of the supposedly 'open' world, as each supplier sought to make its mark on its machines; and darkest of all was the ever-greater dominance of Microsoft.

The man who dispelled these clouds and put free software on the map was a graduate student in Helsinki, Linus Torvalds, who developed his own Unix-style operating system for his PC. In 1991 he released the source code, which he called Linux, on an Internet newsgroup, and persuaded a hundred like-minded hackers to join him. The rapid penetration of the Internet among computer buffs was making mass collaboration easy, and Linux mushroomed.

By 1994 it was clear that Linux was a viable alternative to Windows for running servers on PC networks. Much to Torvalds's surprise, it was immediately embraced by DEC, which put it on its prestigious new Alpha machine, the first of many unexpected endorsements from the corporate world. Other suppliers anxious to curtail Microsoft's hegemony soon followed. The most enthusiastic, ironically, was IBM, which made Linux its favoured operating system for e-business. A growing number of business customers decided that Linux was good enough for them, and, of course, free.

The most indignant at Linux's success has been Microsoft, calling open source in general an 'intellectual property destroyer' and a 'cancer'. In arguments reminiscent of the record industry in the 1980s, faced with the disruptive CD, it even suggested that by reducing the incentive for commercial firms to innovate, open source would cause the industry to stagnate and collapse. The rise of Google, itself an enthusiastic user of Linux and provider of free applications software, gives the lie to this. However, Google also shows that open source is far from being the predominant model – its own code and search algorithms are closely guarded secrets. As Eric Raymond has pointed out, the big breakthroughs invariably start with 'one good idea in one person's head'.

The wisdom of crowds

Craigslist is one of several examples of how the open source philosophy has spread beyond the realm of software. The university world has always believed in freely sharing information and ideas, but it was stunned by MIT's announcement in 2001 that it would make available online the materials used in the teaching of all its courses. Rather than squandering its intellectual property, it felt that Open Courseware would be a showcase for its teaching. By 2007 notes were available on over 1,800 courses, and the initiative had been imitated by several others, notably the Open University in the UK, making publicly accessible a rich selection of its renowned teaching materials.

The largest collaborative project on the Web, and one of the five most popular sites, has been Wikipedia, the free encyclopedia. It is non-profit making, depends entirely on the unpaid efforts of 75,000 anonymous contributors and on financial contributions from donors, the principal one being its founder, Jimmy Wales.

Wales was a former options trader who had made enough money to devote himself to higher things. His first attempt at an online encyclopedia, Nupedia, depended on extensive peer review and made painfully slow progress. When he discovered wikis – software that allows any visitor to add and edit content easily – he realized that a different approach was possible. Starting in 2001, anyone who wanted to could write or contribute to a Wikipedia entry.

Within twenty days, Wikipedia had 600 articles; after six months, 6,000; by the end of its first year, there were 20,000. These were, of course, of distinctly variable quality, but they kept on getting better, as contributors endlessly modified each other's efforts, and they grew at similar rates to the Web itself. By 2005, Wikipedia, with 500,000 articles in English, was twelve times bigger than *Encylopædia Britannica* with 40,000. By 2007 it had 10 million articles, in 250 languages.

Only the most fervent Wikipedians would argue that their articles are as good as *Britannica*'s, but in terms of factual accuracy they do not fall far short. Mischievous errors are corrected in a few minutes, thanks to the army of Wikipedia guardians who watch over their cherished subjects with an eagle eye. Wikipedia is certainly more up to date, particularly on the minutiae of technology and popular culture, which attract hordes of contributors and enormous entries.

This extraordinary commonwealth has so far resolutely refused to accept advertising, though an AdSense deal with Google would make it immensely profitable. Changing the model, however, might dim the ardour of the faithful.

Wikipedia and open source software are outstanding examples not just of a new kind of altruism, but of what James Surowiecki calls *The*

Wisdom of Crowds. Drawing on the ideas of Condorcet, the eighteenth-century philosopher, he marshals a wealth of examples to show how, in certain circumstances, the judgements of groups of people can be sounder than those of individual experts – and always have been.

In 1906, Francis Galton, a pioneer of the science of statistics, analysed 787 votes in a competition to guess the weight of an ox, expecting that most guesses would be wide of the mark. Nobody guessed the actual weight, but to Galton's amazement, the average vote was only out by one pound. So long as a crowd is sufficiently large and diverse, and its judgements are arrived at independently, mistakes tend to cancel each other out and they arrive at pretty good answers. Homogenous groups that deliberate together, on the other hand, are likely to reinforce each other's prejudices – the problem with committees and focus groups.

Mass collaboration is good at correcting errors and works particularly well on Wikipedia and in software development, where eliminating errors and bugs is normally difficult and time-consuming. As Eric Raymond has put it, 'Given enough eyeballs, all bugs are shallow.' Markets that set prices through the uncoordinated actions and interactions of many buyers and sellers and Google's search methods, based on the preferences of millions of Internet users, are further examples of the wisdom of crowds.

Surowiecki sensibly does not construct a grand theory around this shrewd insight. Commentators like Don Tapscott and Anthony Williams are less inhibited: in *Wikinomics: How Mass Collaboration Changes Everything* they claim that peer production is becoming the predominant model in the new economy and suggest that community, collaboration and self-organization are supplanting systems based on hierarchy and control. Millions of people, they claim, are now working together, using 'weapons of mass collaboration', and they hail the 'birth of a new era, perhaps even a golden one, on a par with the Italian Renaissance... A new economic democracy is emerging in which we will all have a lead role.'

The rhetoric is characteristic of the hyperbole that so often greets

the apparently revolutionary. Different forms of networked collaboration were becoming important in business long before the Internet
became a mass medium. The first successful 'groupware' product,
Lotus Notes, caused such a stir that IBM spent $3.5 billion in 1995
to buy the company. While Tapscott and Williams describe some
important social trends, they present no evidence to suggest that
'Wikinomics' will become universal. Most people still prefer to be
paid for their work, however passionately devoted to it they may be,
and self-organization, which is also not entirely new, is likely to
continue to complement formal structures rather than supplant
them.

Hyperbole had its heyday in the closing years of the last century.
Highly intelligent people, not just in Silicon Valley and Wall Street,
became convinced that not only was the world being transformed by
technological miracles but that *everything* was changing, including
the laws of economics and gravity. It became clear that not all crowds
were wise, and that some were distinctly foolish.

Irrational exuberance

The excitement was not just about the Internet. In fact, most
prophets of a digital future scarcely noticed it until about 1995,
despite its exponential growth over the previous ten years. Their eyes
were firmly fixed on two vague and largely imaginary concepts, interactive television and the Information Superhighway. The latter, they
believed, would beam interactivity and all kinds of other exciting
multimedia experiences into people's homes. Louis Rossetto founded
Wired in 1993, with backing from Nicholas Negroponte of MIT's
Media Lab, to chronicle a digital revolution that was creating 'social
change so profound that the only parallel is with the discovery of fire'.
Yet the first editions of this magazine whose pages vibrated with
excitement made virtually no mention of the versatile medium growing like crazy under their noses, almost entirely unplanned. They
soon made up for it.

Three things set in train the explosion of interest in the Net:
Netscape's IPO in 1995, the widespread belief that Internet traffic

was growing at an annual rate of 1,000 per cent, and the publication of *The Internet Report*, by Morgan Stanley in 1996. All three proved somewhat misleading.

Netscape's business model, as we have seen, was flimsy and the company itself quickly marginalized. Internet hyper-growth was actually starting to slow down by 1996 – though it continued to more or less double each year for most of the decade, but the myth of hyper-growth was so powerful that telecoms companies borrowed heavily and invested billions in building vastly more infrastructure than was then needed. When boom turned to bust this meant bankruptcy for many of them.

The *Internet Report* was the first serious attempt to assess the business potential of the Net. It made its author, Mary Meeker, with her ever more bullish pronouncements, one of the seers of the new economy, and not always a wise one. She argued that the Netscape IPO was a 'world-changing event', that a new industrial revolution was underway, and that different valuation models were needed for Internet companies:

> For now, it's important for companies to nab customers and keep improving product offerings: mind share and market share will be crucial... If a company can build subscribers in a small but rapidly growing market with a compelling economic model and maintain that market share when the market gets bigger, it should reap good profit margins.

This was undoubtedly true, though it did not prove to be the case for most of the companies Morgan Stanley was promoting. And the proviso about economic models was soon forgotten in the excitement of astronomically rising share prices. The preferred criteria for valuing companies became mind share and market share. If traffic on a website was rising, then the company must be worth more. The eagerness of investors to buy these stocks seemed to confirm this logic – share prices just kept on rising and sceptical advisers lost credibility. Investment bankers and venture capitalists, desperate not to be left behind, queued up to throw money at almost any business venture involving the Internet.

Magazines like *Business 2.0*, stuffed with advertisements from hi-tech companies, endlessly puffed the merits of Internet businesses and proclaimed themselves 'the oracle of the New Economy'. Kevin Kelly, *Wired's* executive editor, urged readers to forget everything they previously thought they knew. His 'New Rules for the New Economy' included such gems as 'embrace dumb power, more gives more, follow the free, let go at the top, don't solve problems'. Thomas Wurster and Philip Evans of the Boston Consulting Group argued that incumbent businesses were in trouble 'precisely because they are incumbents'. Legacy assets like sales and distribution systems, bricks and mortar, brands and core competences would be an unbearable burden.

It was this kind of thinking that led Time Warner to acquiesce in its acquisition by AOL, the crowning folly of the era. Feeling that it needed a 'digital construction', its management gave away billions of dollars of shareholder value. Almost as bizarre were the valuations of completely unproven companies with preposterous business models. In February 1999, Priceline.com, which described itself as a 'demand collection business' where buyers could name their price, came to the stock market, with virtually no assets and very few customers. It had been trading for less than a year, during which it had lost $114 million, selling $35 million of airline tickets for less than it had paid airlines. Meeker publicly endorsed this get-big-fast strategy. 'This is a time to be rationally reckless. It is a time to build a brand. It really is a land-grab time.'

Her strategic vision and reputation were the decisive factors in Morgan Stanley winning the job of managing the IPO. This valued Priceline at $10 billion, more than the combined value of United Airlines, Continental Airlines and Northwest Airlines. A few weeks later NASDAQ valued it at $21 billion, more than the market capitalization of the entire US airline industry. Between 1995 and 2000 the value of the NASDAQ index rose fivefold, almost twice as much as in the run-up to the 1929 crash. As David Simons put it, each IPO was 'red meat to mad dogs'.

Meeker, to her credit, was starting to get concerned about wildly inflated valuations and looked for a 'correction', but respected technology pundits like the economist George Gilder saw nothing wrong at all. In 1999, he declared:

> I don't think Internet valuations are crazy. I think they represent a fundamental embrace of huge opportunities. Virtually all forecasts estimate something like a thousandfold increase in Internet traffic over the next five years… In ten years, at this rate there would be a millionfold increase.

Several commentators insisted that there was no fundamental reason why share prices could not go on rising indefinitely. 'Reduced to irrelevance are all the conceptual foundations of the computer age,' exulted Gilder. 'A new economy is emerging, based on a new sphere of cornucopian radiance… leaving only the promethean light.'

A year later, the stock market slumped, darkness fell, and thousands of businesses like Priceline were washed away. The AOL part of AOL–Time Warner was soon revealed to be worth very little. The lesson many people drew from the boom and bust was that the Internet was mainly hype and could be safely ignored, which was as foolish as the previous euphoria. The Internet is the most important new communications medium since the invention of television, and its disruptive impact has already been much greater.

Notwithstanding all the 'paradigm shifts', a much-used phrase of the time, the laws of economics had not been repealed: supply still had to be aligned with demand; businesses had eventually to satisfy customers and make a profit; giving products away free was only a sustainable long-term strategy if it helped to generate real revenues; and old-economy companies, with their much-derided 'legacy assets', not to mention their capabilities and customer relationships, were not entirely washed up.

Blown to bits

The Wurster–Evans thesis on the pitiful plight of old-economy businesses bears closer examination. *Blown to Bits* was one of the most plausible and sophisticated interpretations of new paradigms and was greeted with acclaim when it appeared in 2000. *The Economist* felt that 'as an analysis of the impact of the communications revolution on the corporate world, this book is hard to better.' It does indeed contain many perceptive observations, but these are obscured by sweeping assertions and convoluted abstractions. Evans and Wurster argue that separating 'the economics of information from the economics of things' represents a fundamental discontinuity for most industries, since 'most industries are information industries'. 'Information is the glue that holds value chains, supply chains, consumer franchises and organizations together. That glue is melting.'

Information is, of course, only one of the things that bind organizations together and rarely the most important, but according to them it 'accounts for the preponderance of competitive advantage and therefore profitability'. They produced few arguments to support this extraordinary claim but cited as evidence the role of brands in the success of Coca-Cola and Nike, the importance of American Airlines' reservations system, SABRE, and Wal-Mart's use of EDI. Brands are, of course, only peripherally about information and it makes little sense to conflate them with things like IT systems which most companies now have.

Every business, say Evans and Wurster, 'is a compromise between the economics of information and the economics of things. Separating them breaks their mutual compromise and potentially releases enormous value.' They never exactly spell out what these painful compromises consist of or how the value will be released. The main example they use is retailing and the trade-off between display and inventory. Amazon, they suggest, has bypassed this by offering unlimited display without having to hold much in the way of stocks. But, as we have seen, Amazon is not quite as revolutionary as it first appeared. It has not avoided having warehouses or substantial distribution costs, nor has it so far 'released' enormous 'suppressed economic value'. It only showed its first modest profit in 2003, a mere 0.6 per cent of its sales. Amazon is an outstanding success story but, contrary to Evans and Wurster's predictions, it has not yet destroyed the business models of the better conventional bookshops, nor did it ever expect to.

Their most dangerous argument, because it contains a grain of truth and was widely shared during the dot.com era, was that 'Incumbents face massive competitive disadvantage arising precisely because they are incumbents.' This is only true if incumbents face a genuinely disruptive challenge – when the challenger has developed capabilities and propositions they cannot match. Incumbents have advantages over the vast majority of challengers – long-established customer relationships, brands, deeper pockets, economies of scale and scope, experience and know-how. Contrary to what *Blown to Bits* might lead one to expect, these proved decisive in most of the battles between dot.com challengers and conventional retailers. Sears and JC Penney in the US and Tesco in the UK were quick to adopt online methods alongside their traditional stores, and most large retailers now combine online sales with physical shops.

Evans and Wurster make many useful observations, notably that 'deconstruction' can attack a small part of an incumbent's value chain, such as newspapers' classified advertising, and that little can do a lot of damage. However, several are distinctly misleading: 'The vulnerability of a business is proportional to the extent of its embedded compromises: between different activities tied together by information flows, and between its economics of information and its economics of things.' It would be more helpful, and more comprehensible, to say that the most vulnerable businesses are those who assume that they are competitively impregnable and who take no account of how the world is changing.

Airy abstractions

Blown to Bits is a warning of the perils of wallowing in abstractions and metaphors, and not thinking too much about what they really mean. Metaphors are never literally true – they invariably simplify and when taken too far break down. Abstractions like digitization, convergence, content and community are shorthand for many different things – all too often the distinctions and nuances are lost.

Digitization, the conversion of all forms of information into easily transmittable bits and bytes, is not intrinsically disruptive, but, thanks to Nicholas Negroponte, 'Being Digital' became a mantra for a generation. A typical Negroponte pronouncement was his advice to the television industry to 'stop thinking about television as television. TV benefits most from thinking of it in terms of bits.' In other words, forget about the wood, concentrate on the trees. Applying 'digital' indiscriminately to companies, markets and industries only blurs the picture. A 'digital market' or a 'digital strategy'' is as meaningless as the 'digital construction' Gerald Levin burbled about at Time Warner. Vacuous expressions like these lead easily into believing that such markets do not require customer propositions or even paying customers at all.

'Content is King' has been a mantra not just of the media industry, but of naive technologists since the early 1990s, but begs the questions: what content, and for which audiences? Entertainment content has little in common with hard information. A tiny number of titles, like a Harry Potter book or a James Bond film, belong to immensely valuable brands, but these are not remotely typical. More than 90 per cent of books, films and records fail to show a profit at all. The exceptions are often more about merchandising than content as such. The 'product', and certainly the brand, is the celebrity 'author' who in many cases has played little part in producing the 'content'.

The most persistent abstraction of the digital age has been 'convergence', which implies not just that products, industries and technologies are starting to overlap, but that they are melting into one gigantic, digitized whole. As there is clearly considerable convergence between some sectors, it is not difficult to persuade the impressionable that it is an irresistible force, or that more of it must be a good thing. When used to justify momentous moves like mergers and acquisitions this can be disastrous.

The really important dividing lines between companies and industries are based on capabilities – developing entirely new ones is something few businesses accomplish. Sony is a shining exception, but the fact that much of its business in the 1980s depended on music and movies did not mean that it was converging with the entertainment industry, or that it had the capabilities to be effective in it.

The IT and telecommunications industries have some technologies in common, but the capabilities are fundamentally different, as indeed they are between different sectors within those industries. Attempts, like those of AT&T, to straddle such differences have generally been calamitous. While it was comparatively easy for BSkyB to add Internet access and telephony to its television service and for Nokia to make phones that took pictures, this was the equivalent of supermarkets moving into selling clothes and books, rather than a radical transformation. We need to be very clear about what exactly is converging and what is not.

Abstractions like these are seductive because they appear to reduce complexity, but oversimplifying only makes the big picture more obscure.

None so blind

Some psychologists believe that evolution has programmed us to react quickly to sudden events, to reach clear-cut conclusions and stick to them. This ability to think fast enabled our hunter-gatherer ancestors to recognize danger, or to spot their next meal, in the blink of an eye. Careful reflection was probably not a useful attribute – wondering whether their first impression was correct could have had fatal consequences. Intuitive thinking and decisiveness are particularly valuable for managers, who have to be able to size up a new situation quickly and take prompt action, but such snapshot views almost always oversimplify and if never reconsidered are dangerous.

Most of us do less logical, deductive thinking than we would like to believe. We have inherited from our distant ancestors a craving for instant explanations that seem to make sense of new experiences and we quickly construct these 'narratives' or eagerly swallow those of experts spinning plausible stories. When fresh evidence appears that does not fit with the explanation we filter it out, but welcome information that confirms our preconceptions. Everybody does this to some extent – presidents and prime ministers find it virtually impossible to admit that what they proclaimed as certainties yesterday they now know to be untrue.

Executives of all kinds are particularly susceptible, because they have to take decisions quickly, and because previous success reinforces their self-confidence and dislike of criticism. The more successful they have been in the past, the greater the danger. They get out of the habit, if they ever had it, of acknowledging their errors. If those around them never disagree with them or ask a challenging question, they become insulated from reality. All the experience of John Akers and other senior IBM executives confirmed the story they all told themselves in the late 1980s: the mainframe is the heart of the computer industry, the PC is a sideshow, IBM is unassailable. It took years of terrible blows to make them realize that perhaps none of these things was true any more.

Like the frog that stays in water that is slowly brought to the boil till it perishes, few organizations, or people, are good at picking up gradual, subtle alterations in their environments – that their standards have stagnated or slipped while customers' expectations have risen or that competitors have caught up with them. Chronic complacency produces a slow decline in competitive advantage that nobody really notices for years. Suddenly companies find themselves competing on a horribly level playing field. The market may not be literally commoditized – competing solely on price – but there may be little real differentiation left.

The process of creeping commoditization can be masked in markets where customers have little choice, are poorly informed, or are spending someone else's money. Hamel and Prahalad dubbed the mediocrity and homogeneity of these markets a 'lack of genetic diversity'. Customers pay the going rate because they are not aware of, or have little incentive to seek, alternatives. A radically different proposition can have a devastating effect, as it exposes just how little value was really being provided.

Airlines prior to deregulation were a classic example – all offering essentially the same service at much the same price. It was unthinkable that competitors without their networks and routes could have an advantage. Disruptive competitors like Southwest and Ryanair showed that what most customers valued was simply reliable transport at low prices. Likewise retail banks and telephone companies took years to acknowledge that their markets were becoming commoditized.

Most businesses are reluctant to ask themselves how much value they are really providing and why customers should remain loyal, and even more reluctant to listen to the answers. Previously successful, confident organizations cannot imagine that their products might have lost their lustre. Like the Queen in *Snow White*, they cling to the belief that they are still the most beautiful and are reluctant to look hard in the mirror. Sears Roebuck and Marks and Spencer, once the leading retailers in the US and the UK, acted as if their pre-eminence was preordained, took little account of the innovations of competitors, and were rudely dislodged from their thrones. In the worst cases, rigid mental models encourage industry leaders to deny the significance of threats in the face of overwhelming evidence. When they do acknowledge them, like *Encyclopædia Britannica* faced with *Encarta* or ITV up against BSkyB, they can fall apart.

Strong cultures can be a great asset, but they reinforce intellectual rigidity. Tightly knit, self-confident groups develop deeply entrenched beliefs, resistant to alternative points of view, and block out ideas and information that do not conform with the prevailing orthodoxy. Dissenting voices – the very people most likely to spot the significance of something new – are mocked as fools or persecuted as heretics. As Daniel Kahneman has shown, 'When pessimistic opinions are suppressed while optimistic ones are rewarded, an organization's ability to think is undermined'.

Capitalism's staunchest defenders and its fiercest critics would like to believe that hard-headed businessmen take every decision coldly and dispassionately, but, as we have seen, they are ruled by their hearts as much as the rest of us. Executives would be horrified to think that their enthusiasm for a new venture bore any resemblance to romantic impulses, but that only makes them less likely to recognize the symptoms of blind infatuation. Sony with stars in its eyes in Hollywood is not an isolated example of using elaborate strategic arguments to rationalize wishful thinking. Most of us are capable of deceiving ourselves when we really set our heart on something – irrational exuberance plays a big part in making the world go round.

Successful organizations are at the greatest risk from over-confidence. Hubris, the fatal flaw of tragic heroes, leads them to think there is nothing they cannot master, and to refuse to face reality. Most of our heroes were touched by hubris, some fatally. Spectacular success made them feel invincible – no mountain was too high, no challenge too great. Netscape dreamed of toppling Microsoft; AOL thought it owned its customers; Webvan believed it would make supermarkets redundant; IBM and Encyclopædia Britannica refused to believe that the PC could make their mighty products irrelevant.

It is no coincidence that the most dreadful decisions in history – from Philip II's launch of the Spanish Armada and Napoleon's invasion of Russia, to the mass murders of Hitler, Stalin and Mao – were taken by one powerful man, convinced that he was right, and answerable to no-one.

It is hubris that spurs businesses to make acquisitions in markets they do not understand and where they lack the necessary capabilities. AT&T threw away $50 billion in the 1990s on disastrous diversifications in computers and cable television. Marconi, the formerly great GEC created by Albert Weinstock, destroyed itself by ill-considered acquisitions during the nineties stock market boom, as did Vivendi, a former French water utility, recklessly buying media companies to feed the vanity of its chief executive, Jean-Marie Messier.

Without high levels of self-confidence, few businesses would ever get off the ground. Arrogance that would be obnoxious in other circumstances may actually be advantageous to entrepreneurs, but not when it comes to answering the really critical questions: what does it take to succeed in this market and does the company have these capabilities? Supermen without a restraining voice to remind them that they are mortal are not good at facing uncomfortable truths.

Making sense of a constantly changing environment is less a matter of systematic analysis than of being alive to alternative ways of seeing the world: opening minds to new possibilities, abandoning long-held convictions, and resisting the temptation to seek comfort in facile explanations. Proust's maxim about seeing with new eyes is apt. So is Andrew Grove's that 'only the paranoid survive'.

Acute antennae and lively imagination are among the attributes of organizations that manage to hold on to hard-won leadership.

13

THE RIGHT STUFF

There are more things in heaven and earth, Horatio,
Than are dreamt in your philosophy.

William Shakespeare, *Hamlet*

Speculative philosophy, which to the superficial appears a
thing so remote from the business of life and the outward
interest of men, is in reality the thing on earth which most
influences them.

John Stuart Mill, 'Bentham'

Out of the crooked timber of humanity, no straight thing
was ever made.

Immanuel Kant, 'Idea for a Universal History'

Creating a new market is a Herculean accomplishment, but the big prize in business is holding on to it, and that requires all the cunning, versatility and stamina of a Ulysses. The once distinctive capabilities of a market creator easily fade, and will eventually be matched or surpassed. Sooner or later, creative destruction will render them irrelevant – new technologies and business models can completely change the rules and wipe out previous sources of competitive advantage. We can only see these and through a glass darkly: making sense of a volatile world is beyond the capacity of most mortals.

Scarcely any company holds on to industry leadership for more than a few years, and many commentators agree with Foster and Kaplan that the idea of sustainable competitive advantage is now illusory. If sustainable means permanent, they are of course right – no organization keeps it for ever. Nonetheless, some companies, as we have seen, do manage to square the circle, to stay on top of the heap for years. This chapter discusses the qualities shared by those who achieved not immortality but exceptional durability. What is their secret? Is there a formula that others can follow?

One man in particular showed that these are not sensible questions – that there are no universally applicable formulae for human beings and their institutions. But he also suggested a way of thinking about the crooked timber of humanity that helps us to identify the conditions necessary for success.

Foxes and hedgehogs

Isaiah Berlin was a philosopher with no interest in business. He is most famous for dividing thinkers into two kinds: hedgehogs, who have one big idea, an all-embracing organizing principle; and foxes, who know many different things, which cannot be neatly reconciled. Plato, Dante and Marx were hedgehogs; Aristotle, Shakespeare and Montaigne, foxes. Jim Collins hijacked this idea in *Good to Great*, where he argues that great companies are invariably hedgehogs, who simplify complex issues and focus on one clear goal. As *Good to Great* has sold millions of copies, Collins's version is now probably better known than Berlin's. This is a pity as, although his book is a useful account of business success in stable markets, on this issue it is completely wrong.

Where Collins is right is that those implementing strategy need to be focused on a clear goal and not distracted by conflicting ideas. However, there is a fine line between focused hedgehogs and closed-minded conformists. If organizations are to spot significant changes and adapt to them, they need the sharp-eyed, versatile qualities of a fox.

Contrary to what one might conclude from Collins's account, there is nothing about animal behaviour in Berlin's essay, and certainly no preference for hedgehogs. Berlin uses the fox–hedgehog divide as a starting point for a discussion of Tolstoy's theory of history, not a formal categorization. However, there was a serious underlying point to it which ran through most of his work: the difference between those whose minds are made up and those who are open to many points of view, between the technocrats and revolutionaries who pursue big abstract ideas, sometimes blindly and fanatically, and believe that there are scientific answers to most questions, and undogmatic pluralists like Berlin himself who see the world in all its untidy complexity. He was a fox by temperament and by conviction.

Berlin interpreted better than anyone the message of Romantics like Herder and Wordsworth: that our understanding of the world is shaped by the language and metaphors we use, and they are always the product of our particular culture and our own imagination. Great entrepreneurs are creators every bit as much as artists and do what they do primarily to fulfil their unique vision rather than for monetary gain.

Berlin established that in the realm of human affairs, big, all-embracing theories can never be universally applicable. (Science is another matter and sometimes economics is a science. But all scientific systems, as Adam Smith observed, are 'inventions of our imaginations'.) Just as philosophers have debated for centuries how best to lead a good life and the ideal forms of government, without ever reaching agreement, comprehensive answers on business success are likely to remain equally elusive. No single theory can offer a timeless, universal explanation of something as complex and varied, any more than generic strategies can take account of the circumstances of individual firms. Yet businessmen, like most of us, yearn for such answers, ideally simple ones, and there is no shortage of pundits ready to proffer them.

Berlin's big idea was pluralism, and he dedicated his life to the study of many different thinkers, the antithesis of the specialized scholar who ends up knowing more and more about less and less. He challenged the widespread assumption that there must be a right answer to every question, if only we could find it, and that all the right answers must be compatible with each other. He showed that it is simply not possible to reconcile many ideas and beliefs – some of our most cherished values conflict with each other, so we are obliged to make painful choices – between liberty and equality, or justice and mercy, in politics. In the realm of business, managers constantly face dilemmas to which there is no right answer. They are forced to make trade-offs between different goals, different needs, different views of the truth, endlessly adjusting priorities and tactics in the light of changing circumstances and knowledge. If they stuck, hedgehog-like, to a single blueprint, they would eventually be swept away.

As we saw in the last chapter, human beings crave simple explanations for complex problems, single causes, one key factor that governs everything. Business pundits and management consultants are instinctive hedgehogs, proclaiming one big idea, a universal remedy. The starting point of Michael Porter, the doyen of business strategists and a formidable analytical hedgehog, is industry structure and strong barriers to entry. When examining mature, stable markets, his framework is the best starting point, but when things change, new capabilities become critical. Other experts stress scalability as the key to success, while Jim Collins puts powerful arguments for vision and values. Pundits in consumer markets see everything in terms of building great brands. Some even argue now that people are the scarcest resource in the new economy, that what is really crucial is recruiting and retaining the most talented ones. It is not surprising that many managers have concluded that what really counts is not so much strategy as effective implementation.

Murder on the Orient Express offers a clue to the conundrum. As it became clear to Hercule Poirot that each of the surprisingly sympathetic, supposedly unconnected suspects on the train had not just the opportunity but a burning motive to kill the loathsome victim, he realized that the explanation must be that all of them had done it together. In most cases of long-term business success, all of the factors cited above, and several more, play a part. As we have seen, one of them can well prove decisive at a particular juncture, but over time virtually all of them are necessary. As with most complex systems there is always more than one explanation.

Eight essential attributes

What follows is not so much a comprehensive explanation of lasting business success as a set of essential attributes. Eight threads, in addition to the eight attributes of market creators, run through all these stories. It is these that enabled the organizations to maintain, and in some cases enhance, their initial competitive advantage.

The most difficult to achieve, and the most critical when the gales of creative destruction are blowing at full blast, are:

1. The ability to recognize significant changes in the competitive environment and adapt to them.

2. The capacity to renew organizational knowledge and develop new capabilities.

Equally important, however, are four attributes that would have been on most people's list at any time in the last fifty years:

3. A strategy that is distinctive, coherent and realistic.

4. The managerial, marketing and logistical capabilities to penetrate a mass market and to operate efficiently on a large scale.

5. Close, ideally binding, relationships with customers.

6. A strong brand that reinforces differentiation and loyalty.

These qualities were consciously cultivated by all the companies who established industry leadership in the twentieth century, from

Ford to Coca-Cola, Glaxo to British Airways, Michelin to L'Oréal. They were equally important for all our recent industry leaders.

A very different business environment has made two others essential. In many industries, not just the most knowledge-intensive, human capital has become more important than physical assets, while the lifetime career in the same organization has largely disappeared. We have also seen the effective demise of the vertically integrated company that tried to do everything in-house, and its replacement by virtual organizations and collaborative relationships between businesses. In all the cases considered here, we find two further qualities:

7. The ability to attract and retain talent and to nurture human capital.

8. Systematic collaboration with networks of autonomous organizations and individuals.

Most of these eight attributes are closely related to and reinforce each other in a web of feedback loops: adaptability and renewal are inseparable; strong brands are both a reflection of the firm's relationships with its customers and a way of strengthening them; companies with loyal customers tend to have loyal, well-nurtured employees; organizational knowledge and capabilities depend on talented people and human capital; realistic strategies require a broad understanding of the firm's ever-changing competitive position and the ability to react quickly to change; collaborative partnerships help the firm to build new capabilities and to adapt to new circumstances.

They are not, however, always positively correlated – Sony and Apple have scored highly on nearly all of these qualities for much of their history, but at crucial periods have gone astray strategically and in terms of capabilities and collaboration. It is also not possible to make a single generalization about these attributes that encompasses all of them without oversimplifying or missing some out: we need to consider each of them separately. Though some attributes are more equal than others, the evidence strongly suggests that they are all necessary. Certainly any firm with ambitions as an industry leader ignores them at its peril.

They are also to be found among firms whose particular talent is to win leadership in, and to dominate, markets that others have created. Microsoft is the outstanding recent example, but other businesses have also transformed markets through competing aggressively and displacing both pioneers and incumbents. These companies too are innovative, but in a different way from market creators, and possess all these attributes. Market creators who want to remain leaders have to become consistently professional competitors themselves.

We now consider each quality in turn, and why they are critical to lasting success in the networked economy.

Recognize and adapt to change

There is no formula for effective adaptation, because discontinuities are so numerous and unpredictable. But some traits are clearly essential – acute awareness of changes in the environment and their significance, willingness to question cherished beliefs, to sacrifice sacred cows and to anticipate change through exploration and experimentation. This is not easy to combine with running a big business, but somehow our heroes managed it more often than not.

Most firms focus exclusively on their existing market and measure their success largely by their market share and financial performance. In so far as they evaluate their competitive position it is by monitoring rivals, who tend to resemble them in most important respects. In the feeblest cases competitive strategy consists of imitating the initiatives of rivals or retaliating against them. Such a narrow view makes it almost certain that the most important opportunities and threats will not be spotted in time to respond effectively, as they are rarely obvious.

All our market creators identified the original opportunity by taking an unconventional view of customer needs and technological possibilities, and continued to do so long after the initial venture was established. Virtually all of them came to realize that the addressable market was considerably larger than they had first thought, and were quick to adjust their strategies accordingly. Most of them also spotted further opportunities.

Nokia's vision of the market for mobile phones, and of what a mobile phone could be, became progressively bigger and bolder in the course of the 1990s. Cisco constantly redefined its vision of the market for networking technology and of what it had to offer customers, coming a long way from simply selling routers. Ryanair progressively enlarged the scope of its addressable market – from travel between and within Ireland and the UK, to a mass, pan-European market for air transport. It then found ways to make money from passengers paying fares that did not cover its costs. Sony and Apple have endlessly identified new and surprising fields to enter – from the Walkman and PlayStation to the iPod and the iMac. Sky's understanding of the dramatically different television landscape in the 1990s enabled it to evolve from cheeky intruder to dominant player.

Sadly, some attempts to see the bigger picture end in disaster. AT&T fondly imagined it could be successful in computers and cable television. Gerry Levin misread the significance of the Internet for Time Warner and confused AOL's inflated share price with real capabilities. AOL deluded itself that it was a media giant and that a business model which depended on charging dot.coms exorbitant advertising rates could continue indefinitely.

Being aware of other ways of seeing things is immensely difficult to combine with implementing a clear strategic vision and rigorous focus on a market. Nobody can be entirely successful, but these examples show how some organizations met Scott Fitzgerald's test of a first rate intelligence: 'the ability to hold two opposed ideas in the mind at the same time, and still retain the ability to function'. They have what the poet Keats called 'negative capability': they know how to make do with half-knowledge.

The trick, closely related to that of disciplined entrepreneurialism, is to balance iconoclasm and imagination with fidelity to essential values. Most of the people engaged in building or managing a business have to concentrate exclusively on the best ways of doing that. But somewhere in the organization, somebody needs to be doing some divergent thinking, questioning orthodoxy, flirting with heresy, thinking the unthinkable.

'When the facts change, I change my mind. What, sir, do you do?' Keynes's dictum is a great deal easier said than done, since distinguished economists are no more willing than CEOs to admit that they have been mistaken. The obstacles to recognizing how much things may be changing and their implications for the firm are both cognitive and emotional. If a discontinuity represents a threat to deeply held beliefs, a common response is to deny its importance, even when the evidence becomes overwhelming. IBM's senior management in the late 1980s simply could not grasp that the centre of gravity of the computing industry was shifting from mainframes to PCs and from hardware to software.

Dorothy Leonard has put it well: 'Core capabilities are simultaneously core rigidities. The very system that conveys competitive advantage can also disadvantage the company, either when carried to an extreme, or when the competitive environment changes.'

Ken Olsen, the founder of DEC, insisted well into the 1980s that PCs could never challenge the position of the minicomputer – and saw his company shrink into insignificance. Silicon Graphics, despite the warnings of its founder, Jim Clark, failed to take seriously the threat to workstations from the ever more powerful PC, and found its market dwindling. Jim Barksdale, the CEO of Netscape, insisted that Microsoft could never overtake it because the Internet rewarded openness and non-proprietary standards. Even Sony's heroic Masara Ibuka, who had blazed an astoundingly innovative trail with transistor radios, Trinitron televisions and VCRs, insisted that digital recordings could never equal the quality of analogue sound.

Most of our winners were prompter to recognize and deal with discontinuities. When the dot.com economy seemed to be collapsing in 2000, Jeff Bezos saw that this required a different strategy from the breakneck expansion and diversification Amazon had been pursuing. All the energies that had gone into getting big fast were directed into becoming cash-positive. The stock market continued to mark down its share value, but during this period the company was turned round and its market leadership consolidated. Lou Gerstner was quick to see how services could be the salvation of IBM in the 1990s. Apple spotted the commercial potential of digital music long before the record industry. After the VHS/Betamax debacle, Sony recognized the need to win industry-wide agreement on standards. Unfortunately it also drew more questionable conclusions.

Adaptation is not about knee-jerk reactions, as the story of AOL illustrates. Probably its most important facet, and a vital attribute in its own right, is developing organizational knowledge and capabilities and building new ones. This is not something that can be done instantly or by following a simplistic formula. Once again Dorothy Leonard has wise words:

> Successful adaptation is not a chameleon-like response to the most immediate stimuli – a quick switch to a new enterprise or an impulse acquisition. Rather, successful adaptation seems to involve the thoughtful, incremental redirection of skills and knowledge bases, so that today's expertise is reshaped into tomorrow's capabilities.

Extend organizational knowledge and capabilities

Before the twin revolutions of the 1980s, that transformed so many markets, the competitive advantage of most long-standing industry leaders was based on economies of scale and strategic assets like brands, patents and regulatory barriers to new entrants. A tidal wave of new competitors has washed away most of these, though leaders like Microsoft, Cisco and eBay have enjoyed new kinds of strategic assets.

This leaves distinctive capabilities as the bedrock of most companies' competitive advantage. Keeping them distinctive is one of the toughest challenges and learning new tricks, especially for old dogs, even more so. Complacency is always a danger for successful companies, thinking that they have little left to learn. From time to time this has afflicted even our paragons, though few have been as honest in acknowledging it as Jorma Ollila of Nokia. And there is a big difference between aspiration and achievement, yet a few manage to master entirely new capabilities repeatedly over long periods. Arie de Geus, who had some success doing this at Shell, believed that 'the ability to learn faster than your competitors may be the only sustainable source of competitive advantage.'

The challenge is about much more than producing innovative new products – the real innovation is in building entirely new capabilities. The outstanding examples in our set are IBM, Apple, Sony, Cisco, Dell, Sky and Nokia. The first three have been singled out here more for their failings than for their many successes. What saved them from disaster, and took them to new heights, was their remarkable ability to learn and to stretch themselves in new directions. These companies had a culture of enthusiasm for knowledge that ran through large parts of the organization and developed a critical mass of creative brainpower.

Even when IBM was blundering in the boardroom in the late 1980s, another entrepreneurial team was bringing to market a world-beating new minicomputer. While Sony was facing defeat in the VCR market it had set its heart on, it conducted the brilliantly successful launches of the Walkman and the CD. While later making a spectacular fool of itself in Hollywood, another part of the organization was planning the triumphant launch of PlayStation and the creation of another enormous new market. Apple has reinvented not just the idea of a personal computer but that of personalized music, stealing a march on its former role model, Sony.

Cisco took the most radical approach to acquiring new capabilities – acquiring whole companies as a deliberate alternative to developing them internally. Sky in its early years had minimal capabilities in television production but made itself the leader in sports presentation, marketing, subscriber management and pricing, and when digital broadcasting arrived it was the first to master it. Likewise Nokia made itself a master of design and marketing as well as of digital technologies for mobile communications.

These are unusually able companies, but as important as natural ability is the cast of mind described by Dorothy Leonard as replenishing the wellsprings of knowledge – 'the constant rebirth of expertise. That rebirth in turn is determined by the attitude managers and organizations foster towards learning.'

In its short life Google has shown spectacular enthusiasm for learning new things. Over a longer period, Amazon has succeeded in reshaping its business model, becoming the orchestrator of a vast network of affiliates and a provider of computing services. The only survivor as an industry leader we have considered that seems not to have enlarged its capabilities is eBay, which relies, almost certainly dangerously, on its unique strategic assets.

Distinctive, coherent, realistic strategy

Few market creators have explicit, detailed strategies at the outset. Provided that they have a clear, strategic vision and the other qualities described in the first half of this book, this need not matter. Indeed, detailed strategic planning can inhibit entrepreneurialism and adaptation to change. They do, however, have to think strategically.

The single most important challenge for market creators, as the firm and the market grow, is that they soon face new and more formidable competitors and most lose their temporary monopolies. All our winners gave serious thought to something most firms neglect – their strengths and weaknesses compared to those of their competitors, present and future. The three features highlighted – distinctiveness, coherence and realism – relate mostly to this fundamental challenge. They are not the only characteristics of successful strategies in mutating markets, but they are the ones most conspicuously lacking in those that fail.

Lack of realism is probably the commonest failing. This is partly because of the many genuine unknowns about a new market and how it will develop, but also because firms easily delude themselves about their own capabilities and take insufficient account of those of their competitors. Netscape naively believed that speed of movement and first mover advantage would protect it against Microsoft. Webvan took no account of the customer relationships, brands and capabilities that conventional supermarkets enjoyed and gave little consideration to consumers.

Similar criticisms could be made of Apple, AOL and Sony at critical stages in their histories, but their strategies were also incoherent.

Apple believed for a large chunk of the 1980s that it could penetrate and dominate business, professional and consumer markets, even when it had a brand image that put off corporate customers, an inadequate marketing organization for addressing them and underpowered products. Steve Jobs's promotional strategy in 1984 seemed designed to patronize business customers and John Sculley's pricing policy was designed to skim profits rather than to build market share.

AOL became so adaptive and opportunistic in the late 1990s that it ceased to have a serious strategy for the medium to long term. It did not develop real capabilities as the media giant it aspired to be, but exploited its short-term monopoly of Internet advertising so ruthlessly that it made countless enemies.

Sony's strategy of diversifying into movies took little account of the capabilities required and overestimated the potential synergies. This was particularly sad as it was so well endowed with the other attributes – one of the strongest brands in the world, a 'family' company that attracted and nurtured the cream of engineering and marketing talent, and an unequalled ability to learn and to extend its capabilities.

Our success stories developed coherent, realistic strategies to a considerable extent because they were logical extensions of their original strategic visions and because they remained focused on their core markets and capabilities: they did not pursue aspirations that bore no relation to what they did best. Despite the temptations offered by a stratospheric share price, eBay resisted until its purchase of Skype the temptation to acquire businesses unrelated to its core. Ryanair expanded aggressively in continental Europe, but only after firmly embedding its business model and testing the formula thoroughly on the routes between Ireland and Britain. Unlike easyJet, it avoided expensive acquisitions.

The distinctiveness of the strategies of industry leaders, like that of market creators, derives mainly from their capabilities. Much of this chapter is about how they maintain that distinctiveness. However, these companies also made a deliberate strategic choice to put clear distance between themselves and their competitors.

Most of them did this by a combination of unique capabilities and getting big fast – building a market presence that was difficult for others to challenge. This strategy was explicitly followed by Amazon, Nokia and Starbucks, and implicitly by most of the rest. Google and eBay were able to do this particularly quickly because of powerful network effects and feedback loops. Cisco did it by making the range of its networking products very much greater than those of its rivals. Dell, Southwest and Ryanair were distinctive primarily because of their business models, whose robustness was not apparent to their competitors for some years, by which time they had established enormous leads.

Apple and Sony were able to rely for their distinctiveness on their extraordinary ability to innovate and to develop uniquely attractive new products. This is a strategy few others can pursue, as they are likely to be quickly outflanked by competitors who can match them. And even Sony and Apple had to ensure that they had all the other attributes as well.

Operational efficiency and scalability

Some commentators see scaling up to penetrate a mass market as the key challenge in market consolidation. During the dot.com boom, most entrepreneurs and venture capitalists cited scalability – the ability to grow very fast and to achieve economies of scale quickly – as the key feature of their strategies and business models. The initial market may consist largely of early adopters ready to pay a premium for a product or service whose benefits they identify for themselves. Mass markets generally require lower prices, simpler propositions and a different marketing approach.

Getting big fast reinforces first mover advantage, where that applies, and makes it more difficult for competitors to establish strong, defensible positions. However, this is far from being universally applicable, and scarcely any businesses are as naturally scalable as eBay and Google. Most considered here had higher fixed than variable costs, but Amazon and Nokia had to overcome significant logistical obstacles in order to deliver books and manufacture phones on a vast scale.

The ability to manage the transition effectively is essential, and not one that most entrepreneurs, or indeed most large companies, possess. Setting up from scratch the processes that take years for a mature company to establish is a tall order. This quality is partly an extension of disciplined entrepreneurialism, but it also requires the acquisition of new functional capabilities – in marketing and distribution, logistics, finance and administration. Invariably this means recruiting and absorbing many new people, particularly in senior management. Indeed, the norm is for there to be a change of chief executive at this stage.

Amazon and Dell had, or set about building, the qualities, people and mindset necessary for scaling up right from the outset. eBay, Google and Cisco acquired them with new waves of professional management. Nokia, IBM and Sony, of course, started out with cadres of professional managers. Apple sometimes struggled with management disciplines, but John Sculley certainly contributed significant professionalism to an often chaotic outfit.

Achieving scale is basically about efficiency – learning how to produce more and more whilst spending less and less, the central concern of management theorists in the twentieth century, from Frederick Taylor and time-and-motion studies to Michael Hammer and business process re-engineering. According to Hammer, 'the heart of managing a business is managing its processes.' Heart is an interesting word for a mechanistic philosophy. The danger, as the search for ever more efficiency becomes more relentless, is that the business loses its soul, as Apple did under Sculley, and as eBay and Dell seem to have done twenty years later.

When this happens, all the emphasis is on execution rather than exploration and the creative and entrepreneurial juices dry up or are squeezed out. Having highly efficient processes means being governed by rules and becoming bureaucratic. This is often necessary but firms obsessed with efficiency try to reduce spare capacity. Rather than experiment with a number of possible new strategic directions, they eliminate redundancy and diversity. They want to pick a strategy, often long before they can know which would be the most viable.

Unless a business has achieved exceptional economies of scale, efficiency is rarely a solid basis for differentiation, as Dell has discovered recently. Eventually other competitors will catch up. Public companies, with financial markets to please, are under constant pressure to emphasize improved financial performance, particularly in the short term. This often means ignoring longer-term opportunities and threats, and putting cost savings before innovation, customers and people. If carried to extremes, and the company takes too narrow a view of shareholder value, it risks damaging customer loyalty, human capital and the ability to extend capabilities, essential for lasting competitive advantage.

Close customer relationships

The biggest single determinant of long-lasting success for most businesses is the ability to hold on to a loyal body of customers – the stronger the ties that bind them, the more difficult for a rival to tempt them away. Conversely, the strong customer relationships of an incumbent are often the biggest obstacle that a challenger has to overcome.

Achieving binding relationships is to a considerable extent a reflection of other qualities, like strong customer propositions and capabilities, but it is something that all sensible businesses specifically cultivate. Most new market pioneers fail to maintain their initial lead at least partly because they do not have the time to build strong customer relationships.

Young businesses do not have long-standing relationships and are vulnerable to attack from others who may only come close to matching their innovations: most pioneers whose initial success is based on an imitable innovation or low prices are soon marginalized. Customers may like the product or the deal but feel no special bond with the supplier. When another supplier, particularly one they already know, makes an equivalent offer, they are happy to move on.

There are essentially two strategies for binding customers closely. One is to cultivate such levels of satisfaction, trust and loyalty that they do not want to leave – the loving embrace. The other is to use strategic assets to make it difficult for them to prise themselves away – customer lock-in. The two strategies are not, of course, mutually exclusive – indeed relying entirely on one or the other carries risks. Most of our success stories worked hard to create customer loyalty and trust, but few resisted the opportunity to lock them in.

The most convincing advocate of the value of loyalty in business is a surprising one. The Bain Consulting group is a highly analytical firm, not much given to mushy ideas, and strongly emphasizing hard, measurable results. Its research, described by Frederick Reichheld in *The Loyalty Effect*, shows conclusively that, in markets where intellectual capital is important, the most successful businesses are those that pay most attention to encouraging loyalty. In industries like financial services, customer loyalty is as important an indicator of long-term corporate performance as short-term profitability. It is also closely linked to employee and investor loyalty. Companies that invest in human assets (both customers and employees) do better than those who stress short-term earnings as the main measure of corporate performance.

Reichheld argues that as intellectual capital becomes more and more important in our economy, so will loyalty and human assets. The fact that few companies take these ideas seriously in practice in no way invalidates them – very few companies become industry leaders and even fewer stay leaders for long. Companies that treat business as a zero-sum game risk alienating, and eventually losing, both their people and their customers.

These findings are consistent with the conclusions of Maister,

Galford and Green on the related subject of trust. According to them, trust has four key elements – credibility, intimacy, reliability and selfishness (or lack of it). Credibility is the first hurdle, but generally the easiest to overcome, provided that the firm has the necessary professional capabilities. It is essential, but never sufficient in itself. Intimacy with customers is not normally possible for large firms, though many achieve it in their early years. There are also comparatively few opportunities for large companies to demonstrate that they always put the customer's interests ahead of their own, though these can pay enormous dividends.

What really counts for most businesses therefore is demonstrating reliability – consistently fulfilling its promises to customers, time after time. Trust, like loyalty, is nearly always earned over a long period, by repeatedly demonstrating that customers come first. It is invariably actions not words that count, and doing it over and over again.

Time after time in its early days Amazon demonstrated a willingness to exceed customer expectations, to do more, with gestures like supplying a hardcover book when the paperback ordered was out of stock. Google initially grew as fast as it did because the search experience was simply so much better than the alternatives. Starbucks built its brand around the consistency of good customer experiences.

There is one notable exception to this rule – Ryanair, which takes a narrow view of the quality of the customer experience. Like most businesses whose value proposition is primarily about price, it does not inspire love and devotion among its customers. Indeed, many of them complain loudly, even as they book their next flight. In the long run, the lack of deep loyalty could make it vulnerable, particularly if there were big changes in the size or structure of the air transport market in Europe.

AOL also ignored the need to cultivate loyalty, with dire consequences. Its cynical treatment of its advertiser clients at the height of the dot.com boom and the poor service it offered to many consumers left a lasting stain on its reputation.

Customer lock-in

It would be imprudent to rely entirely on trust and loyalty when there are opportunities to bind customers more closely through strategic assets, notably network effects and switching costs.

Pierre Omidyar discovered the power of eBay's magnet serendipitously and made trust between members of 'the community' the cornerstone of eBay's strategy. Under Meg Whitman's less starry-eyed management, network effects were used to exert an iron grip on traders and extract ever higher rents from them.

Google's case is less extreme, but its enormous share of the billions of Internet searches made every day has given it an ever larger share of the ever larger market for search-based advertising. The fact that, like early eBay, it placed great emphasis on ethical principles has perhaps made the locks less irksome to advertisers.

Sky's hold on its customers is mainly due to two exclusive assets – its rights to premium programming, notably football, and its position on Astra – which have kept out competitors and enabled it to ratchet up prices frequently.

The classic lock-in strategy, particularly for new-technology firms seeking industry leadership is to make their technology the standard – the key to JVC's victory over Sony in the titanic videocassette war, much of IBM's initial success with the PC, and, of course, Microsoft's subsequent domination of large parts of the IT industry. The software companies who have resisted the latter's hegemony are those like Adobe and Oracle who have made their products the standards in the narrowly defined markets they dominate – portable document format (PDF) and databases.

In all these cases, customers face significant switching costs if they move to another supplier, mainly because of their initial investment. In some cases the costs of changing are so prohibitive that they are effectively locked in. Microsoft has achieved lock-in with PC operating systems, Xerox did so for many years in photocopying, and IBM in mainframe computing. Cisco, eBay and Sky have achieved customer lock-in through making themselves the industry standard.

The danger is that chains like these tend eventually to corrode genuine loyalty and to be resented. When a disruptive new proposition loosens them, many grateful customers will flee to their liberator, as eventually happened to both IBM and Xerox.

A strong brand that reinforces differentiation

One of the most important expressions of a close relationship with customers is the brand – both a consequence of millions of virtual relationships and a means of reinforcing them. In the modern economy, the brand is what the business means to most consumers, what it stands for. Brands are nearly always among any successful company's most valuable assets – though quantifying that value is hardly an exact science. In several of our cases the brand gradually became a major component in the company's competitive advantage – notably Amazon, Apple, Sony, Nokia, Google and eBay.

Brands are primarily about differentiation, as they have been since cowboys first burned theirs on cattle to discourage rustlers. They are also an expression of how consumers feel about the company, but the brand alone cannot differentiate it – it does not exist in isolation from the company and its products. It only encapsulates something significant about the organization if it is based on differences that mean something to customers. In brand-speak, the company, its products and its people must deliver the promise.

Our market creators succeeded in doing something that the great marketing companies of the mid-twentieth century did, making their brands virtually synonymous with their market. As Coke is to cola, Gillette to razors and Hoover to vacuum cleaners, so is Google to Internet search, Amazon to books online, and Starbucks to coffee shops. Their brands have embedded themselves deeply in the minds of consumers. They achieved this mainly by continuing to do well what made them successful in the first place and by making it difficult for others to imitate them: their differentiation is largely based on their distinctive capabilities.

Amazon was so good at online retailing, Nokia's phones in the 1990s so much better designed than the rest, Phoenix and the OU so dedicated to adult university education that it was difficult for anyone else to get close. During that crucial rapid growth period, when the mass market was created, they imprinted their brands on their customers, and in most cases established an emotional bond with them. The rapid expansion strategies pursued by Amazon, Sky, Starbucks and Ryanair were partly about economies of scale and pre-empting competitors, but also consciously about building their brands as quickly as possible. They then became the semi-automatic first choices of many consumers.

Starbucks famously built its brand one cup at a time, but this was in effect what all the others did too. All of these brands were created primarily by actions rather than words, and reinforced repeatedly. Jeff Bezos's observation that instead of a merchant spending 30 per cent of his time creating a good customer experience and 70 per cent shouting about it, the proportions should be reversed in the online world is actually applicable to most contemporary markets. Amazon and Apple spent heavily on promotion after they had raised substantial venture capital, but far more important were the endorsements of satisfied customers and the associated free publicity.

This is very different from the heavy advertising used for most of the twentieth century to build consumer brands. There the brand was differentiating very similar products, by fashioning an image in potential customers' minds and justifying a price premium. Modern consumers are more sophisticated and resistant to promotional messages that do not match their experience. Heavy promotion has certainly worked for some recently created brands, like The Gap and Nike, but only where customer value corresponded closely with the brand promise. Several other brands created in the last twenty years, like The Body Shop and Häagen-Daz's ice cream also only used advertising at a late stage in their market development.

What our winners have done is build their brands primarily by offering something that their customers valued highly. The brand is to a large extent the expression of that value and of consumer trust. Like size, it is initially a consequence of success, which subsequently becomes an important defensive weapon in its own right.

Attract talent and nurture human capital

'People are our greatest asset' is the hollowest of business clichés in a large field, but it contains an important truth. The quality, skills, knowledge and motivation of the key people in an organization are critical to just about everything it does: levels of innovation, quality of design and production, and effectiveness of marketing. Human capital is essential to sustaining several other critical attributes – strategy, customer loyalty, the brand and, above all, capabilities. A well-organized football team can do well without necessarily having the very best players, but not when competing repeatedly against the best. Many companies now understand that they are competing for talent as much as for customers, and that competition has recently become intense.

There is no general agreement on what constitutes talent, except that most of us think we recognize it when we see it. Sadly, it is not distributed remotely equally, and what really counts for companies is a comparatively small number of people – senior managers, strategists, technologists, marketers and designers. Alan Eustace, Google's vice-president of engineering and research, has declared that one top-notch engineer is worth '300 times or more than the average'. Bill Gates believes that 'if it weren't for twenty key people, Microsoft wouldn't be the company it is today.' Similar power law ratios are probably true of all our stars.

Human capital is a broader concept, though the two are often fudged in public pronouncements to make everyone feel included in both. The calibre of large numbers of front-line staff is of course critical to the quality of the customer experience and to the brands of businesses like Starbucks. Its army of baristas is a crucial part of its human capital, and Starbucks' investment in them has been an important part of its successful growth. However, notwithstanding the warm rhetoric, baristas are fairly easily replaceable foot-soldiers, particularly now that the company has well-established procedures for recruiting, training and motivating them. These processes, however, are something that other suppliers can imitate, and they gradually became enabling, rather than distinctive, capabilities in this market. Generating new capabilities and the continued dynamism of the company depend very much on unusual talent, particularly among senior management.

The concept of human capital was developed in the 1960s by Gary Becker, who subsequently won the Nobel Prize for economics. He applied it mainly to governments investing in education. For businesses, thinking of people as assets rather than liabilities has represented a massive shift – from seeing most employees as a cost to a means of adding value and differentiating the firm.

In a free society human capital ultimately belongs to the individual rather than the firm. As most people can leave whenever they like and competition for the best is intense, firms have two big challenges: attracting, developing and retaining the best people; and ensuring that individuals' knowledge and skills are converted into organizational capabilities and intellectual capital.

The competition for talent has become more intense recently for several reasons. The number of jobs requiring 'tacit' interactions has been growing three times faster than employment in general, according to McKinsey. These are jobs that require high levels of judgement rather than the implementation of rules, and this need is coinciding with the retirement from the workforce of many experienced baby-boomers.

New areas of knowledge, especially in technology, have grown enormously in the networked economy. The supply of even qualified people is limited, and really talented ones even more so. In the past, the promise of lifetime employment and career progression helped to secure high levels of loyalty, and still does in countries like Japan. Now that the promise and the expectation have largely disappeared in the West, so has the automatic loyalty. Consequently competition has concentrated on talented people – most are regularly contacted by head-hunters.

Finding, cultivating and developing good people is, of course, something that companies like IBM, GE, HP, BP and Shell have been doing for decades. The big change has been that responsibility for talent management is no longer seen as something that can simply be left to what is now called the HR department, traditionally the most bureaucratic. The whole company needs to develop what McKinsey calls a talent mindset, a deep conviction that managing talent is critical to competitive advantage. Jack Welch famously spent half of his time nurturing and assessing talent within GE. 'Having the most talented people in each of our businesses is the most important thing. If we don't, we lose.' Many other CEOs give it a similar priority.

It is certainly something that all the industry leaders studied here took seriously. Google, Amazon, Apple and Dell were obsessive from the start about only recruiting the very best people, and the others only slightly less so. Nokia, Sony, Apple and DoCoMo consciously set out to develop a wide range of talents and capabilities, notably in engineering, but also in areas like design and marketing that most technology companies ignored. Meg Whitman's top priority when she became eBay's CEO was professionalizing its workforce and its marketing.

Two of the companies examined here did not appear to pay much attention to cultivating human capital: Ryanair has taken attitudes towards some of its staff as robust as those its customers have experienced: flight attendants are even charged for their uniforms and in-flight coffee. Southwest on the other hand is rated as one of the most desirable employers in the US. AOL never made that list, taking the same short-term, opportunistic approach to its people as it did to so many aspects of its business.

The relentless focus on costs and margins at both Dell and eBay disillusioned many people and made it difficult to attract more. After the Netscape revelations, many software developers were repelled by Microsoft. Google, on the other hand, was the most eagerly sought employer in the world from 2006 to 2008. Apple has always been able to attract outstanding talent: Jony Ive joined in 1992, when many people were writing Apple off. But when Steve Jobs came back five years later, he found many talented people working on the iMac with Ive.

Networked collaboration

By the end of the twentieth century the old command-and-control, vertically integrated company was close to extinction, along with organization man, replaced by networked person. Many mature organizations, from BP to IBM, have 'dis-aggregated' their internal organizations. However, one of the most striking features of the networked economy is the extent to which businesses now systematically collaborate with others, and derive large parts of their competitive advantage from the nature of these relationships.

Collaboration can take several forms and a rich crop of terms has sprouted to describe different facets of the 'extended enterprise' – strategic partnerships, 'coopetition', managing the supply chain, outsourcing, being the platform leader, the keystone of an ecosystem, the orchestrator or hub of a business network. Some of these are specific to technology industries, but there is a common element in all these forms of collaboration or partnering – to form a network of relationships that enable the firm to be both more effective and more efficient than if it tried to do everything in-house or if it tried to match all the capabilities of its partners or suppliers.

The intellectual rationale for this is *The Nature of the Firm*, an article written by a young British economist in 1937. Ronald Coase wondered why it was that, if markets were so efficient at allocating resources, there weren't more small businesses trading with each other, instead of large, hierarchical companies deciding everything internally. The answer, he decided, was that the costs of conducting thousands of individual transactions every day was too high.

Modern information and communications systems have dramatically reduced transaction costs and made possible ways of working that would once have been unthinkable. Value chains have been disaggregated and the boundaries of many firms shifted. Networks of specialized firms can work together closely and effectively, even though geographically as well as organizationally distinct.

Another reason for the trend is the vastly greater complexity of modern economic and industrial systems, which demand more and more specialization. Even exceptionally capable organizations are better at some things than others, and most now concentrate on their core capabilities and markets, while getting others to perform peripheral activities. Tightly controlling every aspect of their business or attempting to do it all is beyond the powers of almost any organization.

The very much lower labour costs of countries like India and China and their integration into the global economy have led to vastly more outsourcing. The most conspicuous cases have been the sub-contracting of manufacturing, but routine administrative functions, call centres and some research and development activities have also moved to Asia.

Building and orchestrating networks of suppliers and distributors requires different capabilities from those required to manage vertically integrated operations. Ironically, one of the earliest adopters of electronically networked collaboration was the automotive industry, which once took vertical integration to such lengths that it used to manufacture the steel for its own cars. For many years now, networks involving dozens of autonomous specialists have designed and proto-typed new models. Sub-contractors now contribute more than half of the value of a typical Ford or Chrysler vehicle.

Companies like Nike, IKEA and Intel have made collaborative networks the cornerstone of their competitive advantage. Nike sub-contracts all its manufacturing and is primarily the orchestrator of a brand. IKEA also sub-contracts its manufacturing to keep costs to a minimum, and indeed persuades its customers to assemble furniture themselves in return for exceptionally low prices. Intel is the hub of a different kind of business network, through being the platform leader. It achieved its dominant position in the PC and related industries, not simply because IBM chose its microprocessor as the (open) standard in 1982, but because it was subsequently able to orchestrate the evolution of the hardware platform. Defining an architecture that allowed its chips to realize more of their potential and stimulating the development of new applications encouraged communities of other suppliers to develop new products and services within that architecture.

Marco Iansiti and Roy Lieven have made the boldest claims about platform leadership in business networks. They argue that the crucial battles are now between networks of firms and that strategy is becoming the art of managing assets one does not own. According to them, the business networks around Intel and Microsoft are analogous to ecosystems, where large numbers of loosely interconnected participants depend on each other for survival. Biological ecosystems generally have active 'keystones' who shape them by 'disseminating platforms' and regulate the health of the system as a whole. 'Dominators' try to extract as much as they can from the system, like the zebra mussels introduced into the Great Lakes who consumed most of the food. Iansiti and Lieven contrast Microsoft's enlightened leadership of the many firms who develop software for the Windows platform with the 'dominator' approach of Apple, which tried to hog as much value as possible for itself.

The fact that Iansiti and Lieven have both consulted extensively for Microsoft inclines one to take this with a pinch of salt. The characterization of Apple as a greedy devourer of other people's vital resources, as compared with wise, benevolent Microsoft, would raise a hollow laugh from some of the firms who have got uncomfortably close to it. However, there is something in these ideas.

Trying to do everything itself was undoubtedly a flaw in Apple's approach in the early 1980s. If it had been prepared to license its operating system, it might have achieved a greater market share, though that would have carried risks. Undoubtedly Apple should have been more consistent in the early 1980s in encouraging and nurturing the software developers who contributed to its success. Its subsequent collaboration with Aldus and Adobe, not to mention Microsoft, was critical to the survival of the Mac. It is significant that two market creators who made no serious attempt to build collaborative networks, AOL and Netscape, failed entirely to establish industry leadership.

Of our set, Cisco, Dell, eBay, Amazon and Google could all be characterised, if not as keystones, certainly as the hubs of business networks. The ways in which they worked with satellite businesses has been central to their business models and strategies. Dell has achieved its tightly coordinated supply chain through the model it calls virtual integration, sharing information with its suppliers as if they were part of its own organization.

Cisco's business model is built around conducting its relationship with suppliers and customers, primarily through the Internet. The Cisco Connection Online helps customers to configure their own networks and decide what specific products and services they need. It passes the leads on to appropriate 'partners' and orchestrates delivery, installation and support, always keeping control of the crucial customer relationship. It has even developed a consulting arm to show its customers how to become more like Cisco themselves.

Of all these businesses eBay is the most virtual – the work of supplying and despatching goods and making payments is done entirely by its customers. Amazon has built several networks of relationships – with other booksellers, with publishers and authors, and with other merchants in Amazon Marketplace; and half of Google's income comes from affiliate sites, like AOL's, where it provides the search capabilities and shares the advertising revenue with the host.

Some have taken a rather instrumental view of collaboration: Nokia has been criticized for its secretiveness and its reluctance to collaborate with network operators and others, but it is, of course, in competition with them for the closest relationship with the customer. Ryanair has rather one-sided relationships with airports, rescuing a number of small European ones from obscurity, but ensuring that it extracted the lion's share of the value from the revenues that its flights generated.

Although the orchestrator of the network normally has the dominant role, in the long run these relationships need to be mutually beneficial in order to be sustainable – as they do with customers and employees. But not all businesses take a long-term view.

The fundamental dilemma

As with the success factors for market creators, these attributes are necessary rather than sufficient conditions for long-term success.

They are no guarantee against the ravages of time or the arrival of a devastatingly disruptive technology or business model, but they do greatly improve the odds. Companies without them will almost certainly lose industry leadership. They are also very rare, particularly in combination. Organizations that achieve all of them, even for a comparatively brief period, can certainly be said to deserve their success.

A case can be made for including other attributes. All these companies, for example, had strong cultures, but they were so different that it would be difficult to generalize about them in a way that could be said to be common to all. The culture of Dell has almost nothing in common with that of Apple, nor eBay's with that of Google. Strong cultures can also be a force for rigidity, as well as for innovation and adaptation, so they are not necessarily a strength in fast-changing markets. The aspects of corporate culture that contribute most to business success are attributes we have considered, like disciplined entrepreneurialism and the ability to learn and to adapt.

Strong leadership can also be a double-edged sword. All these businesses were clearly extremely well led for most the time, and leadership is generally the most important part of the human capital of an organization. In some cases – Lou Gerstner at IBM, Sam Chisholm at Sky, Steve Jobs at Apple, and Jorme Ollila at Nokia – leaders were crucial in turning round a company's fortunes, but in these cases, and every other, they were building on the capabilities and attributes of the organization. The apparent transformation of companies is not a magical process effected by a single superhero.

Technological innovation is likewise critical in many cases but not all, and only benefits the firm if it makes the product or service more valuable to customers, or if it enhances organizational capabilities and differentiation. It certainly did so for Apple, IBM, Cisco, Sony, Nokia and Google, and to a considerable extent for Amazon, Sky and eBay, but it is not universal.

Innovation in a general sense is a thread that runs through both sets of success factors, starting of course with the very first one, radical strategic vision. It applies particularly to recognizing and adapting to change, developing new capabilities, nurturing human capital and developing collaborative networks, which are largely about maintaining and building on the innovativeness of the earlier market creator. The successful organizations described here were for the most part innovative through and through, not just in engineering or product development.

At every stage of a company's development there are trade-offs to be made between innovation and efficiency, entrepreneurialism and management discipline, adaptation and market focus. The central argument of this chapter is that companies have to take account of all eight attributes, not just some of them. They also need to hold on to the qualities that made them successful in the first place. Emphasizing one at the expense of the others risks losing the company's ability to learn, to innovate, to adapt, to attract good people, and ultimately to retain the loyalty of customers.

Companies with powerful strategic assets like switching costs and network effects can be shielded from these imperatives but can easily be lulled into a false sense of security, which makes the shock even worse when the tide turns. In the long run the only way of coping with constant change is repeated reinvention, but that means endless experimentation and exploration, being prepared to try many things and frequently failing, all of which go against the big-company grain of focus, efficiency and consistent financial results. Big companies are optimized for effective execution, not exploration: they concentrate on the core capabilities and markets they know well, on eliminating uncertainty and meeting investors' expectations.

This is the fundamental dilemma – how to resolve the conflicting claims of discipline and entrepreneurialism, execution and exploration, focus and discovery. Most successful companies, for perfectly sensible reasons, choose discipline, execution and focus, as these are critical to operating efficiently on a large scale and to achieving the predictable results demanded by financial markets. Those who make the transition from market creator to large-scale industry leader often lose some of their original radicalism and entrepreneurialism. But when circumstances change, as they undoubtedly will one day, this can leave them high and dry – supplanted or marginalized by the innovations or incursions of others. It is often the biggest winners who suffer the furthest falls and suddenly become losers.

POSTSCRIPT

Oh no, it is an ever-fixed mark,
That looks on tempests and is never shaken;
It is the star to every wand'ring bark,
Whose worth's unknown, although his height be taken.

Shakespeare, Sonnet 116

Despite all the turbulence of the Age of the Internet with its accelerated creative destruction, some things have scarcely changed at all. A torrent of new products and technologies has swept over us, entire industries have disappeared, companies conduct much of their business, and people much of their lives, online or on mobile phones, but there are a few constants, not the least the discontinuities.

Capitalism has mutated many times since its flourishing in late eighteenth-century England and many sub-species have evolved. The version that depended on children working fourteen-hour days has long disappeared in rich countries, where brains have displaced brawn as strategic assets. I have concentrated on the last thirty years, but the era of discontinuities and globalization really started somewhere between 1776 and 1789, with the American, French and Industrial Revolutions, and has proceeded in fits and starts ever since. It was then that the economy started to grow at a rate never seen before. In 1750, when global GDP per head was about $190, only twice that of the hunter-gatherers of thousands of years ago, it was still dominated by landed interests, intent on defending the status quo. By 1800 competitive markets and a new breed of entrepreneurs were disrupting traditional ways of life. This meant painful dislocation for many in the short-term, but the long-term material gains were enormous. By 2000, global GDP per head had grown 36-fold to $6,600 – and in the richest countries exceeded $30,000.

Serendipity and hazard have always played a big part in who benefits from change and who gains. David Landes has argued that amongst the reasons for the rise of Western Europe to wealth and power 500 years ago were two essentially fortuitous developments during the Middle Ages. The ubiquity of church clocks and the widespread observance of the Angelus in rural areas familiarised ordinary people with the measurement of time and helped them appreciate its value. The bitter struggles for power between Church and State and between kings, nobles and cities, led eventually to the idea of pluralistic governance and the gradual acceptance of the value of debate and discussion between many points of view. In other parts of the world single sources of authority were the norm and stifled rational, scientific enquiry and competition. At the time however these divisions, especially religious dissent, were universally deplored as calamities for Christendom.

And one man's disastrous disruption is frequently another's golden opportunity. The appalling plagues which wiped out a third of the population of Western Europe in the fourteenth century, made labour a scarce resource, led to a loosening of the bonds of feudalism and raised wages. Free trade and increases in agricultural productivity in Western countries during the nineteenth and twentieth centuries eliminated tens of millions of jobs. This had tragic consequences for countless individuals and their families, yet many of those forced off the land, often brutally, found better lives in the New World. Many of course ended up in crowded cities, living and working in Dickensian squalor, but the descendants of all these uprooted peasants enjoy vastly higher standards of living than they ever could have in agriculture.

Massive over-investment by telecommunications suppliers in the 1990s led to vast amounts of unused fibre capacity being installed and a vertiginous drop in the price of bandwidth. This wiped billions off the value of debt-ridden telecom companies and pushed many into bankruptcy, but gave a big boost to e-commerce and other Internet-based businesses. It also enabled millions of economic migrants to phone home cheaply and countless small businesses to operate globally.

This example parallels the history of the railway industry in the nineteenth century. Not many investors saw a return on their money, but the enormous improvements in the economic infrastructure made possible the rapid growth of manufacturing industries dependent on cheap transport.

Although the networked economy has several distinctive features, the fundamental principles of supply and demand, economies of scale and scope, and the role of entrepreneurs have not changed. Nor has human nature altered greatly since the days of Homer – the young and vigorous challenge the old and complacent, ingenuity and bold visions move things forward, but ambition and success all too often lead to hubris and self-deception.

Even the success factors are not entirely new. Although their relative importance has altered enormously in the last thirty years, all of them applied to some extent to businesses in earlier times. Distinctive capabilities have been fundamental to the creation of new industries since James Watt and Matthew Boulton commercialized the steam engine, and Richard Arkwright revolutionized the manufacture of textiles 200 years ago. Another eighteenth-century entrepreneur, Josiah Wedgwood, realized that branding and differentiation were as important as technical innovation. Henry Ford's notion of human capital was cruder than today's, but he was as concerned with the principle as he was with revolutionary business models and value propositions. There were collaborative business networks in Europe as early as the Hanseatic League in the twelfth century, and the Medici started a pan-European one in the fourteenth.

What has shuffled the pack in the last thirty years have been the forces that have transformed and integrated the global economy – the lowering of market barriers and the increase in competitive intensity, the diffusion of information and communications technology and the profusion of networks of different kinds. All this has made qualities like adaptability and the ability to extend capabilities imperative for all would-be industry leaders, and the nurturing of human capital and collaborative networks critical for most. And now that fewer companies are able to rely on barriers to entry as their main source of competitive advantage, distinctive capabilities and compelling value propositions have become even more vital to every business.

The pace and tumult of change have prompted an often frantic search for new explanations for new phenomena. In a world where uproar seems the only music, fresh insights are always welcome and we must always be prepared to question our previous assumptions. We should be wary, however, of jettisoning too quickly what we have painfully learned from the past and of seizing on new notions as magical mantras. The key to making sense of constantly shifting patterns is not complex top-down theories or buzzwords. The best way to see the big picture of the networked economy, I have found, is by looking hard at the portraits of the individual organizations that have created it. The more stories one considers, and the more questions one asks, the more patterns one sees.

SOURCES AND BIBLIOGRAPHY

These notes are arranged by chapter and mainly refer to books. In all of the company stories, I have made use of annual reports, where available, and of the excellent online archives of *The Economist* and *Financial Times*. I have used countless other online sources. More detailed notes are available at <www.kieranlevis.com>.

Introduction

This part of the eBay story relies on Adam Cohen, *The Perfect Store: Inside eBay* (Little Brown, 2002), and on *eBoys* (see notes on Chapter 3). The Kodak story draws mainly on annual reports and press coverage.

Joseph Schumpeter and *Capitalism, Socialism and Democracy* (Harper, 1942), where he outlined his theory of creative destruction, are discussed in detail in Chapter 10.

Peter Drucker pioneered the profiling of individual firms with *The Concept of the Corporation* (John Day, 1946). Alfred Chandler was the greatest business historian of the twentieth century. *The Visible Hand: The Managerial Revolution in American Business* (Harvard University Press, 1990) is an outstanding work.

Tom Peters and Robert H. Waterman's *In Search of Excellence* (Harper & Row, 1982) is perhaps the most influential business book of the last thirty years. Its only challengers have been Jim Collins's two. The first he wrote with Jerry Porass, *Built To Last: Successful Habits of Visionary Companies* (Random House, 1994). The second, *Good To Great: Why Some Companies Make the Leap and Others Don't* (Harper Business, 2002), is discussed in Chapter 13.

Phil Rosenzweig in *The Halo Effect… and the Eight Other Business Delusions that Deceive Managers* (Free Press, 2007) savages these books for their scientific pretensions and reliance on stories. I concur with many of his arguments, but not all, for obvious reasons.

Chapter 1

Steven Levy has written two fine books about Apple: *Insanely Great* (Penguin, 1994) and *The Perfect Thing* (Ebury, 2006) – about the iPod. Michael Malone's *Infinite Loop* (Doubleday, 1999) is strong on the early years, as Malone had access to people like McKenna and Scott. It is marred, however, by the author's visceral dislike of Steve Jobs. Leander Kahney's *Inside Steve's Brain* (Portfolio, 2008, Atlantic, 2009) has valuable insights on the new man and his recent achievements.

My account of the early years of Sony relies mainly on John Nathan's excellent *Sony, the Private Life* (Harper Collins, 1999). The Columbia saga is based largely on *Hit and Run: How Jon Peters and Peter Gruber Took Sony for a Ride in Hollywood* (Simon and Schuster, 1996). Alfred Chandler's *Inventing the Electronic Century* (Harvard University Press, 2005) is masterful and influenced several other chapters.

Chapter 2

The words from the *Iliad* are addressed to the young Achilles by his father.

I owe my conviction that organizational capabilities are the most important basis for competitive advantage to John Kay's *The Foundations of Corporate Success: How Business Strategies Add Value* (OUP, 1993). Alfred Chandler's writings and Kay's subsequent columns in the *Financial Times* have reinforced that conviction. Kay disapproves of Hamel and Prahalad's *Competing for the Future: Building and Sustaining the Sources of Innovation* (HBSP, 1996), but their ideas on extending capabilities, non-traditional competition and competitive white space, a phrase they coined, broke important new ground. Dorothy Leonard takes a more rigorous approach to developing new capabilities in *The Wellsprings of Knowledge* (HBSP, 1995).

Niall Ferguson's *The Pity of War* (Basic Books, 1999) discusses German military superiority, but the definitive account is Trevor Dupuy's *A Genius For War* (NOVA, 1977).

Southwest's radical approach and how Ryanair followed in its footsteps are recounted by Siobhan Creaton in Ryanair: *How a Small Irish Airline Conquered Europe* (Aurum, 2004) and Simon Calder in *No Frills: The Truth Behind the Low-Cost Revolution in the Skies*. The ventures of American universities are based on my own *The Business of eLearning* (Screen Digest, 2002). Dell's development of new capabilities is discussed in an interview with Joan Magretta, 'The Power of Virtual Integration', *Harvard Business Review* (March 1988).

Chapter 3

Robert Foster's *Amazon.com: Get Big Fast* (Random House, 2000) is a good, if uncritical, account of the early years. James Marcus's *Amazonia: Five Years at the Epicentre of the dot.com Juggernaut* (The New Press, 2004) is amusing. John Cassidy strikes a more sceptical note in *Dot Con: The Greatest Story Ever Sold* (Harper-Collins, 2002). Nicholas Carr's *The Big Switch: Rewiring the World, from Edison to Google* (W. W. Norton, 2008) shows how Amazon developed new capabilities.

Cassidy gives a good, brief account of the Webvan debacle, but the invaluable source on this was Randall Stross's *eBoys* (Ballantine Books, 2000). Stross spent a year as a fly on the wall at Benchmark Capital when it decided to back both Webvan and eBay, and wrote it before the fate of each was known.

Chapter 4

Hundreds of books have been written on entrepreneurialism and corporate culture. The best is probably still Peter Drucker's *Innovation and Entrepreneurship* (Collins, 1997). Peter Senge's *The Fifth Discipline: The Art and Practice of the Learning Organization* (Doubleday, 1990) is superb.

There is a good account of DoCoMo's approach to i-mode in Takeshi Natsuno's *i-mode Strategy* (John Wiley, 2003). Michael Lewis's *The New New Thing: A Silicon Valley Story* (W. W. Norton, 2000) is a brilliant portrait of Jim Clark.

Chapter 5

Alec Klein's *Stealing Time: Steve Case, Jerry Levin and the Collapse of AOL Time Warner* (Simon and Schuster, 2003) is an enthralling piece of investigative journalism. Julia Wilkinson's *My Life at AOL* (1st Books Library, 2000) gives an insider's perspective. Michael Wolff's *Burn Rate: How I Survived the Gold Rush Years on the Internet* (Touchstone, 1998) has useful insights.

Netscape is well covered both by Lewis (see Chapter 4 above) and by Charles Ferguson's *High Stakes, No Prisoners: A Winner's Tale of Greed and Glory in the Internet Wars* (Three Rivers Press, 1999), a well-written and thoughtful book. Cisco's early story is told in David Bunnell's *The Cisco Connection: The Story Behind the Real Internet Superpower* (John Wiley, 2000).

Chapter 6

Joan Magretta and Nan Stone's *What Management Is: How It Works and Why It's Everybody's Business* (Free Press, 2002) contains, amongst other gems, a brilliant account of how business models work.

Drucker's line on customers comes from his still relevant *The Practice of Management* (Butterworth, 1955). Michael Dell's account of how he built his business, *Direct from Dell* (Harper Business, 1999), is useful.

Chapter 7

Cohen's (see Introduction) is a rather starry-eyed account of eBay's early years, but contains a wealth of useful detail. Eric Jackson paints a darker picture in *The PayPal Wars: Battles with eBay, the Media, the Mafia and the World* (World Ahead Publishing, 2006).

Google has been the subject of at least three good books. David Vise's *The Google Story* (Bantam Dell, 2005) does what it says. John Batelle puts it in a broader context in *The Search: How Google and its Rivals Re-wrote the Rules of Business and Transformed our Culture* (Nicholas Breeley, 2005). Randall Stross provides a more recent perspective in *Planet Google: One Company's Audacious Plan to Organize Everything We Know* (Free Press/Atlantic, 2008).

Google is one of the main characters in Nicholas Carr's *The Big Switch* (see Chapter 3 above), a characteristically astute analysis of how the computing infrastructure is evolving. Google also figures prominently in Gary Hamel's rather less persuasive *The Future of Management* (HBSP, 2007).

Chapter 8

My account of Henry Ford draws on several sources. He told his own story in *My Life and Work* in 1922.

The clearest guide to concepts like switching costs, standards and network effects is *Information Rules: A Strategic Guide to the Network Economy* by Carl Shapiro and Hal Varian (HBSP, 1999). Eric Beinhocker's *The Origin of Wealth: Evolution, Complexity and the Radical Remaking of Economics* (HBSP, 2006) is a masterpiece that prompted me to revise many of my previous assumptions.

Nassim Nicholas Taleb has many interesting, and occasionally infuriating, observations in *The Black Swan, The Impact of the Highly Improbable* (Random House, 2007) and *Fooled by Randomness* (Texere, 2001). *The Fifth Discipline* (see Chapter 4 above) has an excellent discussion of feedback loops.

Malcolm Gladwell's *The Tipping Point: How Little Things Can Make a Big Difference* (Little, Brown, 2000) confines itself to how word-of-mouth sometimes works, without mentioning feedback or virtual networks. Clay Shirky in *Here Comes Everybody* (Penguin, 2008) is an interesting examination of new forms of group activity.

The figures on English language speaking come from David Crystal's contribution to the *Cambridge History of the English Language* (2006).

Chapter 9

Paul Carroll gives a gripping account of the crisis years in *Big Blues: The Unmaking of IBM* (Crown Publications, 1993). Doug Carr describes how it bounced back in *IBM Redux: Lou Gerstner and the Business Turnaround of the Decade* (Harper Business, 1999). Deborah Spar's chapter on Microsoft in *Ruling the Waves: Cycles of Discovery, Chaos and Wealth, from the Compass to the Internet* (Harcourt, 2001) is instructive.

I used several sources for the Encyclopædia Britannica story, notably Philip Evans and Thomas Wurster's *Blown To Bits: How the New Economics of Information Transforms Strategy* (HBSP, 2000) and a personal account by Robert McHenry, a former managing editor, of Britannica's involvement in electronic media, available at <www.howtoknow.com>.

Chapter 10

See 'Introduction' for the Schumpeter reference. The chapter on creative destruction is surprisingly short. Thomas McCraw's biography of the great man, *Prophet of Innovation: Joseph Schumpeter and Creative Destruction* (Belknap Press, 2007) is superb. There is a useful summary of Schumpeter's thinking in Robert Heilbonner's *The Worldly Philosophers* (Simon and Schuster, 1953, 1999).

Tony Judt's *Postwar* (Penguin, 2006) is a magisterial account of the transformation of Europe after 1945. Martin Wolf's *Why Globalization Works: The Case for the Global Market Economy* (Yale, 2004) is a masterly description of how the global economy came about and how it works.

Richard Foster's *Innovation, The Attacker's Advantage*, was published by Summit Books in 1986. His subsequent work, written with Sarah Kaplan, was a direct dig at Jim Collins: *Creative Destruction: Why Companies That Are Built to Last Underperform the Market, and How to Successfully Transform Them* (Currency Doubleday, 2001). Clayton Christensen's seminal work on disruptive technologies is *The Innovator's Dilemma* (HBSP, 1997).

Andy Grove was a skilled practitioner of creative destruction at Intel and showed himself a perceptive observer in *Only the Paranoid Survive* (Doubleday, 1996).

Chapter 11

My account of Sky's rise draws mainly on Matthew Horsman's detailed chronicle: *Sky High: The Inside Story of BSkyB* (Orion, 1997), based on his deep knowledge and interviews with most of the key people. Deborah Spar's *Ruling the Waves* (see Chapter 9 above) adds useful insights. *Out of Time* (Simon & Schuster, 2004), by Alex Fynn and Lynton Guest includes an account of the formation of the Premier League and the first negotiations on television rights.

Nokia commissioned an historian, Martti Haikio, to write *Nokia, The Inside Story* (Edita, 2001; Pearson Education, 2002). It is immensely detailed, but not for the faint-hearted. More instructive is *Mobile Usability: How Nokia Changed the Face of the Mobile Phone* (McGraw Hill, 2003) by Christian Lindholm, Turkka Keinonen and Harri Kiljander. This, too, is an inside view since all them have worked for Nokia.

A more dispassionate and interesting book on the broad phenomenon of mobile telephony is Howard Reingold's *Smart Mobs: The Next Social Revolution* (Perseus, 2003). Reingold had coined the term 'virtual community' long before the arrival of the mass Internet.

Chapter 12

Deductive tinkering is described in Beinhocker (see Chapter 8). For the CDi, see Chandler (Chapter 1.) Steven Weber's *The Success of Open Source* (Harvard University Press, 2004) is a well-written history. Eric Raymond's seminal essay, 'The Cathedral and the Bazaar', is available at <www.catb.org>.

James Surowiecki's *The Wisdom of Crowds: Why the Many are Smarter than the Few* (Doubleday, 2004) is perceptive and well balanced. *Wikinomics: How Mass Collaboration Changes Everything* (Atlantic, 2007), by Don Tapscott and Anthony Williams, is interesting on collaboration but has surprisingly little to say about economics. For *Blown to Bits*, see Chapter 10. The account of the Internet frenzy draws on Cassidy's *Dot Con* (see Chapter 3). Extracts from *The Internet Report* are available at <www.MorganStanley.com>.

The Vivendi disaster has been sharply chronicled by Jo Johnson and Martine Orange in *The Man Who Tried to Buy the World: Jean-Marie Messier and Vivendi International* (Viking, 2003). Dick Martin's *Tough Calls: AT&T and the Hard Lessons Learned from the Telecom Wars* (Amacom, 2004) is a fascinating insider's account.

Nicholas Negroponte's *Being Digital* was published by Vintage in 1995. John Seely Brown and Paul Duguid's *The Social Life of Information* (HBSP, 2000) is an excellent corrective to digerati hype. Mary Midgley discusses the dangers of reductive thinking in *The Myths We Live By* (Routledge, 2003).

The leading authority on behavioural economics and on System 1 and System 2 thinking is Daniel Kahneman, the only psychologist to win the Nobel Prize for economics. His lecture on that occasion is available at <www.nobelprize.org>. The quotation is from 'Delusions of Success: How Optimism Undermines Executives' Decisions', written with D. Lovallo, *Harvard Business Review*, July 2003. Both Beinhocker and Taleb (2007) draw on his work in an accessible way.

Chapter 13

The Hedgehog and the Fox appears in several collections of Isaiah Berlin's writings. The one that best captures the breadth of his thinking is *The Proper Study of Mankind* (Pimlico, 1998), edited by Henry Hardy and Roger Hanscher.

Strategic Safari, by Henry Mintzberg, Bruce Ahlstrand and Joseph Lampel (Free Press, 1998), is a witty, iconoclastic discussion of the many schools of thought on business strategy. *Strategy Bites Back* (FT Prentice Hall, 2005), by the same authors, is even more amusing. John Micklethwait's and Adrian Wooldridge's *The Witch Doctors* (Heinemann, 1996) debunks the pretensions of management gurus. Michael Porter's most influential books are *Competitive Strategy: Techniques for Analyzing Industries and Competitors* (Free Press, 1980) and *Competitive Advantage: Creating and Sustaining Superior Performance* (Free Press, 1985).

Leonard (see Chapter 3) is essential reading on extending capabilities, and Hamel and Prahalad (see Chapter 2) are particularly relevant to recognizing and adapting to change. Gary Hamel's *Leading the Revolution: How to Thrive in Turbulent Times by Making Innovation a Way of Life* (Plume, 2000, 2002) is interesting but comes dangerously close to suggesting that companies can become anything they want. W. Chan Kim and Renee Mauborgne's *Blue Ocean Strategy: How to Create Uncontested Market Space and Make the Competition Irrelevant* (HBSP, 2005) builds on Hamel and Prahalad's idea of 'competitive white space'. *The Only Sustainable Edge: Why Business Strategy Depends on Productive Friction and Dynamic Specialization* (HBSP, 2005), by John Hagel and John Seely Brown, brings out the links between capability building, business networks and human capital.

Costas Markides and Paul Gerosi make an interesting distinction between colonizers and consolidators in new markets. In *Fast Second: How Smart Companies Bypass Radical Innovation to Enter and Dominate New Markets* (Jossey-Bass, 2004) they argue that it is advantageous to let others create the new market.

Frederick Reichheld's *The Loyalty Effect: the Hidden Force Behind Growth, Profits and Lasting Value* (HBSP, 1996) is still relevant, as is *The Trusted Advisor*, by David Maister, Robert Galford and Charles Green, published by Simon and Schuster in 2000.

David Aaker and Erich Joachimsthaler's *Brand Leadership* (Free Press, 2000) is a comprehensive guide to a subject often obscured by airy abstractions. Stan Makin and Simon Knox have interesting ideas on organizational brands, value propositions and customer relationships: *Competing on Value: Bridging the Gap Between Brand and Customer Value* (Financial Times, 1998).

Christensen (see Chapter 10) discusses business networks. Annabelle Gower and Michael Cusulmano's *Platform Leadership: How Intel, Microsoft and Cisco Drive Industry Leadership* (HBSP, 2002) is interesting and persuasive. Slightly less so is Marco Iansiti and Roy Lieven's *The Keystone Advantage: What the New Dynamics of Business Ecosystems Mean for Strategy, Innovation and Sustainability* (HBSP, 2004).

In the vast literature on talent and human capital, *Competing for Global Talent*, edited by Christiane Kuptsch and Pang Eng Fong (International Institute for Labour Studies, 2006) is useful, as is Adrian Wooldrige's survey in *The Economist*, October 2006. *Talent on Demand: Managing Talent in an Age of Uncertainty* by Peter Cappelli (HBSP, 2008) is a fresh look at a perennial problem.

Finally, Richard Bronk's brilliant defence of pluralism and poetry, *The Romantic Economist: Imagination in Economics* (Cambridge, 2009) bolstered my conviction that man cannot live by bread or rationalism alone.

ACKNOWLEDGEMENTS

This book was conceived during a series of conversations with Peter Antonioni early in 2004. Peter has made countless comments on successive drafts and I owe him an enormous debt. Whilst taking full responsibility for any errors, there are many other debts to acknowledge. The most important are to those writers cited in the text, but earlier experiences and people played crucial roles.

Prestel was a brave precursor of the Internet, a solution that never found enough problems, and Alex Reid was the most inspiring leader I ever worked for. Sadly the lessons I learned from him included the limitations of brilliance and charisma, and the dangers of paying insufficient attention to customers. Alex's successor, Richard Hooper, gave me my start as a strategist and taught me that we only really learn from our own mistakes. From John Short I learned that great strengths are invariably accompanied by great weaknesses. Working with all these corporate venturers, and less talented ones subsequently, I learned that entrepreneurial talent cannot just be grafted onto bureaucratic organizations, that capabilities and cultures, like brands, take years to develop.

I also learned as much from three outstanding lieutenants – Tim Kaye, Tom Phillips and Natalie Turner – as they ever did from me.

The other defining learning experiences for me were Supercall, the telephone information business I started for BT Enterprises, and BAe Systems' ventures in satellite television, which I led. There I learned mainly from the many talented people I persuaded to join me, and have continued to do so since from colleagues and clients too numerous to mention. I learned about creative destruction the hard way, not least from Sam Chisholm, who unwittingly destroyed one of my fledgling businesses.

Ironically for someone whose business career had been devoted to starting up new things, I subsequently found myself as a consultant explaining to several clients that the business opportunities they thought golden were not particularly attractive, and generally required capabilities they didn't have. Most accepted this with grace, though they didn't always ask me back again. But when I have warned clients of serious threats to their businesses that they hadn't considered, their invariable reaction has been denial. It gave me little comfort to be subsequently proved right.

Two people prodded me into making this a much better book: David Godwin, my agent, and Sarah Norman, my brilliant editor at Atlantic Books. Many thanks to them, to Toby Mundy for his enthusiastic encouragement, and to Peter Hennessy for introducing me to David and encouraging me to replenish my historical wellsprings.

Many people have commented on drafts, particularly Graham Wheat, my ever-patient wife, Angela, and my not so patient son, Jeremy. My thanks also to Danielle Barr, Steve and Tom Bonnick, Kathleen Brady (and her inspirational father), Michael Cutler, Ketty Dal Lago, Richard Duggleby, John Fielding, Henry Hardy, Tom Hargadon, Gary Heiden, Sophie Hoult, Joan Jackson, John Lawrence, John Madell, Eve Mendelsohn, Christian Michel, Barbara Nokes, Gerald O'Connell, Geoffrey Owen, Marty Perlmutter, Barry Reeves, Peter Slater, Natalie Turner, Pat Whaley, and to Ederyn and Ilona Williams.

Finally I must thank my wonderful doctors, Katja Paeprer, Martin Lowe and Jack McCready, and everyone at the Heart Hospital who restored me to full health and helped me to keep working throughout 2008.

INDEX

@Home 96

A&C Black 257
abstractions 348–50
Accept.com 102
Adair 220
adaptation 361–4
Adobe 37, 373
Adobe Acrobat 37
advertising
    Amazon 93–4, 94, 95
    AOL 147–8
    Apple Computer 29, 34, 41
    Craigslist 275–8
    Encyclopædia Britannica 258
    Google 193, 195–8
    Webvan 108
AdWords 196–8
Akers, John 249–51, 252–4, 256, 351
Alando.de 183
Albertson 115
Allen, Howard 63–4, 65
Allen, Paul 146

AltaVista 190, 191, 193
Alward, Peter 62
Amazon 1, 81, 86–105, 106, 111, 122,
187, 198, 218, 221, 286, 347–8, 363,
366, 369
    acquisitions programme 96, 101–2
    advertising 93–4, 94, 95
    affiliates 102
    assessment 104–5
    auction services 101, 180–1
    branding 374
    business model 92–3
    business networks 381–2
    culture 97–8
    customer care 100, 162, 162–3, 372
    customer experience 98–100
    diversification 95–6
    funding 92–4, 95
    growth 93, 94–6, 104
    and the Internet stock collapse,
    2000 103–4
    launch 90–2
    losses 95, 102, 103–4

market focus 166–7
music sales 96
organizational capabilities 79
origins 87–90
profit margins 175
proposition 115
recruitment 96–7, 378
revenues 93, 95, 102–3, 103–4, 105
scalability 230–1
strategic planning 368
vision 83, 105
Amazon Marketplace 104, 382
Amazon Merchants 104
ambition 83–5
AMD 124
Amdahl, Gene 239, 250, 254
Amelio, Gil 40
American Booksellers Association 92
American Express 161
Andersen, Arthur 301
Andersen, Chris 230–1
Andersen Consulting 87
Andreessen, Marc 125–9, 131, 133,
134–5, 135, 138, 139, 165
Anglo-American Cable Company 330
anti-capitalists 11
AOL 94, 96, 120, 121, 125, 133, 136,
139–59, 196, 199, 202, 203, 221, 226,
231, 234, 346, 353, 362, 364
    acquisition of Netscape 134, 139,
        153
    advertising 147–8
    advertising sales 151–2, 156
    assessment 157–9
    business model 160
    business networks 380, 381, 382
    chat rooms 142–3
    competition 147
    customer loyalty 158, 372
    and eBay 176
    Federal Trade Commission
        investigation 156
    funding 145

growth 142–6, 147–9, 149–51
human capital 378
innovations 143–4
market focus 166
marketing 144–6, 149
merger with Time Warner 153–7,
    345
and Microsoft 146–7
origins 139–42
power 151–3
revenues 150
services 142–4
stock market floatation 144, 149
strategic planning 366–7
Apollo 246
Apple Computer 17–47, 82, 120, 121,
123, 134, 138, 142, 220, 223, 226–7, 269,
360, 362, 365
    achievements 47
    advertising 29, 34, 41
    the Apple II 19, 21, 22, 23–4, 26,
        33–4, 35, 36, 37, 76, 115, 227, 332
    the Apple III 24, 25
    Black Wednesday 25–6
    business networks 381
    the Classic 37
    competition 27, 30, 41–2, 45
    decline 38–9
    the Dynabook 39
    early development 20–3
    financial crisis 35
    first computer 18–19
    human capital 378
    the iMac 40–1
    and innovation 75–6
    iPhone 45
    iPod 14, 42–6, 284
    iPod sales 43, 44–5
    iTunes 272–3
    leadership 46, 165
    licensing agreement with PARC
        24
    the Lisa 24, 27–8, 34

Mac sales 34–5, 37–8
the Macintosh 28–31, 34–5, 36–7, 39, 47
market focus 166
mission statement 36
the Newton 39
organizational capabilities 76
origins 17–19
product compatibility 33
regime change 31–5
research and development spending 24
revenues 23, 27, 28, 34, 38, 41, 43, 45, 45–6
revival 40–2, 84
Sculley's appointment 32–3
Sculley's reforms 35–6
seed capital 21
stock market floatation 24–5
strategic planning 366–7, 368
survival 46–7
take-off 23–5
the Trojan niche 37–8
vision 83, 85
Arkwright, Richard 388
Armstrong, Mike 245, 247–8
Arpanet 279–82, 339
arrogance 353–4
artificial intelligence systems 106
Ask Jeeves 203
ASkyB 310
AT&T 6, 50, 350, 353, 362
Atari 244
Atkinson, Bill 30
attributes, essential 167–9
Auction Universe 177
AuctionWeb (later eBay) 1–3, 170–5
Ayre, Rick 99

BackRub 190
Bain Consulting group 371
Bankers Trust 87
banking crisis, 2008 274–5, 288

Barksdale, Jim 129, 132, 134, 135, 137, 138, 139, 146, 165, 363
Barlow, Frank 309
Barnes & Noble 79, 92, 94
barriers 235
BBC 291, 292, 294–5, 303–4, 312, 313
Beanie Babies 173
Bechtel 108
Bechtolsheim, Andy 191–2
Becker, Gary 377
Beckham, David 232
Beinhocker, Eric 123, 334
Beirne, David 178
Bell, Alexander 330
Bell, Andrew 257
Bell Labs 50
Benchmark Capital 1, 107, 110, 175–6, 178
Benton, William 257–8
Benton Foundation, the 260
Berkowitz, Steve 203
Berlin, Isaiah 10, 356–8
Berlow, Myer 151
Berners-Lee, Tim 40, 126, 280
Betamax VCRs 53–5
Bezos, Jeff 83, 86–90, 91–2, 93, 94, 95, 96–9, 100, 100–1, 102, 103, 105, 157–8, 164, 192, 200, 218, 220, 363, 375
Bezos, Mackenzie 88, 89
Billpoint 185
Bina, Eric 126, 128
Birt, John 305
Blackstone Group 64
Blodget, Henry 102–3
blogs 338
Bloomberg. 338
Blown to Bits (Wurster and Evans) 347–8
Body Shop 375
Books.com 89
Booth, Mark 311
Borders 106
Borders, Louis 106–7, 110, 111, 114

Bosack, Len 120
Boston Consulting Group 345
Boulton, Matthew 388
Bowie, David 273
Brainerd, Paul 36
branding 55–6, 94, 137–8, 193,
198–200, 211, 324, 374–5, 388
Breest, Gunther 62
Brin, Sergy 83, 119, 164, 189–91, 191–3,
194, 196–7, 200–3, 208
Britannica Online 259
British Army 77
British Satellite Broadcasting (BSB)
295, 296–7, 300–1, 332
BSkyB 13, 74, 123, 161, 220, 221,
291–314, 336, 349, 351, 363–4
    acquires rights to Premier League
    302–6
    advertising sales 306
    assessment 313–14
    background 291–5
    and BSB 300–1
    business model 305, 306–9
    Chisholm takes charge 299–302
    Chisholm's final days 309–11
    competition 310–11
    customer lock-in 373
    customer service centre 306
    and the digital TV threat 311–13
    early failure 298–9
    installation business 299
    launch 294–6
    leadership 165
    losses 299, 302
    marketing 298–9
    Movie Channel 307
    multi-channel package 307
    proposition 115
    renewal of Premier League contract
    310
    revenues 307, 308, 314
    spectrum scarcity 293–6
    Sports Channel 304

    stock market floatation 309
    strategic planning 367
    subscribers 302, 305, 308, 314
Buckmaster, Jim 277
budget airlines 79, 161, 283, 352
Buffett, Warren 205
Bulova 51
bureaucracy 120
Burton-Davies, Paul 89, 90, 96
Business 2.0 345
business books 10
business creation 121–2
business judgement 83
business lifespans 268
business models 160–1, 168, 195–8,
218, 283–4, 332, 336, 344
Business Week 33, 100–1, 108
Byte Shop, the 19

cable TV 147, 293–4, 308
Campbell, Bill 38, 202
Cannavino, Jim 251–2
capabilities, distinctive 82–3, 115, 124,
135–6
capital, increased availability of 281–2
capitalism 9, 11, 265–8, 289, 386
Capitalism, Socialism and Democracy
(Schumpeter) 265–6
Cardean University 79, 80
Carlton Communications 311, 312–13,
313
Carp, Daniel 3–4
Cary, Frank 239–40, 241, 256
Case, Dan 140
Case, Steve 136, 140, 142, 143, 144, 146,
147, 149, 153–5, 156–7, 158, 165
Caufield, Frank 140
CBS 55, 57, 58, 61, 61–2, 107
CBS/Sony Records 55, 56, 57
CD-i (Compact Disc Interactive) player
334
CD-NOW 96
CD-ROM 259, 260, 261, 262, 284, 334

Cellnet 317
cellular telephony 81. *see also* Nokia
Chambers, John 188, 335
Chance, David 300, 307, 310, 311
Chandler, Alfred 54
change 8, 361–4, 389
Channel 4 292, 312, 314
Chargeurs 295, 297, 309
charisma 21
Chen, Steve 209
China 207, 281, 380
Chisholm, Sam 165, 299–302, 303–4, 304, 306, 309–11, 383
Christensen, Clayton 124, 268–70
Cisco 6, 20, 120, 121, 122, 138, 188, 221, 230, 335, 364, 365, 368, 369, 373, 381–2
Claris 38
Clark, Jim 116–18, 121, 125, 126–8, 130–2, 134–5, 135, 136, 138, 139, 165–6, 363
classical music 83, 284
CNN 154
Coase, Ronald 379
Coca-Cola 63, 65
Colburn, David 151–2, 156, 157
Cold War, end of 281
collaboration 379–82
Collins, Jim 10, 356–7, 358
Colony, George 94
Columbia Pictures 59–61, 62–71
Columbia Records 61–2
Columbia University 79, 80
commoditization 351
Commodore 142, 244
communications 280–1, 379
compact discs 56–8, 222, 224, 280
Compaq 30, 133–4, 222, 245–6, 247, 248, 252, 255
competency-based education 79
competition 110, 111
competitive advantage 79–80, 177–8, 233, 287, 356, 364, 365, 374
competitive white space 219

complacency 12, 351, 364–5
complementary skills, partnerships between people with 119–20
complex systems 358–9
Compton's Multimedia Encyclopaedia 259–60
CompuServe 133, 141, 142, 144, 150
Computer Literacy Bookshops 89
content 349
Control Video Corporation 139–40
convergence 60, 73–4, 349–50
core competencies 76
corporate control systems 123
costs 9, 336–8
courage 119–20
Craigslist 124, 186, 275–8, 283, 340
Creative Artists Agency 63
creative destruction 7, 116–18, 263–5, 354–5, 384
    acceptance 289–90
    age of discontinuity 279–82
    definition 266
    discontinuities 272–3, 274–82, 287, 361–2
    disruptive competition 263–4, 283–6, 326–7, 355–6, 386–8
    disruptive technology 268–73
    embracing 286–9
    origins 264–8
    role 267–8
creativity 46, 73–4, 83, 118, 120–1, 167
credibility 372
crowds, wisdom of 340–3
culture 383
customer care 100, 162–4, 168, 199, 370
customer loyalty 352, 370–2
customer proposition, the 113–16, 124, 136, 167, 335
customer relationships 137–8, 369–72
customers
    attracting 113–16
    habits 115–16

locking in 222–3, 373–4

D. E. Shaw & Co 87–8

de Geus, Arie 365

DEC 246, 331, 340, 363

decision making 352–3

decisiveness 350–1

deconstruction 348

deductive thinking 349

deductive tinkering 350–1

Dein, David 302, 303

delegation 122

Dell 13, 46, 82–3, 100, 115, 121, 122, 138, 161, 162, 163, 222, 230, 255, 256, 283–4, 365, 368, 369, 378, 381–2

Dell, Michael 45, 121, 122, 163, 164, 200

demand 114

design 55–6

desktop publishing 37

Deutsche Morgan Grenfell 95

Dhuey, Mike 37

Dickens, Charles 267

Digital Hub, the 42, 44, 45

digital photography 4–5, 42, 270

Digital Rights Management 44

digital sound recording 56–8

digital television 311–13, 313–14

digitization 349

Diller, Barry 309

Dillon, Eric 93, 94

direct marketing 82–3

discipline 47, 118–23, 124, 384

discontinuities 272–3, 274–82, 287, 363

Disney 53

disruptive competition 263–4, 283–6, 326–7, 351–2

disruptive technology 268–73

distance learning 80

distinctive capabilities 76, 167, 219, 354, 364–6, 388–9

distribution 105–12

dMarc Broadcasting 209

DoCoMo 119–20, 123, 165, 167, 378

Doerr, John 94, 96, 130, 192–3, 200, 201, 202

Double Click 209

Drucker, Peter 79, 164, 278–9

Drugstore.com 102

Dunlevie, Bruce 175

Dupuy, Trevor 77

Dusseldorf 51

Dutta, Rajiv 187

DVDs 285, 334

Dyke, Greg 295, 304

Dynabook, the 39

dynamic tension 120

Eastman, George 3–5

easyJet 368

eBay 3, 84, 101, 119, 120, 121, 162, 170–89, 220, 221, 226, 230, 283, 286, 364, 366, 369
    acquisition of PayPal 184–6
    acquisition of Skype 186–7, 189
    and AOL 176
    assessment 188–9
    business networks 381–2
    business plan 174–5
    community 171–2, 179–80
    competition 176–7, 180–1, 183–4
    competitive advantage 177–8
    customer lock-in 373
    diversification 181–2
    eBay Motors 181–2
    eBay Stores 185
    Feedback Forum 172–3
    funding 1, 175–6
    Great Collections 181
    growth 2–3, 104–5, 171, 174–9, 188–9
    human capital 378
    international growth 182–3
    leadership 164
    network effects 177, 225, 234–5
    network monopoly 10

origins 1–3, 170–1
profit margins 175
revenues 2, 174, 175, 182, 183
sales 3
stock market floatation 178–9
strategic planning 367
vision 83
eBay Motors 181–2
e-commerce 81, 83
Economist, The 156, 346
Edison, Thomas 330
EDS 124, 254
efficiency 384
Ehrnrooth, Casimir 319
Eindhoven 51
e-learning 79–80, 335
Elstein, David 305–6
email 225–6, 271
e-merchandising 99–100
Encarta 260, 261, 262, 283, 352
Encyclopædia Britannica 6, 167, 257–62,
283, 284, 341, 352, 353
English language 224
e-Niche 102
Enoki, Keichi 119–20, 165
Enron 152, 289
entrepreneurialism 118–23, 123–4, 136,
167, 384–5
equilibrium 265
Ericsson 74, 161, 315–16, 317, 321, 324,
326
    essential attributes 359–61
    adaptation 361–4
    branding 374–5
    customer relationships 370–4
    distinctive capabilities 364–6
    investment in human capital 376–8
    networked collaboration 379–82
    operational efficiency 368–70
    scalability 368–70
    strategic planning 366–8
Estridge, Don 31–2, 120, 240–5, 256
European Union 281

Europolitan 317
Eustace, Alan 3765
Evans, Philip 345, 347–8
Excite 96, 190

Facebook 338
Fairchild 124
Fargo, J. C. 161
Fathom 79
Fax 225
Federal Trade Commission 156
feedback loops 10, 178, 221, 223–4, 225,
225–7, 228, 360
Ferguson, Charles 135–6
Ferguson, Colin 132, 137
file sharing 43, 72, 272–3
Finland 316, 317, 318
Firefox 229
first mover advantage 219, 220–1
First World War 77
Fitel 87
Fleishman, Glenn 96–7
focus groups 99–100
Football League 295, 302–6, 313
Ford, Henry 216–19, 224, 230, 388–9
Ford Motor Company 216–19
Forrester Research 94, 145
Fortune 94, 128
Fortune 500 268
Foster, Richard 268, 287–8, 289, 356
Fox Broadcasting 293
France Telecom 141
Franco-Prussian War, 1870 77
Freeview 314
Friis, Janus 186–7
FTSE index 268
Fujitsu 246, 254
Full Service Network 153, 332
Future of Management, The (Hamel)
207

Galton, Francis 342
Gap 375

Gates, Bill 29, 31, 97, 117, 132, 133, 137, 138, 146–7, 200, 229, 240, 241, 246–7, 251, 339, 376
GDP 281, 282
GE 77, 124, 165
Gear.com 102
GEC 6, 332
General Atlantic Partners 93
General Motors 217
Geometry Machine, the 116–18
German army, organizational capabilities 77
Gerstner, Lou 165, 254–5, 256, 364, 383
Gilder, George 346
Gillette, King C. 161
Gladwell, Malcolm 223
global economy 5–6, 9, 281–2, 389
Global System for Mobile (GSM) 316–17, 317, 320, 333
Gnutella 272
Goldman Sachs 145
Good to Great (Collins) 10, 356–7
Google 14, 119, 121, 189–212, 220–1, 221, 226, 262, 283, 340, 342, 366, 369
  acquisitions 209, 338
  AdSense 198, 206
  advertising revenues 195–8, 207, 231
  AdWords 196–8
  assessment 211–12
  book digitization 210
  Book Search 210
  branding 198–200, 211, 374
  business model 195–8
  business networks 381–2, 382
  and China 207
  competition 199
  customer care 199, 372
  customer lock-in 373
  expansion 206–10
  funding 191, 192, 192–3
  Google Apps 208
  growth 6, 193–5, 197, 230

  leadership 164, 200–3
  marketing 193–4, 198
  and Microsoft 199–200, 229
  organizational capabilities 79
  origins 189–91
  proposition 115
  recruitment 194, 199, 207, 378
  revenues 195, 197, 198, 199, 203, 206
  start-up 191–2
  stock market floatation 203–6
  strategic planning 368
  vision 83
Google AdWords 187
Google Apps 208
Googleplex, the 194
GoTo 195–6, 197
Grade, Michael 296
Granada 292, 295, 297, 312–13, 313
Great Britain
  broadcasting franchises 291–5
  digital television conversion 311–13
  spectrum scarcity 293–6, 312
Griffin, Nancy 60, 67
Griffiths, Jim 172
Gross, Bill 195–6, 197
Grove, Andy 200, 274
Guber, Peter 64–5, 65–8, 69–71
Guber–Peters Entertainment 67–8

Häagen-Daz 375
Half.com 183–4
Hambrecht and Quest 140
Hamel, Gary 207, 289, 351
Hammer, Michael 369
Hanseatic League 389
Hastings, Reed 285
Heidrick & Struggles 32
Hewlett Packard 246
Heymann, Klaus 83
higher education 79–80
history, lessons of 8–9

Hit and Run (Griffin and Masters) 60, 67

Hitachi 30, 246, 250

Hoffman, Steve 181

HomeGrocer.com 102, 103, 109

Homer 10

Homer, Mark 129

Homer, Mike 132

HP 77, 124

Huber, Peter 155–6

human capital 11–12, 371–2, 376–8, 388

Hurley, Chad 209

Hush Puppies 223

Iansiti, Marco 381

iBazar 183

IBM 6, 22, 27, 29, 30, 31, 33, 38, 39, 77, 82, 120, 124, 141, 165, 222, 226, 236–56, 269, 283, 335, 351, 353, 364, 365, 369, 373, 374, 380, 372

    the AS/400 252–3

    assessment 254–6

    competition 245–8, 250, 255

    early development 236–40

    Global Services 254

    leadership 248–51, 256

    losses 253–4

    Management Committee 237–8, 245, 252

    market dominance 238

    market focus 167

    and Microsoft 240, 241, 246–7, 251–2

    OS/2 251–2

    PC development 240–5

    the PC Junior 244

    PC sales forecasts 242

    Prodigy 332

    the PS/2 248

    reinvention 254–5, 256

    research and development 239

    revenues 237, 238, 239, 244, 249, 252

    workforce 249, 250

Ibuka, Masara 48–50, 51, 52–3, 55, 56, 58–9, 70, 76, 164–5, 363

IdeaLab 195–6

Idestam, Fredrik 316

IKEA 380

IMDB 96

i-mode 119–20, 123, 167

iMovie 42

income, inequalities in 232

Independent Broadcasting Authority (IBA) 292, 294–5, 308

India 281, 380

Industrial Revolution 386–7

inequality, increases in 231–5

info-imaging industry 3

information, value of 336–8, 347

Inktomi 194

innovation 75–6, 76, 111–12, 119, 122–3, 123–4, 124, 167, 218, 265, 287, 365, 383–4

Innovation, the Attacker's Advantage (Foster) 268

Innovator's Dilemma, The (Christensen) 269

Intel 46, 87, 124, 138, 223, 241, 274, 330–1, 334–5, 380

inter-connections 10

Internet 9, 39, 72, 83, 98, 271, 276

    boom 116

    broadband 158

    excitement about 343–6

    free content 335–7

    growth 87, 141–2, 280, 332, 388

    online learning 79–80

    security 90

    stock collapse, 2000 103–4, 346

    structure 190

    valuations 345–6

    Web 2.0 338–40

Internet Explorer 131–4, 136, 147, 229

Internet Report (1996) 145, 343, 344

intuitive thinking 350

inventory handling 106–7

iPhoto 42
iPod 14, 42–6, 284
Iridium 333
iTunes 42, 43, 44, 272–3
iTunes Music Store 44
ITV 291–2, 292, 295, 297, 302, 302–4, 312, 313, 352
ITV Digital 313
Ive, John 42, 378

Jackson, Eric 184
Jagger, Mick 34
Japan 48, 58
Japan Measuring Instruments Company 48
Java 133
Jobs, Steven 17–19, 20, 21–3, 24, 26, 28, 29, 31–3, 34, 34–5, 36, 37, 40–2, 42–4, 45–6, 46, 121, 129, 165, 367, 378, 383
Johnson and Johnson 124
Junglee Corp 96
JVC 53, 226, 373

Kagle, Bob 175–6
Kahneman, Daniel 352
Kakkonen, Kari 316
Kaphan, Sheldon 88–9, 91, 96, 164, 356
Kaplan, Sarah 287–8, 289
Kapor, Mitch 243
Kelly, Kevin 345
Keynes, John Maynard 266–8, 363
Kijiji 186
Kimsey, Jim 139, 140, 142, 146
Kirschbaum, Laurence 210
Klein, Alex 140
Kleiner Perkins 94, 130, 131, 139, 192–3
Kleiner Perkins Caufield & Byers 140
Kleinwort Benson 333
K-Mart 106, 244
Knight-Ridder 107
Kodak 3–5, 11, 226

Kopelman, Josh 183–4
Kordestani, Omid 192, 203
Korhonen, Pertti 321
Kruse International 181

Landes, David 387
Lands End 81
LaserWriter 36–7
leadership 46, 136, 164–6, 168, 200–3, 248–51, 256, 383
Lehman Brothers 103
Leonard, Dorothy 363, 364, 366
Leonsis, Ted 146, 147, 165
Lerner, Sandra 120
Levin, Jerry 153–5, 156, 156–7, 349, 362
Lieven, Roy 381
Linux 229, 276, 339–40
London Weekend Television 292
long tail, the 230–1
L'Oréal 124
Lotus 1-2-3 227, 228, 243
Lotus Notes 343
Lowe, Bill 240, 241, 245, 247–8, 251
loyalty 49
Loyalty Effect, The (Reichheld) 371
Lucas, George 39, 117
LVMH 107

McCracken, Ed 117, 121
Macfarquhar, Colin 257
McKenna, Regis 20, 23, 26, 31, 35, 39, 46
McKinsey 320, 377, 378
MacWorld Expo trade shows 41
Maginot Line, the 235
Malone, John 147
managers, professional, conflict with creatives 120–1
Maples, Mike 228
Marconi 353
Marconi, Giuseppe 271, 330
Marcus, James 97–8, 99

market creation 216–23
market evolution 82–3
market focus 166–7, 168
market research 99–100, 110
marketing 18, 81, 198, 258
    AOL 144–6, 149
    BSkyB 298–9
    Google 193–4
    Nokia 323–4
Markkula, Mike 20–1, 25, 26, 31–2, 165
Marks and Spencer 352
mass collaboration 340–3
Masters, Kim 60, 67
Matsenuga, Mari 119–20
Matsushita 53, 54
mature companies, entrepreneurialism 123–4
Mayfield Fund 1, 117, 175
Meeker, Mary 95, 145–6, 344, 345–6
Merrill Lynch 79, 145, 333
Messier, Jean-Marie 353
MGM 63
Micro Channel Architecture 248
Microsoft 14, 42, 46, 47, 82, 117, 131–2, 138, 220–1, 221, 223, 269, 284, 338, 361, 364
    and AOL 146–7
    business model 261
    business networks 381
    customer lock-in 373
    customer proposition 136
    domination of 227–9
    Encarta 260, 261, 262, 283, 352
    and Encyclopædia Britannica 259–60
    Excel 228
    and Google 199–200, 229
    Hotmail 161
    human capital 376
    and IBM 240, 241, 246–7, 251–2
    Internet Explorer 131–4, 136, 147, 229
    and Linux 340
    MSN 132, 146–7
    and Netscape 131–2, 138–9, 147, 166, 229
    network effects 228–9, 234–5
    network monopoly 10
    Office 228
    power 229
    revenues 227, 228, 252
    source codes 339
    stock market floatation 247
    Vista 42
    Windows 227, 229, 247, 251–2
    Windows 3.0 251–2
    Windows 95 131, 146
    Windows sales 39
    Word 228
Minitel 141
MITI 50, 340
Miyoake, Senri 52–3
mobile telephony 45, 81. *see also* Nokia
monopolies 10, 233, 282
Moore, Gordon 241, 330–1
Moore's Law 9, 30, 42, 330
Morgach, John 120
Morgan Stanley 95, 130–1, 145, 343, 344, 345
Morita, Akio 48–52, 52, 54, 55, 58–9, 60, 61, 62, 64, 65, 67, 68, 70, 75, 76, 84, 164–5
Moritz, Michael 192–3, 200
Mosaic 125–7
Mosaic Communications 128
Motorola 161, 316, 317, 321, 324, 326, 327, 333
moveable type 270–1
Mozilla 134
MP3 280, 285
MSN 132, 146–7
MTV 148
Mueller, Glen 117
Murdoch, Elizabeth 310
Murdoch, James 311

Murdoch, Rupert 13, 155, 292–3, 294, 295, 298, 299, 301, 303–4, 304, 306, 309–10, 311
Museum of Modern Art 28
music industry 43–4, 57, 58, 72, 272–3
MySpace 338

Napster 43, 272
NASDAQ 103, 109, 156, 194, 345
Nathan, John 60
National Center for Supercomputing Applications (NCSA) 125, 128
Natsuno, Takeshi 119–20
Nature of the Firm, The (Coase) 378
Naxos 83, 115, 284
need 218
Negroponte, Nicholas 343, 349
Neil, Andrew 298, 309
Netflix 285
Netscape 96, 118, 125–39, 147, 221, 226, 280, 337, 353, 363, 381
  AOL acquires 134, 139, 153
  assessment 134–9
  business model 137, 160, 344
  competition 131–4, 138–9
  customer proposition 136
  customer relationships 137–8
  distinctive capabilities 135–6
  entrepreneurialism 136
  funding 128
  growth 146, 226
  human capital 378
  leadership 136, 165–6
  market focus 166
  and Microsoft 131–2, 138–9, 147, 166, 229
  Netscape Navigator 1.0 130
  origins 125–7
  pricing 129
  revenues 130
  start-up 127–30
  stock market floatation 130–1
  strategic planning 366–7

vision 134–5
network effects 221, 222, 223, 224–5, 228–9, 234–5, 337
network monopoly 233–4
networked collaboration 379–82
networks 9–10
new markets 14, 78–9
New York 223
New York Times 69, 203
New York University 79
New Yorker 194
Newmark, Craig 275–8
News Corps 299, 309, 338
News Datacom (NDC) 307
News International 292–3, 298, 306
NeXT 35, 40, 126
Nike 375, 380
Nintendo 71–4
Nokia 45, 123, 221, 289, 314314–28, 350, 362, 365, 369
  assessment 326–8
  branding 324, 374
  breaks Ericsson's monopoly 315–16
  business model 161
  business networks 382
  competition 321–2, 326–7
  corporate culture 325–6
  design philosophy 322–4
  digital camera production 5
  first mobile handsets 316
  funding 319
  growth 6, 320–1, 324–5
  human capital 378321–1
  leadership 165
  losses 318
  Ollila saves 317–21
  origins 315
  revenues 324
  strategic planning 368
Novell 201
NTL 313
NTT 119–20

Nuovo, Frank 323
NYU Online 79, 80

Office of Fair Trading (OFT) 308
Ohga, Norio 55–8, 58, 60–2, 62, 64, 65, 67, 68–9, 70
O'Leary, Michael 120, 163–4
Ollila, Jorma 165, 317–21, 322, 324, 324–6, 326–7, 365, 383
Olsen, Ken 331, 363
Omidyar, Pierre 1, 83, 119, 162, 164, 170–2, 174, 175, 176–7, 178, 179, 180, 185, 186, 188, 373
ONdigital 313
online learning 79–80
ONnetwork 313
ONrequest 313
OnSale 177
Opel, John 240, 249, 253, 256
Open Courseware 340
open source software 229, 338–40, 341
Open University 80, 162, 165, 331, 340–1
operational efficiency 368–70
opportunity 361–2
Oracle 373
Orange 81, 317
organizational capabilities 12, 76–9
organizational knowledge 364–6
Origin of Wealth, The (Beinhocker) 334
originality 118
outsourcing 4
over-confidence 353
Overture 197, 199, 203
Ovitz, Michael 63, 64

Page, Larry 83, 119, 164, 189–91, 191–3, 194, 196–7, 200, 201–2, 203, 204–6, 208
Pagemaker 36, 37
PageRank 190, 191
Pareto, Vilfredo 232
Parry, Rick 303, 303–4, 310–11

Parsons, Dick 157
Pathfinder 154
PayPal 184–6, 187, 225
PC industry 82–3. see also Apple Computers; IBM
PC Magazine 192
Pearson 295, 297
Pensare 79, 80
Perez, Antonio 5
perfect competition 265
Perry, Walter 165
Peters, Jon 64–5, 66–8, 69–71, 70
Pets.com 102, 103
Philips 51, 56–7, 66, 334
photography 3–5
Pittman, Bob 148–9, 149, 157, 165
Pixar 40
Pizza Hut 140
PlanetAll 96
pluralism 358
Polygram 66–7
Porter, Michael 358
PostScript 37
power laws 232
preconceptions 350–1
Premier League, the 302–6, 310, 314
pre-recorded video cassettes 54
Priceline.com 345, 346
Princeton University 86
Prodigy 141, 142, 144, 145
professionalism 135–6
profit margins 106, 111
progress 265
Proust, Marcel 330, 354

Quantum Computing Services 140, 142
Quattrone, Frank 130–1

radio 271
Radiolinja Oy 317, 320
railways 267–8, 388
Raskin, Jeff 28, 34, 37
Raymond, Eric 340, 342

RCA 52
Reaganism 281
realism 366–7
Reed 295
Reichheld, Frederick 371
reinvention 124
Reith, John 292
reliability 372
research and development 24, 122–3,
239
Reuters 338
revenues
    Amazon 102–3, 103–4, 105
    AOL 150
    Apple Computer 23, 27, 28, 34, 38,
    41, 45, 45–6
    BSkyB 307
    Craigslist 277
    eBay 2, 3, 174, 175, 182, 183
    Google 195–8, 199, 206, 207, 231
    IBM 237, 238, 239, 244, 249, 252
    Kodak 4
    Microsoft 227, 228, 252
    Netscape 130
    Nokia 324
    Sony 58, 71
    Tesco 114
    Webvan 108, 109
rewards, inequalities in 231–5
Rheingold, Howard 141–2
rigidities 363
Risher, David 97
Robinson, Gerry 309
Rosen, Benjamin 23
Rosen, Dan 131–2
Ross, Steve 63
Rossetto, Louis 343
Russell, Sir George 305, 308
Ryan, Tom 120
Ryanair 79, 120, 122, 163–4, 283, 337,
352, 362, 367–8, 372, 375, 378, 382
Safari 229
Safra, Jacob 260

Samsung 326
San Diego 52–3
San Francisco Chronicle 277–8
SAP 124
satellite broadcasting 294–6. *see also*
BSkyB
SATV 294
scalability 93, 138, 230–1, 325, 368–70
Schein, Harvey 52
Schmidt, Ed 119
Schmidt, Eric 137, 200–3
Schulhof, Michael 60–2, 62, 65, 67,
70, 70–1
Schultz, Howard 121, 321
Schumpeter, Joseph 7, 9, 264–8, 278,
290
Scott, Mike 21–2, 23, 25–6, 31, 165
Scott, Ridley 29
Sculley, John 28, 32–5, 35–6, 37, 38–9,
120, 121, 165, 367, 369
Search Engine Watch 194
Sears Roebuck 141, 257, 332, 352
Seattle Computer Products 240
Second World War 77, 235, 264
seed capital 21
self-confidence 353–4
self-improvement 12
Sequoia Capital 107, 192–3, 200
Seybold, Patricia 100
Shakespeare, William 10
Shamir, Adi 306
Sheehan, George 108–9, 109, 111
Shell 77, 365
short-term financial performance 11
Siemens 66, 318
Siino, Rosanne 128
Silicon Graphics, Inc 27, 39, 116–18,
121, 125, 126, 130, 246, 363
Simmons-Gooding, Anthony 296–7
Simons, David 345–6
sincerity 199
Skoll, Jeff 1–3, 164, 171, 172, 174, 174–5,
176–7, 178, 179, 188

Sky. *see* BSkyB
Skype 186–7, 189
Sloan, Alfred 217
SmartMarket Technology 174–5
social networking 338
Société Européenne des Satellites (SES) 295
solutions looking for problems 331–6
Sonafon 317
Sondheim, Stephen 34
Sony 31, 48–74, 123, 220, 222, 226, 350, 353, 360, 362, 363, 365, 369
    acquisition of Columbia Pictures 59–61, 62–71
    acquisition of Columbia Records 61–2
    acquisition of Guber–Peters Entertainment 67–8
    assessment of synergy strategy 71–4
    Betamax VCRs 53–5
    brand 55–6
    breaks into US market 51–2
    CBS/Sony Records 55, 56, 57
    CD sales 58
    competition 72
    design 55–6
    development of the compact disc 56–8, 280
    first ever loss 70
    first products 49–52
    first US manufacturing facility 53
    human capital 378
    and innovation 75–6
    leadership 164–5
    organizational capabilities 76–7
    origins 48–52
    the PlayStation 71, 74
    problems 73
    revenues 58, 71
    Sony Pictures Entertainment losses 70
    strategic planning 366–7, 368
    transistor licence 50

    Trinitron television 52–3
    vision 83
    the Walkman 58–9, 75, 84, 284, 285
Sony Corporation of America 51, 58, 61
Sony Pictures Entertainment 70–1
Sony Walkman 58–9, 75, 84, 284, 285
Sony-Ericsson 327
Sotheby's 101
Sothebys.com 181
Southwest Airlines 79, 283, 352, 368, 378
Soviet Union, collapse of 281, 318
spam 195–6
Spiegel, Joel 97
Spielberg, Steven 66, 117
Spindler, mike 40
Springer, Howard 73
Spyglass 130–1
Starbucks 100, 115, 121, 122, 162, 221, 283, 321, 368, 372, 374, 375, 376
strategic assets 220, 221, 235, 384
strategic planning 279–80, 366–8
style 45
sub-contractors 380
Sullivan, Danny 194, 206–7
Sun, the 292–3
Sun Microsystems 39, 133, 192
Sunday Times 293, 309
supermarkets 284
Suria, Ravi 103
Surowiecki, James 341–2, 342
sustaining innovation 269
switching costs 222–3

Taleb, Nassim 231
Tapscott, Don 342–3
Taylor, Frederick 369
teamwork 119–22
technological innovation, economic consequences of 9
technology
    disruptive 268–73
    first impressions of 330–1

practical applications 331–6
telephones 330
Tesco 111, 114–15
Tett, Gillian 275
Thatcher, Margaret 293, 297
Thatcherism 281
Theory of Economic Development
(Schumpeter) 265
Thiel, Peter 184–5, 225
Thomson Learning 80
3M 124
Time 103, 147, 194
Time Warner 63, 68, 117, 148, 149,
153–7, 210, 332, 345, 349, 362
The Times 293
Times Mirror 175, 177
tipping points 223
Tisch, Laurence 61–2
Tivola, Mika 318, 319
Torvalds, Linus 339–40
Toshiba 42
Totsuko 48–51
transformational technology 270–1
transistors 50
traveller's cheques 161
Turner, Ted 34
Twain, Mark 9
20th Century Fox 293

uncertainty 6–7
UNext Inc 79
unique resources 221
uniqueness 115, 116
United News and Media 310
United States of America
    CD sales 58
    first Japanese manufacturing facility
    53
    GDP 282
Universal City Studios 53
universities, online learning 79–80
Unix, 339
unknown unknowns 274–5

Valentine, Don 20, 23, 24, 120
value chains 379
Vanity Fair 92
VCRs 53–5, 284–5
venture capital 93, 94, 107, 117, 130,
139, 140, 168, 321, 344–5, 377
Viacom 149
video rental stores 54, 285
videoconferencing 334–5
virtual community 141–2
Visicalc 227
VisiCalc spreadsheet program 23–4,
332
vision 83–5, 105, 119–20, 121, 124,
134–5, 167
Vivendi 353
Vodafone 81, 317
Voice Over Internet Protocol (VOIP)
186–7
von Maister, Bill 139–40
Vonnegut, Kurt 34

Wales, Jimmy 341
Wall Street crash, 1929 224
Wall Street Journal, The 93–4, 155–6,
248
Wal-Mart 106, 108, 110, 111, 115, 163,
203
Wang 246
Warhol, Andy 34
Warner, Harry 330
Warner, John 37
Warner Bros 67, 68, 69
Watson, Thomas 'Old Tom' 236–7
Watson, Thomas 'Young Tom' 236–9
Watt, James 388
wealth, distribution of 232
Web 2.0 338–40
websites
    operation 81
    popularity 190
Webvan 105–12, 114–15, 160, 226, 332,
353, 367

Wedgwood, Josiah 388
Weinstock, Albert 353
Welch, Jack 165, 378
Well, The 275
Western Electric 50
Western Governors University 79, 80
Westly, Steve 176, 177
What You See Is What You Get 36
Wheeler, David 306
Whitman, Meg 119, 178, 179–80, 181, 183, 184, 185, 186, 187, 188–9, 373, 378
Wikinomics: How Mass Collaboration Changes Everything (Tapscott and Williams) 342–3
Wikipedia 261, 338, 341, 3
wikis 341
Williams, Anthony 342–3
Wilson, Mike 174, 176, 180
winners attributes 11–12, 13

Wired 143, 145, 343, 345
Wisdom of Crowds, The (Surowiecki) 341–2
Woods, Tiger 232
WordPerfect 228
World Wide Web 3, 40, 140, 280
WorldCom 150–1
Wozniak, Stephen 17–19, 20, 22, 25, 28, 35, 76
Wurster, Thomas 345, 347–8
Xerox PARC 24, 29, 82, 122–3

Yahoo 1, 91, 96, 107, 159, 180, 183, 191, 194, 230
Yellow Pages 234
Yetnikoff, Walter 61–2, 63, 64, 67, 70
YouTube 209, 338

Zennström, Niklas 186–7